Environmental science

B. J. Smith

B.Sc., Ph.D., F.C.I.O.B., F.C.I.B.S., C.Eng., M.Inst.P., F.I.O.A., F.I.M.A. F.B.I.M.

G. M. Phillips

B.Sc., M.Sc., M.C.I.O.B., M.C.I.B.S., M.I.O.A.

M. E. Sweeney

B.Sc., M.Sc.

Environmental science

Longman
Scientific &
Technical

Longman Scientific & Technical
Longman Group UK Limited,
Longman House, Burnt Mill, Harlow,
Essex CM20 2JE, England
and Associated Companies throughout the world.

First published 1983
Reprinted by Longman Scientific & Technical 1987, 1988, 1989, 1990

British Library Cataloguing in Publication Data

Smith, B.J.
   Environmental science. – (Longman technician
series. Construction and civil engineering)
   1. Human ecology
   I. Title   II. Sweeney, M. E.   III. Phillips, G. M.
   333.7   GF41
ISBN 0-582-41620-5

Printed in Malaysia
by Percetakan Jiwabaru Sdn. Bhd.,
Bangi, Selangor Darul Ehsan

Contents

Preface

In writing this book the authors aimed to cover the requirements for the Business and Technician Education Council Level Three & Four Units in Environmental Science. At the same time it was recognised that there was a real need for a suitable book for those studying Building, Quantity Surveying, Architecture and Environmental Health at 1st degree level. This book should therefore form a useful first year introductory text for both 'A' level and BTEC entrants.

The book contains a large number of worked examples in the text as well as many student questions at the end of each chapter. Experiments have been included, not with the intention of being exhaustive, but to give ideas. Some areas of work lend themselves to student practical work better than others so that some inbalance is inevitable.

'Environmental Science' should give students an introduction to the environmental problems in construction and the methods which may be used to provide a satisfactory and economic solution.

Acknowledgements

We are grateful to the following for permission to reproduce copyright material: British Standards Institution for our Fig. 9.18 and our Table 9.15 from *CP 3* Chapter 1, Part 1, 1964; our Table 3.1 from *BS 5250*, 1975; Building Establishment, Department of the Environment for our Fig. 9.12 from Fig. 1 *BRS Daylight Factor Protractors*; our Fig. 9.17, our Table 9.4 from Table 1, our Table 9.5 from Table 2 and our Table 9.10 from Table 3 *BRS Digest*, 42 by permission of the Controller of Her Majesty's Stationery Office; The Chartered Institution of Building Services for our Fig. 8.15 from Fig. 8 *CIBS (IES) Technical Report No 2*; our Fig. 9.3 from Fig. 11, our Fig. 9.13(a) and (b) from Fig. 3, our Figs. 9.7, 9.8 and 9.9 from Table 7, our Table 10.4 from Table 1, our Table 10.5 from Table 2 and our Fig. 10.9 from *CIBS (IES) Technical Report No. 4*; our Table 10.2 from Table 1.8, our Table 10.1 from Table A1 and our Fig. 10.8 from Fig. 2 *CIBS (IES) Technical Report No. 10*; our Figs. 9.1 and 9.2 from Fig. 1 *CIBS (IES) Lighting Guide*; our Table 2.1 from *CIBS Guide*, Section A1; our Fig. 8.11 from Fig. 32, our Fig. 10.7 from Fig. 16, our Fig. 10.14 from Fig. 3, our Fig. 10.6 from Fig. 7 and our Tables 8.2 and 9.14 from *CIBS (IES) Code for Interior Lighting* 1977; our Fig. 10.17 from Fig. 14 *CIBS (IES) Code for Interior Lighting* 1977 and Fig. 16 *CIBS (IES) Code for Interior Lighting* 1973; our Figs. 10.12 and 10.13 reproduced from the paper 'Eyestrain: the environmental causes and their prevention' by J. H. Goacher *CIBS National Lighting Conference* 1980; Lighting Industry Federation Ltd for our Fig. 10.4 from fig p 30 *Interior Lighting Design*; Longman Group Ltd for our Table 9.2 from Table 6.6, our Table 9.3 from Table 6.8, our Table 9.6 from Table 6.7, our Table 9.11 from Table 6.5, our Fig. 9.10 from Fig. 6.6, our Fig. 8.22 from Fig. 6.8 and our Fig. 10.2 from Fig. 1.6 *Lighting* (2nd edn) by D. C. Pritchard, Environmental Physics Series; A. M. Marsden and Thorn Lighting Ltd for our Fig. 10.5 from Fig. 2.5 *Lamps and Lighting* by Henderson and Marsden published by Edward Arnold.

List of units

There are three classes of SI Units (Système International d'Unités, or International System of Units):

1. Base units
2. Supplementary units
3. Derived units

1 Base units

Quantity	Name of unit	Symbol
length	metre	m
mass	kilogram	kg
time	second	s
electric current	ampere	A
thermodynamic temperature	kelvin	K
luminous intensity	candela	cd
amount of substance	mole	mol

2 Supplementary units

Quantity	Name of unit	Symbol
plane angle	radian	rad
solid angle	steradian	sr

3 Derived units

Quantity	Name of unit	Symbol	Relationship to base or supplementary unit
frequency	hertz	Hz	$1 \text{ Hz} = 1 \text{ s}^{-1}$
force	newton	N	$1 \text{ N} = 1 \text{ kg m s}^{-2}$
pressure or stress	pascal	Pa	$1 \text{ Pa} = 1 \text{ N m}^{-2}$
work, energy, quantity of heat	joule	J	$1 \text{ J} = 1 \text{ N m}$

Quantity	Name of Unit	Symbol	Relationship to base or supplementary unit
power	watt	W	$1\ W = 1\ J\ s^{-1}$
quantity of electricity	coulomb	C	$1\ C = 1\ As$
electrical potential, potential difference, electromotive force	volt	V	$1\ V = 1\ W\ A^{-1}$
electrical resistance	ohm	Ω	$1\ \Omega = 1\ V\ A^{-1}$
electrical capacitance	farad	F	$1\ F = 1\ AsV^{-1}$
inductance	henry	H	$1\ H = 1\ VsA^{-1}$
luminous flux	lumen	lm	$1\ lm = 1\ cd\ sr$
illumination	lux	lx	$1\ lx = 1\ lm\ m^{-2}$

Multiples

Name	Symbol	Factor
tera	T	10^{12}
giga	G	10^{9}
mega	M	10^{6}
kilo	k	10^{3}
hecto	h	10^{2}
deca	da	10
deci	d	10^{-1}
centi	c	10^{-2}
milli	m	10^{-3}
micro	μ	10^{-6}
nano	n	10^{-9}
pico	p	10^{-12}
fempto	f	10^{-15}
atto	a	10^{-18}

It is preferable to express all values so that a number between 0.1 and 1000 can be written down for the quantity.

Examples of units

Quantity	SI units	Multiples
plane angle	rad	m rad, μ rad
length	m	km, cm, mm, μm, nm
area	m^2	km^2, dm^2, cm^2, mm^2
volume	m^3	dm^3, cm^3, mm^3
time	s	ks, ms, μs, ns

Examples of units (*Cont.*)

Quantity	SI units	Multiples
angular velocity	rad/s	
velocity	m/s	
acceleration	m/s^2	
mass	kg	Mg, g, mg, μg
density	kg/m^3	
momentum	kgm/s	
moment of inertia	kgm^2	
force	N	MN, kN, mN, μN
moment of force	Nm	
pressure	Pa	GPa, MPa, kPa, mPa, μPa
stress	Pa, N/m^2	GPa, MPa(N/mm^2), kPa, mPa, μPa
surface tension	N/m	mN/m
heat, energy, work	J	MJ, kJ
power, heat flow rate	W	MW, kW, mW, μW
Celsius temperature	°C	
temperature interval	K	
linear expansion coefficient	K^{-1}	
thermal conductivity	W/(mK)	
heat capacity	J/K	kJ/K
specific heat capacity	J/(kgK)	kJ/(kgK)
specific latent heat	J/kg	MJ/kg, kJ/kg
electric current	A	kA, mA; μA, nA, pA
electric charge quantity of electricity	C	kC, μC, nC, pC
potential difference	V	MV, kV, mV, μV
capacitance	F	mF, μF, nF, pF
resistance	Ω	GΩ, MΩ, kΩ, mΩ, $\mu\Omega$
resistivity	Ωm	GΩm, MΩm, kΩm, Ωcm, mΩm, $\mu\Omega$m, nΩm
wavelength	m	nm, pm
radiant energy	J	
luminous intensity	cd	
luminous flux	lm	
quantity of light	lm s	
luminance	cd/m^2	
illuminance	lx	
sound power level	dB	(Reference level 10^{-12} W)
sound pressure level	dB	(Reference level 2×10^{-5} N/m^2)
sound reduction index	dB	
reverberation time	s	

Section I

Heat and thermal effects

Chapter 1

Thermal transmission

Temperature

Heat is associated with the continual random motion of molecules within all states of matter – solid, liquid and gas. Temperature is a measure of this molecular energy.

The transfer of heat energy to or from a body is dependent upon temperature. If a cold body at a low temperature is placed in contact with a hotter body at a higher temperature, heat is transferred from the hotter body to the cold body until thermal equilibrium is established and both bodies are at the same temperature. If no heat flows upon contact, there is no difference in temperature.

The establishment of a temperature scale requires fixed points at which the temperature is always the same and reproducible. Any property of a substance which changes with temperature, such as length, volume, resistance or thermoelectric electromotive force can be used for temperature measurement.

Thermodynamic temperature scale

The basic temperature is the thermodynamic temperature (T) in kelvin (K).

The unit of thermodynamic temperature, the kelvin, is the fraction $1/273.16$ of the thermodynamic temperature of the triple point of water (approximately $0\,°C$). The triple point of water is the temperature at which ice, water and water vapour coexist in equilibrium at a pressure of $610\,N/m^2$.

Celsius temperature scale

Practical temperature measurements are usually made in degrees Celsius (°C). On the Celsius scale of temperature the lower fixed point is the temperature of melting ice at standard atmospheric pressure ($101.325 \, kN/m^2$) and is defined as $0 \, °C$. The upper fixed point is the temperature of steam above water boiling at standard atmospheric pressure and defined as $100 \, °C$.

Measurements show that

$$0 \, °C = 273.15 \, K$$

$$100 \, °C = 373.15 \, K$$

Hence an interval or change of $1 \, °C$ on the Celsius scale has the same magnitude as an interval or change of $1 \, K$ on the kelvin scale. θ or t denote temperature on the Celsius scale (°C), and T denotes temperature on the thermodynamic and absolute scales (K).

Since $0 \, °C = 273.15 \, K$

$$T = 273.15 + \theta$$

so that

$$21 \, °C = 273.15 + 21$$
$$= 294.15 \, K$$

Sufficient accuracy is usually obtained by writing 273 for 273.15, so that approximately,

$$21 \, °C = 273 + 21$$
$$= 294 \, K \text{ (Fig. 1.1)}$$

Quantity of heat

Heat is a form of energy. Quantity of heat is measured in terms of the fundamental energy unit, the joule (J). The joule is defined as the work done when the point of application of a force of 1 newton (N) is displaced through a distance of 1 metre (m) in the direction of the force. A newton is that force which applied to a mass of 1 kilogram (kg) gives it an acceleration of 1 metre per second per second (m/s^2).

The application of heat to a body may result in a rise in temperature or a change in state. Sensible heat is a term sometimes used in heating and cooling to indicate any quantity of heat which changes only the temperatures involved. The heat energy used to change the state of a substance is known as the latent heat.

Sensible heat

The specific heat capacity (c) is the heat required to raise the temperature of unit mass (1 kg) of a substance by $1 \, K$.

4

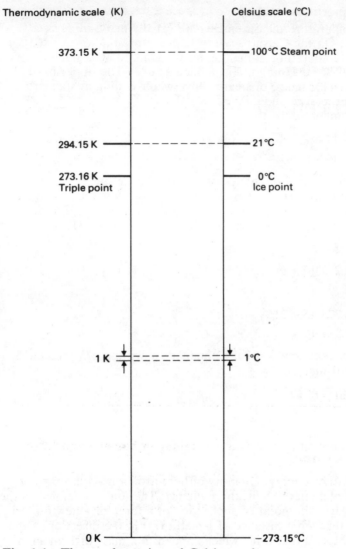

Fig. 1.1 Thermodynamic and Celsius scale

The term heat capacity or thermal capacity of a body is defined as the quantity of heat required to raise the temperature of that body by 1 K. The thermal capacity, often useful as a measure of the heat storing capacity of an element or structure, is obtained from the product of mass and specific heat capacity:

thermal capacity = mass × specific heat capacity
(J/K)   (kg)   (J/kg K)

The quantity of heat required to produce a temperature change θ in a body of mass m with specific heat capacity c is given by

quantity of heat $= mc\theta$

where, $m =$ mass   (kg)

  $c =$ specific heat capacity   (J/kg K)

and $\theta =$ temperature difference   (K)

Example 1.1 Calculate the quantity of heat required to raise the temperature of 0.2 kg of water from 15 °C to 90 °C if the specific heat capacity of water is 4.2 kJ/kg K.

$m = 0.2$ kg

$c = 4.2$ kJ/kg K

  $= 4.2 \times 1000$ J/kg K

  $= 4.2 \times 10^3$ J/kg K

Temperature rise required, θ

  $= 90 - 15$

  $= 75$ °C

Quantity of heat $= mc\theta$

  $= 0.2 \times 4.2 \times 10^3 \times 75$

  $= 63 \times 10^3$ J

Quantity of heat $= 63$ kJ

Latent heat

When heat is supplied to a body at its melting or boiling point there is no temperature change.

The latent heat is the quantity of heat required to change the state of a substance without change of temperature.

The heat required to change a solid to a liquid at the melting point is called the latent heat of fusion. The heat required to change a liquid to a gas at the boiling point is called the latent heat of vaporisation.

The specific latent heat is the quantity of heat required to change the state of unit mass (1 kg) of a substance, without change of temperature.

Transmission of heat energy

Heat energy transmission takes place from a higher temperature region to one at a lower temperature. This heat energy transfer may be by one or more than one of three modes: conduction, convection, and radiation.

Conduction

Conduction of thermal energy may take place in all three states of matter – solid, liquid, gaseous. The kinetic energy of molecules within a body may be associated with the temperature of that body. An increase in temperature corresponds to an increase in the energy and hence the activity of molecules within the body. These molecules are not isolated but interact with one another. The transfer of thermal energy takes place as a result of interactions between higher energy molecules and ones of lower energy. Thermal energy transference by conduction is in the direction of decreasing temperatures and involves a continuous gradient of temperature throughout the body.

In the solid state molecules are closely packed together at fixed positions within the solid. Mose molecules do not move throughout the solid but vibrate and rotate at their own position in the solid. If thermal energy is supplied to one region of the solid, the molecules in that region become hotter, more energetic, vibrate and rotate more rapidly and interact with neighbouring molecules so that this excess energy is soon shared with the neighbouring molecules, which in turn share their increase in energy with their nearest neighbours. The heat energy is transmitted from molecule to molecule throughout the whole solid.

Liquid molecules are closely packed together, but do not have the fixed positions of solid molecules. The molecules move in random fashion in the liquid. In this continuously changing arrangement of molecules thermal energy transfer and hence heat conduction take place less rapidly than in most solids.

The molecules in a gas are relatively far apart and make random collisions with each other and the walls of their container. The wide separation between gas molecules leads to fewer molecular collisions than in liquids, and hence less efficient thermal energy transfer than in liquids or solids.

Conduction is the transfer of thermal energy by molecular interactions, from one part of a body to another under the influence of a temperature gradient, without appreciable disturbance of the usual molecular arrangements within the body.

Those materials in which heat transfer takes place most rapidly, usually solids, and especially metals, are known as good conductors. In poor conductors, such as gases and insulating materials heat transfer takes place more slowly.

Thermal conductivity

Thermal conductivity (k) is a term used to assess the ability of a material to transfer heat by conduction.

The heat flow rate (q) represents thermal energy transferred in unit time and is measured in watts (joules/second).

$$q = \frac{\text{quantity of heat}}{\text{time}}$$

$$\text{watt} = \frac{\text{joule}}{\text{second}}$$

In Fig. 1.2 a steady heat flow rate q is shown through a block of material of area A and thickness L with face 1 maintained at temperature θ_1 and face 2 maintained at temperature θ_2. Assume the block is lagged so that no heat escapes through the sides of the block and heat flow is along parallel straight lines normal to faces 1 and 2. The temperature difference divided by the thickness L is known as the temperature gradient.

$$\text{Temperature gradient} = \frac{\theta_1 - \theta_2}{L}$$

Experiment has shown that the rate of heat transfer (q) is directly proportional to the area (A) and temperature difference ($\theta_1 - \theta_2$), but inversely proportional to the thickness (L).

$$q \propto \frac{A(\theta_1 - \theta_2)}{L}$$

In equation [1.1] the proportionality is made an equality by introducing the thermal conductivity (k) as a constant dependent upon the nature of the block material.

$$q = \frac{kA(\theta_1 - \theta_2)}{L} \qquad [1.1]$$

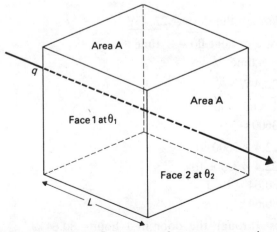

Fig. 1.2 Heat flow normal to two opposite faces of a homogeneous block

To define thermal conductivity consider unit area ($A = 1\,m^2$), unit temperature difference ($\theta_1 - \theta_2 = 1\,K$) and unit thickness ($L = 1\,m$) in equation [1.1]

$$q = \frac{k \times 1 \times 1}{1}$$

Thermal conductivity (k) is the rate of heat transfer (watts), through unit area of a slab of a uniform homogeneous material of unit thickness, when unit difference of temperature is maintained between two opposite surfaces, the other surfaces being insulated. The units of thermal conductivity are W/m K.

Example 1.2 A wooden door $2\,m \times 0.8\,m$ and 40 mm thick has an inside surface temperature of 13 °C and an outside surface temperature of 9 °C. If the thermal conductivity of wood is 0.14 W/m K, calculate:

  (a) the rate of heat flow through the door;
  (b) the quantity of heat lost through the door in 1 hour.

(a) Rate of heat flow $(q) = \dfrac{kA(\theta_1 - \theta_2)}{L}$

$$k = 0.14\,W/m\,K$$
$$A = 2 \times 0.8 = 1.6\,m^2$$
$$\theta_1 - \theta_2 = 13 - 9 = 4\,K$$
$$L = 40\,mm\ (0.040\,m)$$
$$q = \frac{0.14 \times 1.6 \times 4}{0.040}$$
$$= 22.4\,W$$

Rate of heat flow through the door = 22.4 W

(b) Quantity of heat = rate of heat flow × time
$$= q \times time$$
$$q = 22.4\,W$$
time = 1 hour $= 1 \times 60 \times 60$
$$= 3600\,s$$
quantity of heat $= 22.4 \times 3600$
$$= 22.4 \times 3.6 \times 10^3$$
$$= 80.64 \times 10^3\,J$$
$$= 80.64\,kJ$$

Quantity of heat lost through the door in 1 hour = 80.64 kJ

The thermal conductivity values in Table 1.1 show that k is relatively high for good thermal conductors such as metallic solids and low for liquids, gases and insulating materials. In general the thermal conductivity of liquids is greater than the thermal conductivity of gases.

Table 1.1 *Some typical thermal conductivity values*
Values refer to normal temperatures unless a temperature is specified.

	Thermal conductivity (k) (W/m K)
Copper (0 °C)	403
Lead (0 °C)	36
Steel, carbon (0 °C)	50
Concrete, ballast 1:2:4	1.5
Concrete, aerated blocks	0.2
Glass, sheet window	1.05 ⎱ average values
Plaster	0.13 ⎰
Polystyrene, cellular	0.035 ⎱ average values
Vermiculite granules	0.065 ⎰
Ice (−5 °C)	2.3
Water (0 °C)	0.561
(80 °C)	0.673
Air	0.025

The thermal conductivity value depends upon temperature, but in the case of most building and insulating materials k may be taken as constant unless large temperature ranges are involved.

Thermal conductivity varies with density, porosity and moisture content of materials. The k value of an isotropic material is the same for all directions of heat flow. In an anisotropic material k may depend upon the direction of heat flow. Most building and insulating materials may be regarded as isotropic although in wood a directional preference for heat flow is caused by the fibrous structure. In many building materials, density, porosity, composition (for example in mixtures such as concrete and brick) and moisture content lead to thermal conductivity variation from sample to sample.

Thermal conductance (C) (W/m² K)

Thermal conductivity is a property of a material. It is sometimes convenient to introduce thermal conductance (C) to express the heat conducting capacity of a structural component or structure.

Thermal conductance is the heat flow rate through unit area of a uniform structural component or structure of thickness L, per unit temperature difference between two opposite surfaces, the other surfaces being insulated. It is assumed that heat flows along straight parallel lines normal to the two opposite surfaces

$$q = A \times C \times \text{temperature difference}$$

where, $q =$ heat flow rate (W)

$\quad\quad A =$ area (m²)

and $\quad C =$ thermal conductance (W/m² K)

The temperature difference is measured between the two opposite surfaces and $C = k/L$, where k is the thermal conductivity and L is the thickness of the structure.
(C replaces k/L in equation [1.1].)

Example 1.3 A 25 mm (0.025 m) thickness of expanded polystyrene ($k = 0.034$ W/m K) has a thermal conductance

$$C = \frac{k}{L} = \frac{0.034}{0.025} = 1.36 \, \text{W/m}^2 \, \text{K}$$

Thermal resistivity (r) (mK/W)
In calculation it may be convenient to use the thermal resistivity (r) of a material. Thermal resistivity is the reciprocal of thermal conductivity with units mK/W. $r = 1/k$

Example 1.4 For expanded polystyrene $k = 0.034$ W/m K, therefore resistivity

$$r = \frac{1}{k} = \frac{1}{0.034} = 29.4 \, \text{mK/W}$$

Thermal resistance (R) (m² K/W)
Thermal resistance (R) is a measure of the resistance to heat flow (through unit area) of a material of any thickness or of a combination of materials. For unit area, thickness L and thermal conductivity k,

$$R = \frac{L}{k}$$

Example 1.5 A 25 mm (0.025 m) thickness of expanded polystyrene ($k = 0.034$ W/m K) has a thermal resistance

$$R = \frac{L}{k} = \frac{0.025}{0.034} = 0.735 \, \text{m}^2 \, \text{K/W}$$

Many practical situations arise when heat flows through two or more layers of different materials, as for example heat flowing through a typical wall construction.

Heat transfer through a composite plane wall
Assume inside and outside *surface* temperatures are known but the temperatures of the interfaces between two different materials

are not. The heat flow rate (q) through each layer of the composite wall is the same, if conditions are steady and there is no lateral heat flow between the various layers (Fig. 1.3). For unit area of wall ($A = 1\,\text{m}^2$),

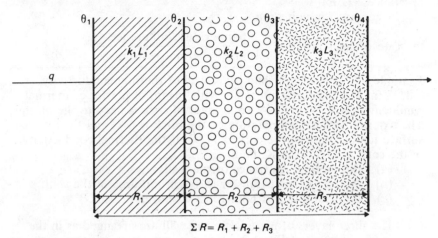

Fig. 1.3 Heat transfer through a composite plane wall

$$q = \frac{k_1}{L_1}(\theta_1 - \theta_2)$$

$$q = \frac{k_2}{L_2}(\theta_2 - \theta_3)$$

$$q = \frac{k_3}{L_3}(\theta_3 - \theta_4)$$

These equations may be rewritten

$$\theta_1 - \theta_2 = qR_1 \tag{1.2a}$$

$$\theta_2 - \theta_3 = qR_2 \tag{1.2b}$$

$$\theta_3 - \theta_4 = qR_3 \tag{1.2c}$$

where

$R_1 = \dfrac{L_1}{k_1}$, $R_2 = \dfrac{L_2}{k_2}$ and $R_3 = \dfrac{L_3}{k_3}$ are the thermal

resistances of the three layers. Adding equations [1.2a], [1.2b], [1.2c],

$$\theta_1 - \theta_4 = q(R_1 + R_2 + R_3),$$

so that for the composite wall in Fig. 1.3

$$q = \frac{\theta_1 - \theta_4}{R_1 + R_2 + R_3}$$ [1.3]

The denominator of equation [1.3] is the sum of the three thermal resistances. (ΣR)

In general for unit area,

the rate of heat flow $(q) = \dfrac{\text{temperature difference}}{\text{sum of thermal resistances}}$

Example 1.6 A 220 mm solid brick wall of thermal conductivity 0.84 W/m K is plastered internally with 16 mm of plaster of thermal conductivity 0.50 W/m K. A 5 mm layer of cork is fixed to the plaster. The external surface temperature of the brick is 6 °C and the internal surface temperature of the cork is 17 °C. If the thermal conductivity of the cork is 0.05 W/m² K, find:

(a)   the thermal resistance of the wall;
(b)   the rate of heat transfer through unit area of the wall;
(c)   the heat loss in 1 hour for a wall area of 9 m².

(a)   The three layers of material in the wall are arranged as in the composite wall of Fig. 1.3.
The thermal resistance of the wall is equal to the sum of the thermal resistances of the three layers.

Cork $k_1 = 0.05$ W/m K

$\qquad L_1 = 5$ mm $(0.005$ m$)$

$\qquad$ Thermal resistance of cork $(R_1) = \dfrac{L_1}{k_1}$

$$= \frac{0.005}{0.05}$$

$$= \underline{0.100\,\text{m}^2\,\text{K/W}}$$

Plaster $k_2 = 0.50$ W/m K

$\qquad L_2 = 16$ mm $(0.016$ m$)$

$\qquad$ Thermal resistance of plaster $(R_2) = \dfrac{L_2}{k_2}$

$$= \frac{0.016}{0.50}$$

$$= \underline{0.032\,\text{m}^2\,\text{K/W}}$$

Brick $k_3 = 0.84$ W/m K

$\qquad L_3 = 220$ mm $(0.220$ m$)$

$$\text{Thermal resistance of brick } (R_3) = \frac{L_3}{k_3} = \frac{0.220}{0.84}$$
$$= 0.262 \, \text{m}^2 \, \text{K}/\text{W}$$

∴   Thermal resistance of the wall $= R_1 + R_2 + R_3$
$$= 0.100 + 0.032 + 0.262 \, \text{m}^2 \, \text{K}/\text{W}$$
$$= 0.394 \, \text{m}^2 \, \text{K}/\text{W}$$

It is often convenient to represent data in table form:

Material	Conductivity (k) (W/m K)	Thickness (L) (m)	Resistance (R) (m² K/W)
Cork	0.05	0.005	0.100
Plaster	0.50	0.016	0.032
Brick	0.84	0.220	0.262

sum of resistances 0.394

The thermal resistance of the wall is $0.394 \, \text{m}^2 \text{K}/\text{W}$

(b)   For unit area,

$$q = \frac{\text{temperature difference}}{\text{sum of thermal resistances}}$$

temperature difference $= 17 - 6 = 11 \, \text{K}$

$$q = \frac{11}{0.394}$$

$$= 27.92 \, \text{W}$$

The rate of heat transfer through unit area of the wall is 27.92 W.

(c)   Heat loss through area A in time $t = A \quad \times \quad q \quad \times t$

area × heat loss rate per unit area × time

(J)   (m²)   (W/m²)   s

$A = 9 \, \text{m}^2$, $t = 1 \, \text{hour} = 3600 \, \text{s}$

Heat loss $= 9 \times 27.92 \times 3600$
$$= 9 \times 27.92 \times 3.6 \times 10^3$$
$$= 904.6 \times 10^3 \, \text{J}$$
$$= 904.6 \, \text{kJ}$$

The heat loss in 1 hour for a wall area of $9 \, \text{m}^2$ is 904.6 kJ.

This composite wall example uses surface temperatures, but in most buildings only the air temperatures are known. Heat is transferred from the inside air to the structure and finally from the structure to the outside air. The surface resistance (R_s), the resistance to the flow of heat between a surface and the adjacent air, involves the other heat transfer modes, convection and radiation, which are considered in the next sections.

An analogy is often constructed between the conduction of heat and electricity. For steady heat conduction through a series of layers, as in Fig. 1.3 the total thermal resistance is the sum of the resistances of the components. In electrical conduction resistances in series are added.

Convection

Convection is an important mode of heat transfer in gases and liquids, which involves the movement of the gas or liquid molecules throughout the fluid. This bulk movement of fluid takes place because temperature gradients within the fluid give rise to density variations of the fluid.

Consider heat applied to one region of a simple system such as a fluid in a container (Fig. 1.4). Heat transfer takes place through the container wall and to the adjacent fluid molecules by conduction. Transferring heat energy through the container wall causes an increase in temperature, expansion and hence a density reduction of the fluid molecules in the vicinity of the heat source. This small region of hotter, less dense fluid tends to rise within the fluid and is replaced by a quantity of cold fluid. The new quantity of cold fluid in the region of the heat source receives thermal energy and undergoes the same process of temperature rise, density reduction and movement upwards through the fluid. In this way convection currents

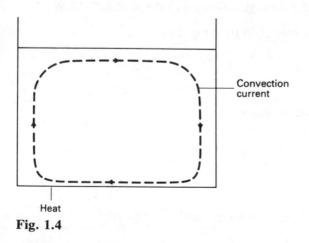

Convection current

Heat

Fig. 1.4

are set up and thermal energy is transported through the fluid by the movement of the fluid itself.

When the main mass of the fluid is stationary and fluid motion is caused by density differences arising from temperature differences, the convection process is termed free or natural convection. Heat transfer by conduction is not significant in a fluid because once a temperature gradient exists in the fluid, natural convection currents will occur from density differences. If the fluid movement arises from external causes such as a pump, fan or wind blowing across the surface of a building so that the main mass of the fluid is in motion, the convection is said to be forced convection.

When the surface of a solid such as a wall, floor or roof is warmer or colder than the adjacent air convective heat transfer occurs. Prediction of the heat transfer to or away from a solid surface in contact with a fluid is complex, for although the transfer in the immediate vicinity of the surface is by conduction, movement or velocity of the surrounding fluid is usually an important variable in the heat transfer by convection.

Natural convection

For the small temperature differences usually encountered in the built environment, the rate of heat transfer by natural convection from a plane surface is given by

$$q_c = Ah_c(\theta_1 - \theta_2) \qquad [1.4]$$

where q_c = rate of heat transfer by natural convection   (W)

A = surface area   (m^2)

θ_1 = surface temperature   (°C)

θ_2 = air temperature   (°C)

and h_c is the convection conductance (W/m^2 K)

It is an acceptable approximation to regard h_c as constant for small temperature differences. For unit area,

$$q_c = h_c(\theta_1 - \theta_2) \qquad [1.5]$$

The convection conductance (h_c), is defined as the rate of transfer of heat to or from unit area of a surface in contact with air or other fluid due to convection, per unit difference between the temperature of the surface and the temperature of the neighbouring air or other fluid. Typical values of h_c, which may also be known as the convective heat transfer coefficient, convection coefficient or unit film conductance are 3.0 W/m^2 K for walls, 4.3 W/m^2 K for upward flow to ceilings and 1.5 W/m^2 K for downward flow to floors.

Forced convection

Newton's law of cooling, an experimental law, states that the rate of loss of heat from a body is directly proportional to the excess of the temperature of the body above that of its surroundings. This law is found to be true for quite large temperature differences, provided the body cools under the influence of a strong draught of air (forced convection).

Therefore equation [1.5]

$$q_c = h_c(\theta_1 - \theta_2)$$

may also be used to account for heat transfer by forced convection. Because heat transfer by forced convection is dependent upon the velocity of the bulk of the fluid past the surface, the value of h_c, the convection conductance, must be modified.

Radiation

Heat transfer by radiation is the transport of energy by electro-magnetic waves. Radio waves, infra-red, visible, ultra-violet, X-rays and γ-rays are all electromagnetic waves, travelling at the speed of light $(3 \times 10^8 \, \text{m/s})$ but differing in wavelength (Fig. 1.5). The sensation of warmth is produced in the human body by the infra-red region of the electromagnetic spectrum.

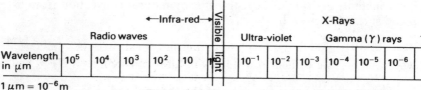

1 μm = 10⁻⁶ m

Fig. 1.5   The electromagnetic spectrum

Electromagnetic radiation originates from materials at high temperatures when excited molecules return to lower energy states and thermal energy is transferred to the lower temperature surrounding surfaces. The sun, our most important source of infra-red radiation, also provides visible and ultra-violet light.

Black body radiation

At the beginning of the nineteenth century experimenters observed that some of the thermal energy falling upon a surface was absorbed by that surface. The amount of absorption depended upon the nature of the surface with matt dark surfaces absorbing most infra-red radiation and highly polished silvered surfaces absorbing least. When the temperature of a surface was raised above the surrounding temperatures the dark matt surfaces were the best emitters of radiation while the highly polished silvered surfaces were poor emitters.

In order to eliminate this dependence upon surface characteristics when dealing with the basic laws of thermal radiation physicists after Kirchhoff postulated a black body as one that absorbs all the energy in every wavelength and conversely emits every wavelength. The black body was chosen as a convenient concept and should not be confused with an actual body having a surface colour black. Not even lampblack possesses this ideal property when examined over a wide range of wavelengths. A good approximation to blackbody radiation is obtained from a small aperture in a constant temperature cavity constructed of opaque absorbing material.

Black body radiation depends only on the temperature of the body and not on the nature of the surface.

Stefan–Boltzmann law

The thermal energy emitted in unit time (1 s) from unit area (1 m^2) of a black body radiator is proportional to the fourth power of the absolute temperature.

$$W_b = \sigma T^4$$

where W_b is the total emitted thermal radiation leaving the surface per unit time and per unit area, Stefan's constant $\sigma = 5.7 \times 10^{-8}$ W/m^2 K^4 and T is the absolute temperature of the surface in kelvin (K). W_b, known as the total emissive power (or emissive power) of a black body radiator, has units of W/m^2.

Emissivity (ϵ)

Real surfaces, unlike the black body are not ideal radiators, perfect emitters, but it is convenient to compare real surfaces with black body radiators by introducing the term emissivity (ϵ).

The emissivity (ϵ) is the ratio of the thermal radiation from unit area of a surface to the thermal radiation from unit area of a black body (full radiator) at the same temperature.

Since by definition a black body has the maximum possible emissive power for radiation of all wavelengths, for a black body radiator $\epsilon = 1$. Real surfaces have emissivity values less than unity and also dependent upon wavelength.

The sun radiates energy as a black body radiator having a surface temperature of about 6000 K. As the solar spectrum in Fig. 1.6 shows, most of the solar energy is concentrated in the visible portion of sunlight and in the infra-red region. Absorption of radiation in the atmosphere, especially by water vapour, carbon dioxide and ozone modifies the black body characteristic of the solar energy supply to the earth. Surfaces in the built environment at around 300 K emit longwave radiation in the infra-red region of the electromagnetic spectrum (Fig. 1.7).

Absorptivity (α)

Thermal radiation incident upon a surface may be:

(a) absorbed by the surface;

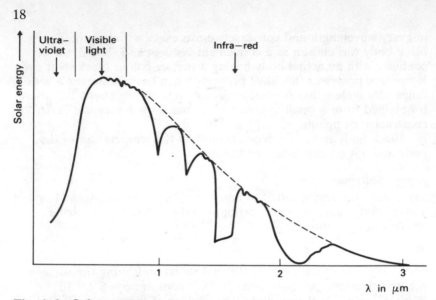

Fig. 1.6 Solar spectrum

(b) reflected away from the surface;
(c) transmitted through the body.

   The absorptivity (or total absorptivity) of a surface is the fraction of incident thermal energy of all wavelengths which is absorbed. The ideal black body is a perfect absorber and so for a black body $\alpha = 1$. For real surfaces α is less than unity and varies with the wavelength of the incident radiation. When the radiation incident upon a surface is of solar origin it is convenient to define *solar absorptivity* (α_s) as

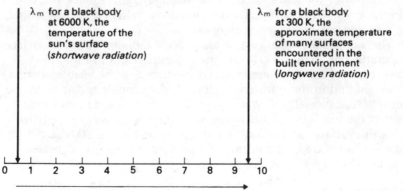

Fig. 1.7 λ_m, the wavelength at which maximum energy is emitted for two surface temperatures

Table 1.2 Some typical values of emissivity (ϵ) and solar absorptivity (α_s).

Surface		Emissivity of surfaces at 10–40 °C	Solar absorptivity
Black non-metallic		0.90–0.98	0.85–0.98
Concrete		0.85–0.95	0.65–0.80
White paint, whitewash		0.85–0.95	0.3–0.5
Aluminium	dull	0.2 –0.3	0.4–0.65
	polished	0.02–0.05	0.1–0.4

the fraction of solar radiation, incident upon a surface, which is absorbed.

For real surfaces both emissivity and absorptivity are functions of surface temperature and surface nature such as roughness and, if metallic, degree of oxidation of surface. Absorptivity is also dependent upon the spectrum of the incident radiation so that absorptivity and emissivity are not always equal (see Table 1.2). However, when the radiation emission is predominantly in the infra-red region of the electromagnetic spectrum, as it is for many materials involved in radiation exchange at normal temperatures, surfaces behave so that approximately,

$$\epsilon = \alpha$$

Because the sun's radiation is produced at such high temperatures (about 6000 K), the solar spectrum is significantly different from that of radiation produced by most other surfaces encountered in practice. The solar absorptivity (α_s) depends upon surface colour, being low for white and high for dark colours. White and light colours may be used externally to reflect solar radiation.

Emissivity values at normal temperatures are important when calculating heat losses from buildings.

Radiation exchange between surfaces

If two surfaces within visual range of each other are separated by a medium which does not absorb radiation, and are at different temperatures above absolute zero, there is a net transfer of heat energy from the warmer surface to the colder surface as a result of reciprocal processes of radiation emission and absorption. If both surfaces are at the same temperature the net heat exchange is zero.

For all but the simplest geometries calculation of the net rate of heat transfer by radiation exchange becomes complex. The net rate of heat transfer by radiation exchange between an object of surface area

A_1 and temperature T_1 and its surroundings at a uniform temperature T_2 may be obtained approximately from equation [1.6]

$$q_r = A_1 E h_r (T_1 - T_2), \tag{1.6}$$

where q_r = heat transfer rate via radiation   (W)

   A_1 = surface area of object   (m^2)

   T_1 = surface temperature of object   (K)

   T_2 = temperature of surrounding surfaces   (K)

   h_r = radiative heat transfer coefficient   (W/m^2 K)

and E an emissivity factor, allows for the emissivity and geometrical relationship of both emitting and receiving surfaces on radiant heat transfer. $(T_1 - T_2)$ represents a temperature difference and may be expressed in °C.

   For unit area,

$$q_r = E h_r (T_1 - T_2) \tag{1.7}$$

The radiative heat transfer coefficient (h_r) is the net quantity of heat radiated per unit time to or from unit area of a black body divided by the difference between the mean temperature of the radiating surface and that of the surrounding surfaces with which it is exchanging radiation. h_r may also be known as the radiation conductance or the black body radiation coefficient. For a mean surface temperature of 20 °C, $h_r = 5.7$ W/m^2 K.

Example 1.7   A small object of temperature 25 °C and surface area 0.25 m^2 is situated towards the centre of a large room of average surface temperature 18 °C. Calculate the net rate of heat transfer from the object to its surroundings. Assume an emissivity factor of 0.9 and a radiative heat transfer coefficient of 5.7 W/m^2 K.

$q_r = A_1 E h_r (T_1 - T_2)$

$A_1 = 0.25$ m^2, $E = 0.9$, $h_r = 5.7$ W/m^2 K

$T_1 = 25$ °C, $T_2 = 18$ °C, $T_1 - T_2 = 7$ °C

$q_r = 0.25 \times 0.9 \times 5.7 \times 7$

   $= 8.98$ W

The net rate of heat transfer from the object to its surroundings is 8.98 W.

   This approximate calculation method gives an answer close to the exact Stefan–Boltzmann law. It is sufficiently accurate for most calculations involving the relatively small temperature differences encountered in buildings.

Surfaces and air cavities

Surface conductance (h_s) and surface resistance (R_s)

The total heat transfer rate (q) from unit area of a surface via convection and radiation may be obtained by the addition of q_c (equation [1.5]) and q_r (equation [1.7])

$$q = q_c + q_r = h_c(\theta_1 - \theta_2) + Eh_r(T_1 - T_2)$$

$\quad\quad$ convection $\quad\quad\quad\quad$ radiation

(W/m^2) $\quad\quad\quad\quad$ (W/m^2) $\quad\quad$ (W/m^2)

Both T_1 and θ_1 represent the surface temperature. T_2 is the temperature of the surrounding surfaces, which in many practical cases may be close to θ_2, the air temperature.

$\quad$ Hence *approximately*,

$$q = h_c(\theta_1 - \theta_2) + Eh_r(\theta_1 - \theta_2)$$

$$q = h_s(\theta_1 - \theta_2)$$

$\quad$ $(W/m^2\ K)\ (K)$

where $h_s = h_c + Eh_r$, is the surface conductance or the surface heat transfer coefficient.

$\quad$ *The surface conductance* (h_s) is defined as the rate of transfer of heat to or from unit area of a surface in contact with air or other fluid due to convection and radiation, per unit difference between the temperature of the surface and the temperature of the neighbouring air or other fluid.

$\quad$ *The surface resistance* (R_s) is the reciprocal of the surface conductance

$$\text{surface resistance} = \frac{1}{\text{surface conductance}}$$

The surface resistance may be calculated from

$$R_s = \frac{1}{h_c + Eh_r}$$

where $\quad R_s$ = surface resistance $\quad (m^2\ K/W)$

$\quad\quad\quad h_c$ = convective heat transfer coefficient $\quad (W/m^2\ K)$

$\quad\quad\quad h_r$ = radiative heat transfer coefficient $\quad (W/m^2\ K)$

and E is the emissivity factor which varies with the emissivity (ϵ) of the surface.

Emissivity	Surface
0.9	Ordinary building materials including glass
0.2	Unpainted and untreated metallic surfaces such as unpainted galvanised steel and dull aluminium
0.05	Polished aluminium

Table 1.3 Internal surface resistance

Building element	Heat flow	R_{si} (m² K/W)	
		High emissivity surface	Low emissivity surface
Walls	Horizontal	0.12	0.30
Ceilings or roofs, flat or pitched; floors	Upward	0.10	0.22
Ceilings and floors	Downward	0.14	0.55

(These internal surface resistance values are calculated from CIBS data; see Question 6.)

The surface resistances occurring at the inside and outside surfaces of building elements are referred to as R_{si}, the inside surface resistance or internal surface resistance, and R_{so}, the outside surface resistance or external surface resistance.

Typical values of R_{si} as shown in Table 1.3 assume that the radiation coefficient $(h_r) = 5.7 \, \text{W/m}^2 \, \text{K}$ as for a mean surface temperature of 20 °C and the air adjacent to the surface is moving only under the influence of convection currents which results in very low air speeds.

The standard outside surface resistances (R_{so}) in Table 1.4 are estimated for normal exposure to wind. Normal exposure corresponds to a wind speed of 3.0 m/s at roof surfaces and a wind speed of two-thirds the roof value at wall surfaces. For sheltered and severe

Table 1.4 Standard outside surface resistance

Building element	R_{so} (m² K/W)	
	High emissivity surface	Low emissivity surface
Wall	0.06	0.07
Roof	0.04	0.05

exposure wind speeds at roof surfaces of 1.0 m/s and 9.0 m/s respectively are used.

In the evaluation of external surface resistance, the severity of exposure is classified by location and height of building. The exposure of buildings in city centres is considered to be sheltered below the fourth floor, normal from the fourth to the eighth floor and severe at the ninth floor level and above. Most suburban and country premises experience normal exposure, but above the fifth floor the exposure is severe. The exposure of buildings on the coast or on exposed hill sites is also severe.

A change in the exposure of a well insulated structure has little effect upon its thermal resistance and the values of R_{so} for normal exposure (Table 1.4) may be used in calculations. For structural elements of low thermal resistance such as glazed areas, a variation in exposure has a noticeable effect upon the thermal resistance. The value of R_{so} appropriate to the severity of the exposure may be obtained from tables (*CIBS Guide*, Section A3) or by calculation.

Example 1.8   Calculate the external surface resistance of a roof for wind speeds of (i) 1.0 m/s and (ii) 9.0 m/s.
For air movement over a surface, the convection coefficient, $h_c = 5.8 + 4.1v$; where v = velocity of the air flow (wind speed) in m/s. Assume a high emissivity surface with an emissivity factor of 0.81, a mean surface temperature of 0 °C and a radiation coefficient, $h_r = 4.6\,\text{W/m}^2\,\text{K}$.
The external surface resistance (R_{so}) is given by

$$R_{so} = \frac{1}{h_c + Eh_r},$$

where E is the emissivity factor.

High emissivity surface ($E = 0.81$)	Sheltered $v = 1.0\,\text{m/s}$	Severe $v = 9.0\,\text{m/s}$
$h_c = 5.8 + 4.1v$	9.9	42.7
Eh_r	3.7	3.7
$h_c + Eh_r$	13.6	46.4
R_{so}	$\text{m}^2\,\text{K/W}$	$\text{m}^2\,\text{K/W}$
$\dfrac{1}{h_c + Eh_r}$	0.07	0.02

External surface resistance: (i) 0.07 m^2 K/W
                                         (ii) 0.02 m^2 K/W

Cavity and air space resistance (R_a)

Air cavities may form an integral part of the construction of floors, walls and roofs. The thermal resistance of a cavity (R_a) includes the resistance of the air space and the surface resistances of cavity boundaries. Conduction, convection and radiation contribute to the heat flow from surface to surface across the cavity.

Factors which influence convection and radiation, the important heat transfer modes within a cavity, alter the thermal resistance of the cavity. The thermal resistance of a tall vertical air space such as occurs in a typical cavity wall construction is mainly affected by (a) cavity thickness, (b) surface emissivity, and (c) ventilation rate.

(a) For unventilated cavities the thermal resistance increases with increase in cavity thickness up to a width of about 25 mm. Convection within the cavity prevents further improvement and for cavity thicknesses of 25 mm or more a constant value of resistance is used.

(b) About 60 per cent of the heat transfer in an air cavity takes place by radiation from commonly used building material surfaces with high emissivity. Low emissivity materials such as aluminium foil may be used to line the cavity walls in order to reduce this radiative heat transfer and so increase the thermal resistance of the cavity.

(c) The additional heat flow path provided by ventilation of the air space reduces the thermal resistance of the cavity.

Table 1.5 shows some standard thermal resistance values for unventilated air spaces.

Table 1.5

Thickness of air space	Surface emissivity	Thermal resistance (R_a) $(m^2 K/W)$	
		Heat flow horizontally or upwards	Heat flow downwards
5 mm	High	0.10	0.10
	Low	0.18	0.18
25 mm or more	High	0.18 (horizontally) 0.17 (upwards)	0.22
	Low	0.35	1.06

In horizontal cavities the resistance to heat flow depends upon whether the direction of heat transfer is downwards or upwards. A horizontal air space offers greater resistance to downward heat flow because downward convection is small.

Thermal transmittance (U value) (W/m² K)

The ability of a material to transfer heat by conduction is measured by its thermal conductivity (k). The thermal resistance of a layer of the material of thickness L is given by L/k. For normal heat conduction through a series of layers the total thermal resistance is the sum of the resistances of the components (equation [1.3]).

In a building construction when convective and radiative heat transfer processes occur at surfaces and in air spaces it is still convenient to calculate the total thermal resistance. The component resistances are added together to produce a total thermal resistance from which the total heat flow (air to air) may be calculated if the temperature difference is known.

Total resistance, $\Sigma R = R_{si} + R_{so} + R_a + R_1 + R_2 + \cdots$

where ΣR = sum of all resistances

R_{si} = inside surface resistance

R_{so} = outside surface resistance

R_a = air space or cavity resistance

$R_1, R_2 \ldots$ thermal resistances of construction elements

The steady heat flow rate through area A is given by

$$q = \frac{A}{\Sigma R}(t_i - t_o)$$

where t_i is the temperature of the inside environment suitably defined and t_o is the outside temperature.

The thermal transmittance or U value of any component of a structure, such as the floor or wall or roof, measures its ability to transmit heat under steady flow conditions. The thermal transmittance is calculated from the reciprocal of the total resistance (air to air) of the component.

$$U = \frac{1}{\Sigma R} = \frac{1}{R_{si} + R_{so} + R_a + R_1 + R_2 + \cdots}$$

where ΣR represents the sum of all the resistances of the component. The quantity of heat flowing through the component in unit time (q) is given by

$$q = AU(t_i - t_o) \tag{1.8}$$

heat flow rate = area × U value × temperature difference
(W)   (m²)   (W/m² K)   (K)

where q = heat flow rate   (W)
        A = area   (m²)
        U = thermal transmittance   (W/m² K)
        t_i = inside temperature   (°C)
        t_o = outside temperature   (°C)

The inside temperature (t_i) may be assumed to be equal to the inside air temperature (t_{ai}) when calculations are approximate or when the structure is lightweight and well insulated. The *CIBS Guide* recommends the use of the inside environmental temperature (t_{ei}) to allow for differences between inside surface temperatures and inside air temperature (Appendix). The closeness of the internal environmental temperature to the internal air temperature depends upon the type of heating and the nature of the surface.

The outside temperature (t_o) may be assumed to be equal to the outside air temperature (t_{ao}) in heat loss calculations under winter conditions. If heat loss over any length of time is to be estimated a value of external air temperature which is the average for the time length considered must be used. A design outside temperature (often −1 °C for the United Kingdom) is appropriate for plant size estimation.

To define thermal transmittance (U) it is convenient to rearrange equation [1.8] to obtain U,

$$U = \frac{q}{A(t_i - t_o)}$$

and substitute $A = 1\,\text{m}^2$, unit area
                 $t_i - t_o = 1\,\text{K}$, unit temperature difference

whence $U = q$

where q is the steady heat flow in unit time (1 s). A decrease in U value results in smaller heat flow and improved insulation.

The *thermal transmittance (U value)* is defined as the quantity of heat that will flow through unit area, in unit time per unit difference in temperature between inside and outside environment.

The practical importance of the U value is in the calculation of heat loss through the fabric of a building and in the comparison of the thermal insulation values of different floor, wall or roof constructions.

Since the U value depends upon the resistances of the component parts which vary with moisture content of materials, convective and radiative heat transfer at surfaces and within air spaces and other factors, standard U values may be calculated from resistances evaluated for given standard conditions. Standard U values (*BRE Digest*, 108) are required for comparing different constructions on a common basis and to meet values specified by a client or by regulations.

Example 1.9 Calculate the thermal transmittance (U value) of a cavity wall (unventilated cavity) with 105 mm brick outer leaf, 100 mm lightweight concrete block inner leaf and with 16 mm dense plaster on the inside face (Fig. 1.8). Assume the following values for thermal conductivity and resistance:

brickwork, $k = 0.84\,\text{W/m K}$

lightweight concrete block, $k = 0.19\,\text{W/m K}$

dense plaster, $k = 0.50\,\text{W/m K}$

internal surface resistance, $R_{si} = 0.12\,\text{m}^2\,\text{K/W}$

external surface resistance, $R_{so} = 0.06\,\text{m}^2\,\text{K/W}$

air space in cavity wall construction, $R_a = 0.18\,\text{m}^2\,\text{K/W}$

Since $U = 1/\Sigma R$, where ΣR represents the sum of all the resistances of the wall it is convenient to construct a table to evaluate and

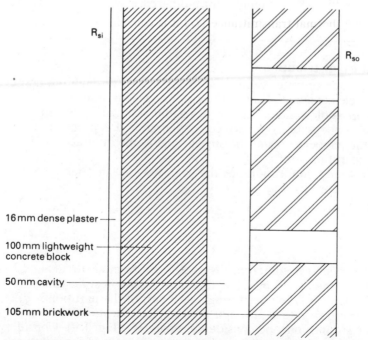

16 mm dense plaster

100 mm lightweight concrete block

50 mm cavity

105 mm brickwork

Fig. 1.8

sum these resistances. The resistance of any solid material equals L/k, where L is the thickness.

Construction element	Thickness L (m)	Conductivity k (W/m K)	Resistance L/k (m² K/W)
R_{si}			0.12
R_{so}			0.06
R_a			0.18
Brickwork	0.105	0.84	0.125
Concrete block	0.100	0.19	0.526
Dense plaster	0.016	0.50	0.032
			$\Sigma R = 1.043$

$$U = \frac{1}{1.043} = 0.96 \, \text{W/m}^2 \, \text{K}$$

Thermal transmittance (U value) of the cavity wall = $0.96 \, \text{W/m}^2 \, \text{K}$; 0.96 J is the quantity of heat flowing through unit area of the wall in 1 second when the temperature difference is 1 degree. In general, the heat flow rate through area A of the wall,

$$\begin{array}{ccccc} q & = A & \times U & \times & \text{temperature difference} \\ \text{(W)} & \text{(m}^2\text{)} & \text{(W/m}^2\text{K)} & \text{(K)} \end{array}$$

Temperature distribution

The transmission of water vapour on to surfaces or through the fabric of building elements may lead to the risk of condensation. In order to indicate possible condensation zones within a building element (interstitial condensation) or surface condensation it is necessary to calculate the temperatures at the surface and at the interfaces of a composite partition.

Under steady state conditions the heat flow rate through a composite structure is given by

$$q = AU(t_i - t_o)$$
$$= \frac{A}{R_t}(t_i - t_o)$$

where the temperature difference is established from air to air, and R_t represents the total resistance (air to air). If lateral heat flow may be neglected, the heat flow rate through each part of the partition is q, the same heat flow rate as through the complete partition.

Each resistance may be considered separately (Fig. 1.9). Consider R_{si}, the internal surface resistance, so that t_1 the surface temperature

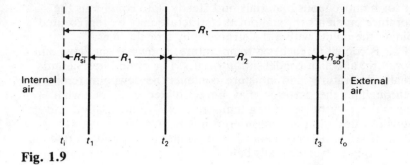

Fig. 1.9

may be obtained

$$q = \frac{A}{R_t}(t_i - t_o) = \frac{A}{R_{si}}(t_i - t_1)$$

$$\frac{t_i - t_o}{R_t} = \frac{t_i - t_1}{R_{si}}$$

$$\frac{t_i - t_1}{t_i - t_o} = \frac{R_{si}}{R_t}$$

$$\frac{\text{temperature difference (room air to wall surface)}}{\text{temperature difference (air to air)}} = \frac{\text{surface resistance}}{\text{total resistance}}$$

$$t_1 = t_i - \frac{R_{si}}{R_t}(t_i - t_o)$$

Similarly, . $$\frac{t_1 - t_2}{t_i - t_o} = \frac{R_1}{R_t}$$

$$\frac{t_2 - t_3}{t_i - t_o} = \frac{R_2}{R_t}$$

$$\frac{t_3 - t_o}{t_i - t_o} = \frac{R_{so}}{R_t}$$

Alternatively temperatures at any of the interfaces may be obtained directly. For example to find t_2 combine R_{si} and R_1,

$$\frac{t_i - t_2}{t_i - t_o} = \frac{R_{si} + R_1}{R_t}$$

$$t_2 = t_i - \frac{(R_{si} + R_1)}{R_t}(t_i - t_o)$$

The temperatures calculated at these interfaces may be plotted at the appropriate distance through the partition. Since equation [1.1] is linear for homogeneous materials and steady state conditions the temperature gradients throughout the structure may be represented by joining the points within the structure by straight lines.

If air is cooled it reaches a temperature, known as the dewpoint temperature, at which condensation occurs. For given internal and external temperatures and moisture contents the dewpoint temperature throughout the partition may be calculated in a somewhat similar manner to the preceding temperatures. (Condensation and methods of calculation are considered in Chapter 3.)

Condensation occurs whenever the temperature of any surface or position within a structure falls below the dewpoint temperature at that surface or position.

Example 1.10   A solid wall, 220 mm brickwork, is lined with 12.5 mm of insulating board. Calculate the temperature throughout the wall and indicate any condensation zone. Assume that the temperatures of the internal and external environment are 21 °C and 3 °C respectively and that,

$k = 0.84$ W/m K for brickwork; $R_{si} = 0.123$ m^2 K/W

$k = 0.05$ W/m K for insulating board; $R_{so} = 0.055$ m^2 K/W

$$U = \frac{1}{\Sigma R}, \quad t_i = 21\,°C, \quad t_o = 3\,°C$$

Construction element	Thickness (L) (m)	Conductivity (k) (W/m K)	Resistance (L/k) (m^2 K/W)
R_{si}			0.123
Insulating board	0.0125	0.05	0.250 (R_1)
Brickwork	0.220	0.84	0.260 (R_2)
R_{so}			0.055
			$\Sigma R = 0.688$

$$U = \frac{1}{0.688} = 1.45 \text{ W/m}^2\text{ K}$$

For unit area ($A = 1$ m^2),
$q = 1 \times U \times$ temperature difference
$\quad = 1 \times 1.45 \times (21 - 3) = 26.1$ W

Neglecting lateral heat flow, the heat flow rate through any part of the wall is q, the same heat flow rate as through the complete wall.

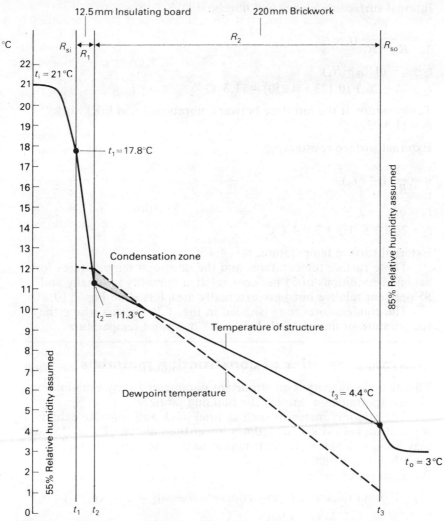

Fig. 1.10 Section showing temperature of structure and dewpoint temperature through wall. (See Chapter 3 for estimation of dewpoint temperature.)

Internal surface resistance (Fig. 1.10)

$$q = \frac{1}{R_{si}}(t_i - t_1) \quad (A = 1\,\text{m}^2)$$

$$t_1 = t_i - qR_{si}$$
$$t_1 = 21 - 26.1 \times 0.123 = 17.8\,°\text{C}$$

Internal surface temperature, $t_1 = 17.8\,°\text{C}$

Internal surface resistance and fibreboard

$$q = \frac{1}{R_{si} + R_1} (t_i - t_2)$$

$$t_2 = t_i - q(R_{si} + R_1)$$
$$t_2 = 21 - 26.1 (0.123 + 0.250) = 11.3\,°C$$

Temperature at the interface between fibreboard and brickwork, $t_2 = 11.3\,°C$

External surface resistance

$$q = \frac{1}{R_{so}}(t_3 - t_o)$$

$$t_3 = qR_{so} + t_o$$
$$t_3 = 26.1 \times 0.055 + 3 = 4.4\,°C$$

External surface temperature, $t_3 = 4.4\,°C$

These surface temperatures and the dewpoint temperatures for assumed conditions of 55 per cent relative humidity internally and 85 per cent relative humidity externally are plotted on Fig. 1.10.

The condensation zone (shaded in Fig. 1.10) occurs where the temperature of the wall falls below the dewpoint temperature.

Thermal properties of construction materials

Thermal conductivity is an important parameter in any examination of heat transmission through the building fabric.

The masonry materials such as brickwork and concrete exhibit thermal conductivities many times lower than metals (Table 1.6) and may provide a moderate contribution to the thermal resistance of most building structures.

For example consider:

(1) 100 mm thickness of cast concrete (density = 2100 kg/m³)

$$\text{Resistance} = \frac{L}{k} = \frac{0.100}{1.4} = 0.07\,m^2\,K/W$$

(2) 100 mm thickness of lightweight concrete block
(density = 600 kg/m³)

$$\text{Resistance} = \frac{L}{k} = \frac{0.100}{0.19} = 0.53\,m^2\,K/W$$

The highest thermal conductivity and therefore minimum resistance is given by the dense cast concrete. The lowest thermal conductivity of the lightweight concrete blocks produces the maximum thermal resistance.

Table 1.6 Typical values for thermal properties at 20 °C

Material	Density (ρ) (kg/m³)	Conductivity (k) (W/m K)	Specific heat capacity (c) (J/kg K)
Aluminium	2790	164	883
Copper	8960	386	389
Steel (1% carbon)	7800	43	473
Marble	2600	2.8	808
Brick (outer leaf)	1700	0.84	800
(inner leaf)	1700	0.62	800
Cast concrete	2100	1.4	897
Lightweight concrete block	600	0.19	1000
Glass	2500	1.05	
Timber	650	0.14	
Water	998 (1000)	0.603	4182 (4200)
Glass wool (lightweight mats, quilts)	25	0.04	
Woodwool slabs	450	0.09	
Expanded polystyrene board	25	0.033	
Air	1.203	0.025	1004
Plaster (dense)	1300	0.5	
(lightweight)	600	0.16	
Plasterboard (gypsum)	950	0.16	
(perlite)	800	0.18	
Fibre insulating board	260	0.050	
Corkboard	160	0.043	

Window glass has a thermal conductivity and density of the same order as brickwork and dense concrete. However the thickness of glass in a typical window contributes negligibly to the thermal resistance of the window. The resistance of the window glass is calculated from the two surface resistances R_{si} and R_{so}, 0.18 m² K/W for normal exposure, so that the thermal transmittance $U = 5.6$ W/m² K. In double glazing an air cavity of 25 mm may improve the total thermal resistance to 0.36 m² K/W, a U value of 2.8 W/m² K. Window frames may be metal or timber. Since metal has a high thermal conductivity (low resistivity), the U value of the metal-framed window may be greater than that of the glazing. The thermal conductivity of glass may be seven or eight times that of timber. A timber frame increases the thermal resistance of the window by an amount dependent upon the proportion of window occupied by the frame. The thermal transmittance is reduced.

The next group of materials listed in Table 1.6 provide thermal insulation. The low thermal conductivities and low densities of these high void materials are associated with high thermal resistances. The ultimate minimum values for thermal conductivity and density would be those of still air in most situations. Because of their mechanical properties and finish these materials will usually be found in cavities and roof spaces or in some way protected by external cladding or internal linings.

The final group of materials in Table 1.6 is normally used for internal finishes and lining boards. Notice that the dense plaster has a thermal conductivity approximately three times that of lightweight plaster or plasterboard, and therefore provides only about 30 per cent of the thermal resistance for the same thickness of material. The low conductivities of the corkboard and fibre insulating board compare favourably with the values of insulating materials. These lightweight lining materials may make a useful contribution to the thermal resistance of a wall and hence a reduction in its thermal transmittance.

In building construction metals may be used for example as metal framing where the high thermal conductivity may lead to thermal bridges unless care is taken to increase the thermal resistance in the metal areas with additional insulation.

The thermal behaviour of a building is also much influenced by its thermal or heat storing capacity. The amount of heat stored in an element depends largely on the mass of the material and its specific heat capacity. The more dense materials can absorb more heat and therefore will take longer to heat up or to cool. Lightweight linings in massive buildings or lightweight structures require less heat and less time to warm or cool. Lightweight structures or linings are considered by some to be especially advantageous when intermittent heating is used in the building.

Effect of moisture content

The presence of moisture is of importance in the consideration of heat transfer through building materials. Many of these, such as brick, concrete and insulation material, are porous.

The thermal conductivity of water is many times greater than that of air (Table 1.6). The replacement of some or all of the air within the pores by water and the diffusion of water vapour across the building material produce an increase in thermal conductivity.

Thermal insulating material in a building must be kept dry if its low thermal conductivity value is to be maintained. If the material becomes wet, whether by rain penetration from the outside or by condensation from within, the value of the insulation is very much reduced.

Table 1.7 shows the effect of different moisture content upon the average thermal conductivity values of brickwork and concrete. External walls with a moisture content of 5 per cent and internal walls with 1 per cent moisture content for brick and 3 per cent moisture content for concrete are taken as standard for the calculation of the standard U value.

Table 1.7 Typical thermal conductivity values of brickwork and concrete at standard moisture content

	Standard moisture content % by volume		Bulk dry density (ρ) (kg/m^3)	Thermal conductivity (k) $(W/m\ K)$
Brickwork (Fletton average density bricks)	Protected from rain	1%	1500–1800	0.48–0.71
	Exposed to rain	5%		0.65–0.96
Concrete	Protected from rain	3%	1400–1700	0.51–0.77
	Exposed to rain	5%		0.57–0.85

Moisture factors (BRE Current Paper 1/70) may be used to estimate the thermal conductivity of masonry materials for other moisture contents within the range 1% to 20% (% by volume).

Thermal bridges

A component of relatively high thermal conductivity may extend partly or completely through the thickness of a building element. The higher thermal conductivity component provides an easier path for heat flow and is referred to as a thermal bridge or a cold bridge. Lower temperatures occur in the region of a thermal bridge because of the increased heat flow through it. Examples of thermal bridges include lintels over openings, ties for securing concrete and brickwork through insulation material, curtain wall construction in which panels with good insulating properties are set in metal framing and insulation material bridged by timber battens in framed construction.

Lower temperatures may lead to increased condensation risk (both surface and interstitial) and pattern staining. These effects may be lessened by the application of local insulation to reduce the heat flow through the thermal bridge.

Thermal insulation of construction materials

Thermal insulation is designed to obstruct the flow of heat between an enclosure and its surroundings.

While the performance of a high void insulating material is characterised by a low thermal conductivity value, a reflective insulation material such as a aluminium foil depends for its effectiveness upon the low emissivity and absorptivity of its surface.

High void insulation

The low thermal conductivity of a high void insulating material, usually of a porous or fibrous nature, is due to the large number of air filled voids or interstices within its structure. Air is a poor conductor.

The basic types of high void thermal insulation are as follows:

(a) Flake, small particles or flakes which finely divide the air space, such as vermiculite and expanded mica.
(b) Fibrous, slender fibres which finely divide the air space, for example, glass fibre and rock fibre.
(c) Cellular, small individual cells sealed from each other as in the closed-cell structure of expanded PVC. This term is also applied to the open-cell structure of some foamed plastics in which the cells are mostly interconnected. (*BRE Digest*, 224).
(d) Granular, small nodules which contain voids or hollow spaces such as cork. It is not considered a true cellular material since air can be transferred between the individual spaces.

The heat flow rate (q) through a solid for a temperature difference ($\theta_1 - \theta_2$) is proportional to the cross-sectional area (A) and inversely proportional to the thickness or conduction path through the solid (L).

$$q = \frac{kA(\theta_1 - \theta_2)}{L} \qquad [1.1]$$

where k, the thermal conductivity of the solid, should be as low as possible to provide thermal insulation.

In a high void insulating material heat transfer by solid conduction is reduced by the introduction of many voids or interstices containing air or some other gas, so that the cross-sectional area of solid available for conduction is much reduced and the solid conduction path is replaced by a longer tortuous one, either a continuous solid path around the pores or cells in a cellular material or an even more difficult solid path in the case of a fibrous material. The contribution to the overall thermal conductivity from solid conduction is very small.

Heat may also be transferred across a void by conduction, convection and radiation, but if the void remains small enough conduction is the chief heat transfer mode. The thermal insulation characteristics of a solid improve as the proportion of air in it is increased. A reduction in the bulk density of the majority of these materials lowers the thermal conductivity. For some materials an optimum density is reached below which k increases, because the air voids have become sufficiently large to allow an increase in heat transfer by convection and radiation.

For a given value of bulk density the thermal conductivity tends to decrease if the particle size or fibre size is reduced. Reduction in particle or fibre size leads to smaller air-filled voids, so that convective and radiative heat transfer is negligible and the thermal conductivity of the material approaches closer to that of still air.

If the fibres in a fibrous material are arranged in layers the heat flow is anisotropic. Resistance to heat flow perpendicular to the plane of the layers is greatest.

High void insulating materials are supplied in the following forms:

(a) **Loose fill**, mineral fibre (glass or rock) as pellets, granulated cork, polystyrene beads, exfoliated vermiculite.
(b) **Quilt, blanket, batt, mat or felt** of mineral fibre.
(c) **Rigid board or slab**, expanded polystyrene board, expanded PVC board, corkboard, concrete block, insulating board, wood wool slab.
(d) **Formed *in situ***, urea formaldehyde or polyurethane foamed *in situ*.

Reflective insulation

Reflective insulating materials reduce the radiative heat transfer from surfaces. The application of metallic foil to a surface reduces the emissivity of about 0.9 which occurs with most construction materials to the 0.05 of highly polished aluminium.

The surface resistance, $R_s = \dfrac{1}{h_c + Eh_r}$

where

h_c = convection coefficient

h_r = radiation coefficient

E = emissivity factor for normal temperature radiation.

At internal surfaces and wherever convection is small (h_c small), the surface resistance is much increased by reducing Eh_r from a typical value of $0.8 \times 5.7 = 4.6\,\text{W/m}^2\,\text{K}$ for a high emissivity factor to a value of $0.05 \times 5.7 = 0.285\,\text{W/m}^2\,\text{K}$ for low emissivity. (Examination of the values of h_c in Example 1.8 shows that when convection (h_c) is large, changes in Eh_r make little difference to R_s.)

A bright metallic foil on either the warm side or the cold side of

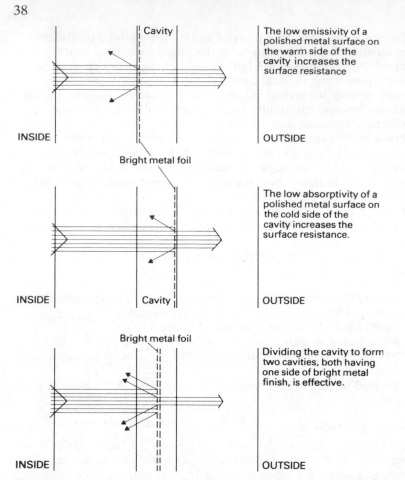

The low emissivity of a polished metal surface on the warm side of the cavity increases the surface resistance

The low absorptivity of a polished metal surface on the cold side of the cavity increases the surface resistance.

Dividing the cavity to form two cavities, both having one side of bright metal finish, is effective.

Fig. 1.11 Application of reflective insulation material in lining a cavity

an unventilated cavity increases the surface resistance of the cavity and hence reduces the thermal transmittance of the structure. Lining both sides of a single cavity is not appreciably better than lining only one side, but dividing the cavity to form two separate cavities, both having one side of bright metallic foil is effective (Fig. 1.11).

A film of water from condensation or a layer of dust on the metal surface will increase the emissivity and absorptivity and reduce the effectiveness of reflective insulation material.

Thermal insulation of structural components

In a typical dwelling in winter heat loss is usually greatest through the walls and roof. Windows, ground floors and ventilation make further contributions to this loss. Insulation may be applied to reduce this heat loss and to improve condensation and mould problems.

In existing housing wall construction depends upon the age of the house (Table 1.8).

Table 1.8 Typical thermal transmittance values

Construction	U value for normal exposure $(W/m^2 K)$
(i) Solid brick wall (pre-1930)	1.9–2.1
(ii) Cavity wall, 50 mm cavity between two leaves of brick	1.3–1.5
(iii) Cavity wall, inner leaf of lightweight block (1970s)	0.96

The *Elemental Approach* to the insulation of the building fabric (Building Regulations, 1985, Part L) allowed a maximum U value of 0.6 W/m²K for external walls in new dwellings. In view of the need for conservation of fuel and power this maximum permissible U value was reduced to 0.45 W/m²K for external walls in dwellings and other buildings (1990 Edition, Part L).

Nearly half the existing houses have solid masonry walls, and while heavy walls moderate the extremes of climate, they have the disadvantages of high thermal transmittance, rain penetration and because of low surface temperatures, surface condensation.

Solid brick wall

Insulation may be applied either to the internal surface or the external surface of a solid wall. Internally applied insulating wall linings are usually lightweight with a low thermal capacity and so may be heated up more quickly, an advantage especially with intermittent heating. The incidence of surface condensation will be reduced because of the higher surface temperatures, but the risk of interstitial condensation may be increased because the brick wall remains colder. All internal insulation methods should incorporate a vapour barrier at the warm side of the structure.

Externally applied insulation must be accompanied by an external render which keeps the insulating material dry, is durable, capable of resisting impact and has a surface finish with an acceptable appearance. External insulation keeps the whole structure warmer so lessening the interstitial condensation risk. An improvement in surface temperatures occurs because of the reduced U value. External insulation of a solid wall may be considered worth while when existing damp problems are remedied at the same time as thermal improvement.

Examples of thermal linings include the following:

1. Internally: (a) Aluminium foil backed plasterboard fixed to treated timber battens.

(b) Plasterboard fixed to treated timber battens with high void insulation filling the cavity (fibreglass, rock wool).

(c) A composite board of plasterboard laminated to an insulant and usually fixed to the wall with an adhesive.

2. Externally: (a) Sheet of water-repellent insulant (fibreglass) laminated to a mechanical fixing device and waterproof render.

The following examples illustrate the improvement in thermal transmittance which may be obtained by various insulation treatments, the thermal transmittance (*U* value) is improved when its value is reduced.

Example 1.11 A 220 mm solid brick wall ($k = 0.84$ W/m K) is plastered internally with 16 mm of dense plaster ($k = 0.5$ W/m K). Assume normal exposure so that $R_{si} = 0.12$ m^2 K/W and $R_{so} = 0.06$ m^2 K/W. Calculate the thermal transmittance of this solid brick wall. The sum of internal and external surface resistance $R_{si} + R_{so}$ will be written as 0.18 m^2 K/W.

Basic solid wall

Element	Thickness (L) (m)	Conductivity (k) (W/m K)	Resistance (L/k) (m^2 K/W)
$R_{si} + R_{so}$			0.18
Dense plaster	0.016	0.5	0.032
Brickwork	0.220	0.84	0.26
			$\Sigma R = 0.472$

Total resistance $\Sigma R = 0.472$ m^2 K/W

$$U = \frac{1}{\Sigma R} = \frac{1}{0.472} = 2.1 \text{ W/m}^2 \text{ K}$$

The thermal transmittance of the basic solid wall = 2.1 W/m^2 K.

Example 1.12 An internal lining of highly polished aluminium foil backed plasterboard fixed to treated timber battens is applied to the basic solid wall of Example 1.11. The resistance of the 25 mm cavity formed with the plastered wall (R_a) is 0.30 m^2 K/W. The plasterboard is 12.7 mm thick with a thermal conductivity of

0.16 W/m K. Calculate the thermal transmittance of the wall (Fig. 1.12).

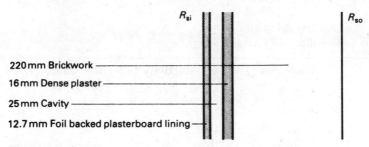

Fig. 1.12

Basic solid wall + internal lining

Element	Thickness (L) (m)	Conductivity (k) (W/m K)	Resistance (L/k) (m² K/W)
$R_{si} + R_{so}$			0.18
Plasterboard	0.0127	0.16	0.08
R_a			0.30
Plaster	0.016	0.5	0.03
Brickwork	0.220	0.84	0.26
		$\Sigma R =$	0.85

$$U \text{ value} = \frac{1}{\text{total resistance}} = \frac{1}{0.85} = 1.2 \text{ W/m}^2\text{K}$$

Thermal transmittance of solid wall with foil backed plasterboard lining = 1.2 W/m² K.

Example 1.13 A composite board of plasterboard laminated to an insulant (expanded polystyrene) is attached to the basic solid wall of Example 1.11 with adhesive and partial mechanical fixing (Fig. 1.13).

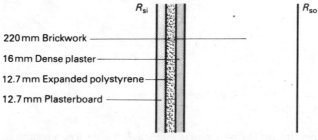

Fig. 1.13

Plasterboard, thickness = 12.7 mm, $k = 0.16$ W/m K.
Expanded polystyrene, thickness = 12.7 mm, $k = 0.037$ W/m K.

Calculate the improved U value.
Basic solid wall + internal lining

Element	Thickness (L) (m)	Conductivity (k) (W/m K)	Resistance (L/k) (m² K/W)
$R_{si} + R_{so}$			0.18
Plasterboard	0.0127	0.16	0.08
Insulation (expanded polystyrene)	0.0127	0.037	0.34
Plaster	0.016	0.5	0.03
Brickwork	0.220	0.84	0.26
			$\Sigma R = 0.89$

$$U \text{ value} = \frac{1}{\text{total resistance}} = \frac{1}{0.89} = 1.1 \text{ W/m}^2 \text{ K}$$

Thermal transmittance of solid wall with composite board lining = 1.1 W/m² K.

In Examples 1.12 and 1.13 the thermal resistance has been completely recalculated in order to obtain the U value. The U value may be obtained by correcting the total resistance of the basic solid wall for any additions to (or subtraction from) the basic structure.
Consider Example 1.12 again.
Basic solid wall + internal lining

Element	Thickness (L) (m)	Conductivity (k) (W/m K)	$+/-$ Resistance (L/k) (m² K/W)
Basic wall			+ 0.47
Plasterboard			+ 0.08
Foil lined cavity (R_a)			+0.3
			$\Sigma R = 0.85$

$$U = \frac{1}{\text{total resistance}} = \frac{1}{0.85} = 1.2 \text{ W/m}^2 \text{ K as previously.}$$

Neither of these internal linings produces a U value low enough to satisfy current regulations.

Example 1.14 Using data from Example 1.13 calculate what thickness of polystyrene insulation is required to improve the thermal transmittance to (i) $1.0\,\text{W/m}^2\,\text{K}$ (ii) $0.6\,\text{W/m}^2\,\text{K}$.

(i)　A U value of $1.0\,\text{W/m}^2\,\text{K}$ corresponds to a total thermal resistance $R_T = 1/U = 1.0\,\text{m}^2\,\text{K/W}$.

If the required amount of insulation resistance $= R_I$, $R_I = L/k$, where L = insulation thickness and k = thermal conductivity.

Basic wall + plasterboard + insulation

Element	Thickness (L) (m)	Conductivity (k) (W/m K)	+/− Resistance (L/k) (m² K/W)
Basic wall			+ 0.47
Plasterboard	0.0127	0.16	+ 0.08
Polystyrene	L	0.037	+ R_I
			R_T

$R_T \quad = 0.47 + 0.08 + R_I = 1.0\,\text{m}^2\,\text{K/W}$

$\therefore \quad R_I = 1.0 - 0.47 - 0.08 = 0.45\,\text{m}^2\,\text{K/W}$

$L = kR_I = 0.037 \times 0.45 = 0.0166\,\text{m}$

$L = 16.6\,\text{mm}$

16.6 mm of polystyrene insulation is required to improve the thermal transmittance of the solid brick wall to $1.0\,\text{W/m}^2\,\text{K}$.

(ii)　A U value of $0.6\,\text{W/m}^2\,\text{K}$ corresponds to a total thermal resistance $R_T = 1/U = 1.67\,\text{m}^2\,\text{K/W}$

$R_T \quad = 0.47 + 0.08 + R_I = 1.67\,\text{m}^2\,\text{K/W}$

$\therefore \quad R_I = 1.67 - 0.47 - 0.08 = 1.12\,\text{m}^2\,\text{K/W}$

$L = kR_I = 0.037 \times 1.12 = 0.041\,\text{m}$

$L = 41\,\text{mm}$

41 mm of polystyrene insulation is required to improve the thermal transmittance of the solid brick wall to $0.6\,\text{W/m}^2\,\text{K}$. This thickness of insulation provides much thermal improvement but internal fixing problems with radiators, sinks, skirting boards, electric sockets and detailing around windows and doors will require careful attention, and the solid wall outside this lining will remain colder.

Example 1.15 A solid brick wall ($U = 2.1\,\text{W/m}^2\,\text{K}$) is externally insulated and rendered (Fig. 1.14). Calculate the new thermal transmittance if the combined thermal resistance of the exterior wall insulation and render is $0.82\,\text{m}^2\,\text{K/W}$.

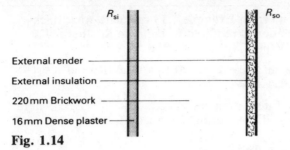

R_{si} R_{so}

External render
External insulation
220 mm Brickwork
16 mm Dense plaster

Fig. 1.14

Basic solid wall + external insulation and render

Element	Thermal resistance (m² K/W)
Solid wall	0.47
Render and insulation	0.82
	$\Sigma R = 1.29$

$$U = \frac{1}{\text{total resistance}} = \frac{1}{1.29} = 0.78 \, \text{W/m}^2\,\text{K}$$

Thermal transmittance of externally insulated and rendered solid wall = $0.78 \, \text{W/m}^2\,\text{K}$.

The U value measures the heat transmission rate through unit area of a structure for unit temperature difference. The U values for the internally improved solid walls considered are listed in Table 1.9 together with the percentage reduction in heat transmission rate through these walls compared to the basic solid wall under the same environmental conditions.

Assume a temperature difference of 1 K
Heat flow rate through unit area of the basic solid wall = 2.1 W
(= 100 per cent)
Heat flow rate through unit area of the improved wall = U value
Reduction in heat flow rate = $2.1 - U$

Percentage reduction in heat transmission rate = $\dfrac{(2.1 - U)}{2.1} \times 100$

Cavity wall

The chief purpose of the cavity in a cavity wall is to prevent damp passing from the exposed external leaf of brickwork into the house. The cavity does provide some additional thermal resistance and is also another method of wall thermal improvement. The U value of a cavity wall may be reduced by insulation applied

(a) internally, (b) externally, (c) within the cavity.

Table 1.9 Percentage reduction in heat transmission through a wall compared to the basic solid wall for the same environmental conditions

Wall construction	U value $(W/m^2 K)$	Reduction in heat transmission (%)
Basic solid	2.1	—
Basic solid + foil backed cavity	1.2	43
Basic solid + 12.7 mm insulation	1.1	48
Basic solid + 16.6 mm insulation	1.0	52
Basic solid + external insulation and render	0.78	63
Basic solid + 41 mm insulation	0.6	71

In many existing houses complete cavity fill is the most convenient method of thermally upgrading the walls. However filling a cavity is a contravention of the Building Regulations (in England and Wales) dealing with the prevention of damp. (The local authority should be advised of details before any work is carried out.) In situations where walls are particularly exposed to driving rain, cavity fill may lead to rain penetration problems (*BRE Digest*, 236).

The cavity is completely filled with the appropriate insulant which is injected into the cavity through holes drilled in the outside brickwork. Insulation commonly used with the standard existing cavity wall is one of the following:

(a) urea formaldehyde foam* (formed *in situ*, hardens to bubble-filled plastic);
(b) blown mineral fibre;
(c) expanded polystyrene beads.

The outer leaf of the wall will remain colder and wetter with cavity fill. The outer leaf should be checked for susceptibility to frost damage in this situation and also when the wall (solid or cavity) is internally insulated.

Example 1.16 A cavity wall is constructed with 105 mm brickwork, 16 mm lightweight plaster and a 50 mm cavity (Fig. 1.15). The sum

---

* Concern exists about the extent to which formaldehyde vapour does penetrate into the interior of buildings insulated with urea formaldehyde.

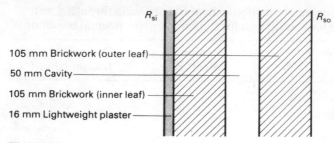

105 mm Brickwork (outer leaf)
50 mm Cavity
105 mm Brickwork (inner leaf)
16 mm Lightweight plaster

Fig. 1.15

of the surface resistances $R_{si} + R_{so} = 0.18\,\text{m}^2\,\text{K/W}$ and the cavity resistance $R_a = 0.18\,\text{m}^2\,\text{K/W}$.

(a) Calculate the thermal transmittance of:
  (i)   the cavity wall;
  (ii) the cavity wall with urea formaldehyde completely filling the cavity.
(b) Calculate the percentage reduction in heat transmission obtained by introducing the cavity fill.

Thermal conductivity values are:
Brickwork (external leaf, 5 per cent moisture content)
$k = 0.84\,\text{W/m K}$
Brickwork (internal leaf, 1 per cent moisture content)
$k = 0.62\,\text{W/m K}$
Lightweight plaster $k = 0.16\,\text{W/m K}$
Urea formaldehyde foam $k = 0.04\,\text{W/m K}$

(a)   (i)   Cavity wall

Element	Thickness (L) (m)	Conductivity (k) (W/m K)	Resistance (L/k) (m^2 K/W)
$R_{si} + R_{so}$			0.18
Lightweight plaster	0.016	0.16	0.1
Brickwork (inner leaf)	0.105	0.62	0.17
Cavity (R_a)	0.050		0.18
Brickwork (outer leaf)	0.105	0.84	0.125
			$\Sigma R = 0.755$

$$U = \frac{1}{\text{total resistance}} = \frac{1}{0.755} = 1.3\,\text{W/m}^2\,\text{K}$$

The thermal transmittance of the cavity wall = $1.3\,\text{W/m}^2\,\text{K}$.

(ii) The cavity resistance (R_a) is replaced by the foam resistance.

Cavity wall + urea formaldehyde fill

Element	Thickness (L) (m)	Conductivity (k) (W/m K)	+/− Resistance (L/k) (m² K/W)
Cavity wall			0.755
Urea formaldehyde	0.050	0.04	+ 1.25
Cavity (R_a)	0.050		− 0.18
			$\Sigma R = 1.83$

$$U = \frac{1}{\text{total resistance}} = \frac{1}{1.83} = 0.55 \, \text{W/m}^2 \, \text{K}$$

The thermal transmittance of the foam filled cavity wall = 0.55 W/m² K.

(b) Assume unit temperature difference
Heat flow rate through unit area of cavity wall = 1.3 W.
Heat flow rate through unit area of cavity filled wall = 0.55 W.
Reduction in heat flow rate obtained by cavity fill = 1.3 − 0.55 = 0.75 W.
Percentage reduction in heat transmission obtained by introducing the cavity fill

$$= \frac{0.75}{1.3} \times 100 = 58 \text{ per cent.}$$

Percentage reduction in heat transmission = 58 per cent.

In the more recently built house the inner leaf of brickwork is replaced by lightweight concrete block ($L = 100$ mm, $k = 0.19$ W/mK) to produce a U value of 0.96 W/m² K, when the inner face is plastered with dense plaster ($L = 16$ mm, $k = 0.5$ W/mK). This structure may be thermally improved by any of the previous methods.

In new construction the cavity may be insulated by building in water-repellent batts as building takes place (glass fibre or rock fibre water-repellent batts).

Partial cavity fill (in new construction) may be considered when a cavity is required in order to prevent damp penetration. The insulation in rigid sheets is attached within the cavity to the inner leaf. A cavity (preferably at least 40 mm) is maintained between the outer face of the insulation boards (fibreglass or expanded PVC boards) and the outer leaf of brickwork.

48

Example 1.17 A wall with partial cavity fill (Fig. 1.16) has the following elements:

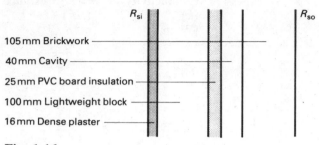

Fig. 1.16

16 mm plaster, $k = 0.5$ W/m K
Inner leaf of 100 mm lightweight concrete block, $k = 0.19$ W/m K
25 mm PVC board insulation, $k = 0.035$ W/m K
40 mm cavity with resistance $R_a = 0.18$ m² K/W
Outer leaf of 105 mm brickwork, $k = 0.84$ W/m K
Sum of internal and external surface resistances
$R_{si} + R_{so} = 0.18$ m² K/W.

What is the U value of the wall?

Wall with partial cavity fill

Element	Thickness (L) (m)	Conductivity (k) (W/m K)	Resistance (L/k) (m² K/W)
Surface resistances ($R_{si} + R_{so}$)			0.18
Plaster	0.016	0.5	0.032
Concrete block	0.100	0.19	0.526
PVC board	0.025	0.035	0.714
Cavity (R_a)			0.18
Brickwork	0.105	0.84	0.125
			$\Sigma R = 1.757$

$$U \text{ value} = \frac{1}{\text{total resistance}} = \frac{1}{1.757} = 0.57 \text{ W/m}^2 \text{ K}$$

The U value of the wall $= 0.57$ W/m² K

Roof

A traditional timber framed pitched roof with tiles has a U value of about $1.5\,\text{W/m}^2\,\text{K}$. Insulation of the roof space should be one of the first tasks undertaken in any thermal improvement programme. Current Building Regulations (1990, Part L) allow a maximum U value of $0.25\,\text{W/m}^2\text{K}$ for house roofs. The following are typical insulation methods available.

(a) Aluminium foil may be used as a lining with low emissivity, but this low emissivity increases once dust obscures the polished surface.

(b) Loose fill insulation materials such as vermiculite can be spread straight from the sack and levelled to the depth required.

(c) Glass fibre blanket may be obtained in rolls in a suitable thickness. The glass fibre is placed between the joists and is not easily disturbed by draughts. It has good insulating properties and is perhaps the most commonly used method of roof insulation.

Example 1.18   The U value of an uninsulated traditional pitched roof is $1.5\,\text{W/m}^2\,\text{K}$. What thickness of fibreglass blanket ($k = 0.04\,\text{W/m K}$) is required to improve this U value to (i) $0.6\,\text{W/m}^2\,\text{K}$ (ii) $0.4\,\text{W/m}^2\,\text{K}$?

$$\text{Thermal resistance of the uninsulated roof} = \frac{1}{1.5}$$

$$= 0.667\,\text{m}^2\,\text{K/W}$$

(i) Thermal resistance equivalent to a U value of $0.6\,\text{W/m}^2\,\text{K}$ is $1/0.6 = 1.667\,\text{m}^2\,\text{K/W}$
Therefore the resistance to be provided by the insulation $= 1.667 - 0.667 = 1.0\,\text{m}^2\,\text{K/W}$

$$\text{Resistance of insulation} = \frac{\text{thickness}}{k}$$

Thickness of fibreglass insulation

$$= k \times \text{resistance of insulation}$$

$$= 0.04 \times 1.0 = 0.04\,\text{m}$$

$$= 40\,\text{mm}$$

(ii) Thermal resistance equivalent to a U value of $0.4\,\text{W/m}^2\,\text{K}$ is $1/0.4 = 2.5\,\text{m}^2\,\text{K/W}$
Therefore the resistance to be provided by the insulation $= 2.5 - 0.667 = 1.833\,\text{m}^2\,\text{K/W}$

$$\text{Resistance of insulation} = \frac{\text{thickness}}{k}$$

Thickness of fibreglass insulation

$$= k \times \text{resistance of insulation}$$
$$= 0.04 \times 1.833 = 0.073 \, \text{m}$$
$$= 73 \, \text{mm}$$

40 mm of fibreglass blanket is required to improve the U value to $0.6 \, \text{W/m}^2 \, \text{K}$
73 mm of fibreglass blanket is required to improve the U value to $0.4 \, \text{W/m}^2 \, \text{K}$

Frequently the fibreglass blanket will be available in standard thicknesses. Thicknesses of 50 mm and 100 mm will improve the U value of this roof to $0.52 \, \text{W/m}^2 \, \text{K}$ and $0.32 \, \text{W/m}^2 \, \text{K}$ respectively.

	Insulation resistance (L/k) $(\text{m}^2 \, \text{K/W})$	Total resistance $(\text{m}^2 \, \text{K/W})$	U value $(\text{W/m}^2 \, \text{K})$	Reduction in heat flow rate
Uninsulated roof		0.667	1.5	
Roof + 50 mm fibreglass	$\dfrac{0.050}{0.04} = 1.25$	1.917	0.52	65%
Roof + 100 fibreglass	$\dfrac{0.100}{0.04} = 2.5$	3.167	0.32	78%

Low U values mean a colder roof space with an increased risk of condensation and its possible effect on the roof timber. Care should be taken to ensure that ventilation gaps are kept clear of insulation material and that adequate loft ventilation is provided. Materials used within the roof space should not be flammable.

Floors

The thermal transmittance of a ground floor is dependent upon the floor size (*BRE Digest*, 145). Some values are given in Table 1.10. Edge insulation at the construction stage may reduce the U values shown in Table 1.10 by about 25 per cent if the insulation extends to a depth of 1 m.

If a floor is provided with overall insulation the thermal resistance of the insulating layer should be added to that of the floor slab so that the new U value may be calculated.

Table 1.10 Thermal transmittance of floors

Solid floor in contact with the earth with four exposed edges		Suspended timber floor directly above the ground	
Size (m)	U value (W/m^2 K)	Size (m)	U value (W/m^2 K)
15 × 7.5	0.62	15 × 7.5	0.61
7.5 × 7.5	0.76	7.5 × 7.5	0.68

Windows

The thermal transmittance values of 5.6 W/m^2 K for single glazing and 2.8 W/m^2 K for double glazing are for the glass only and do not account for the effect of the surrounding frame and glazing bars. These values indicate the 50 per cent reduction in heat transmission through the window which is achieved by double glazing. This improved insulation leads to an increase in the surface temperature of the glass facing the room. Comfort may be improved and the risk of surface condensation reduced.

Example 1.19 Calculate the thermal transmittance of single glazing and double glazing (Fig. 1.17).

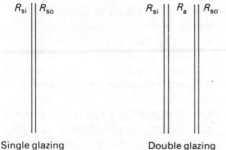

Single glazing   Double glazing

Fig. 1.17

The thermal resistance of perhaps 4 mm of glass is negligible.

Resistance of glass $= \dfrac{L}{k} = \dfrac{0.004}{1.05}$, where $k = 1.05$ W/m K for glass.

Calculations are based on the surface resistances (R_{si} and R_{so}) and the insulating effect of introducing an air cavity (R_a) for double glazing.

Assume: $R_{si} = 0.12$ m^2 K/W, $R_{so} = 0.06$ m^2 K/W for normal exposure, $R_a = 0.18$ m^2 K/W for an air cavity at least 25 mm wide.

Thermal resistance of single glazing $= R_{si} + R_{so}$

$$= 0.12 + 0.06$$
$$= 0.18 \, \text{m}^2 \, \text{K/W}$$

Thermal transmittance of single glazing $= \dfrac{1}{\text{total resistance}}$

$$= \frac{1}{0.18} = 5.6 \, \text{W/m}^2 \, \text{K}$$

Thermal resistance of double glazing $= R_{si} + R_a + R_{so}$

$$= 0.12 + 0.18 + 0.06$$
$$= 0.36 \, \text{m}^2 \, \text{K/W}$$

Thermal transmittance of double glazing $= \dfrac{1}{\text{total resistance}}$

$$= \frac{1}{0.36} = 2.8 \, \text{W/m}^2 \, \text{K}$$

For single glazing $U = 5.6 \, \text{W/m}^2 \, \text{K}$
For double glazing $U = 2.8 \, \text{W/m}^2 \, \text{K}$ (25 mm air space)

Many insulating materials and methods of application are available. Correctly applied insulation will reduce heat transmission and may ameliorate condensation problems. Energy will be saved but some of the benefits of insulation may be taken in improved comfort. The choice of which insulation material to use cannot be based entirely upon its thermal conductivity value. Resistance to rot, vermin, damp, fire hazards, manufacturers' advice on fixing and durability, mechanical strength, influence of dust or air movement on its efficacy, thermal expansion, ventilation requirements and cost are some of the factors to be considered.

The insulation examples described illustrate the reduction of heat transmission from the interior of an enclosure to the outside in a cool climate. Insulation methods are also available to protect enclosures from overheating in hot climates.

Heat loss

Method of heat loss

Buildings lose heat by transmission through the external fabric (floor, walls, roof) and by ventilation.

Heat transfer through the fabric
The rate of heat loss through the fabric of a building is determined by the U values and areas of the various elements of the external fabric, and by the expected temperature conditions. For one element of a

building under steady conditions, the rate of heat loss (q) is given by equation [1.8] in which t_i is the internal temperature suitably defined and t_o is the external temperature suitably defined.

$$q = AU(t_i - t_o) \qquad\qquad [1.8]$$

heat loss = area × U value × temperature difference
per second
  (W)   (m²)   (W/m² K)   (K)

The fabric heat loss rate from a building Q_u may be obtained by summing the contributions from all the elements.

$$Q_u = A_1 U_1 (t_i - t_o) + A_2 U_2 (t_i - t_o) + \ldots$$
$$Q_u = \Sigma AU(t_i - t_o)$$

total   sum of $AU(t_i - t_o)$
fabric   for all elements
heat
loss per
second

Example 1.20 (a) Calculate the rate of heat transfer through the fabric of a wall ($U = 2.1$ W/m² K, solid wall area $= 13$ m²) if the wall contains three single glazed windows ($U = 5.6$ W/m² K) and a glazed door ($U = 5.6$ W/m² K). Assume an area of 12 m² is occupied by windows and door and the difference in temperature between the internal environment and the external air is (i) 1 K (ii) 16 K.

    (b) How much heat is lost through the wall in 6 hours?

(a)

Building element	A (m²)	U (W/m² K)	AU (W/K)
wall	13	2.1	27.3
windows + door	12	5.6	67.3
			$\Sigma AU = 94.5$

ΣAU represents the rate of heat loss through the fabric of the wall for a temperature difference of 1 K.

$Q_u = \Sigma AU \times$ temperature difference

  (i) temperature difference $= 1$ K
      $Q_u = 94.5 \times 1 = 94.5$ W

 (ii) temperature difference $= 16$ K
      $Q_u = 94.5 \times 16 = 1512\text{W} = 1.512 \times 10^3$ W
                 $= 1.5\text{kW}$ ($1\text{kW} = 1 \times 10^3$ W)

(b) q and Q_u represent heat flow rates, the quantities of heat transferred in 1 second. Quantity of heat (energy) may be obtained by multiplying heat flow rate (power) by time (expressed in seconds)

quantity of heat = heat flow rate × time
(energy)   (power)   × (time)
(J, kJ, MJ)   (W, kW)   (s)

(i) $Q_u = 94.5$ W, time = 6 hours = $6 \times 60 \times 60$ seconds
$$= 21\,600\,\text{s} = 21.6 \times 10^3\,\text{s}$$

Quantity of heat lost = $Q_u \times$ time
$$= 94.5 \times 21.6 \times 10^3 = 2041 \times 10^3$$
$$= 2.041 \times 10^6\,\text{J}$$
$$= 2\,\text{MJ}$$
$$(1\,\text{MJ} = 1 \times 10^6\,\text{J})$$

(ii)   Quantity of heat lost = $Q_u \times$ time
$$= 1.512 \times 10^3 \times 21.6 \times 10^3$$
$$= 32.66 \times 10^6\,\text{J}$$
$$= 32.7\,\text{MJ}$$

The joule is the basic energy unit, but the use of the kilowatt hour is also common.

1 kilowatt hour = 1 kilowatt × 1 hour
(energy)   (power)   × (time)
(kWh)   (kW)   (h)
$$= 1000\,\text{W} \times 3600\,\text{s} = 3.6\,\text{MJ}$$
(energy)

In this example:

(i) $Q_u = 94.5$ W = 0.0945 kW
0.0945 kW for 6 hours ≡ $0.0945 \times 6 = 0.57$ kWh
(kW) × (h)

(ii) $Q_u = 1.5$ kW
1.5 kW for 6 hours ≡ $1.5 \times 6 = 9$ kWh
(kW) × (h)

(a)   The rate of heat transfer is (i) 94.5 W (ii) 1.5 kW
(b)   The quantity of heat lost is (i) 2 MJ or 0.57 kWh
                                  (ii) 32.7 MJ or 9 kWh

Ventilation heat loss
Ideal ventilation is the replacement of vitiated air by sufficient fresh air for health and comfort, without causing draughts and with as little loss of heat in the process of air change as possible. Controlled ventilation is not supplied in most dwellings but air infiltration occurs through gaps in and around doors and windows. In many older buildings natural ventilation often greatly exceeds the minimum ventilation requirements.

In winter, air which has been warmed to make a room comfortable leaks and so transfers heat outside. Fresh outside air must be heated to the room air temperature. The heat transfer rate during ventilation (Q_v) is the quantity of heat required per second to heat the air infiltrating from outside to the temperature of the inside air.

$$Q_v = mc(t_{ai} - t_{ao}) \qquad [1.9]$$

Ventilation heat transfer per second	mass of air infiltrating per second	specific heat capacity of air	temperature difference
=	×	×	
(W)	(kg/s)	(J/kg K)	(K)

Ventilation rate may be described in terms of the number of air changes per hour. In an enclosure of volume v with 1 air change per hour (1 ach), a volume of air v is heated in 1 hour from outside air temperature (t_{ao}) to inside air temperature (t_{ai}). With N air changes per hour the volume of air heated in 1 hour is Nv. A volume Nv per hour is equivalent to a volume $Nv/60 \times 60$ per second or a mass $Nv\rho/60 \times 60$ per second.

(mass = volume × density, where ρ = density)

Therefore substituting for the mass of air infiltrating per second in equation [1.9]

$$Q_v = \frac{Nv\rho c}{60 \times 60}(t_{ai} - t_{ao})$$

For dry air at 20 °C, $\rho = 1.203 \, \text{kg/m}^3$, $c = 1004.4 \, \text{J/kg K}$

$$Q_v = \frac{Nv \times 1.203 \times 1004.4}{60 \times 60}(t_{ai} - t_{ao})$$

$$\frac{1.203 \times 1004.4}{60 \times 60} = \frac{1208.3}{3600}, \text{ which is approximately } \tfrac{1}{3} \text{ or } 0.33$$

$$Q_v = \tfrac{1}{3}Nv(t_{ai} - t_{ao}) \qquad [1.10]$$

Ventilation heat transfer per second	$\tfrac{1}{3}$ × number of air changes per hour	× volume of enclosure	× temperature difference
=			
(W)	(ach)	(m³)	(K)

Example 1.21 Calculate the rate of heat loss through ventilating air in a room in which there are 2 air changes per hour, the internal air temperature is 18 °C, and the external air temperature is 2 °C. Room volume = 40 m³.

$$Q_v = \tfrac{1}{3}Nv(t_{ai} - t_{ao})$$

$N = 2$ ach, $v = 40 \, \text{m}^3$, $t_{ai} - t_{ao} = 18 - 2 = 16 \, \text{K}$

$$Q_v = \tfrac{1}{3} \times 2 \times 40 \times 16 = 427 \, \text{W}$$
$$= 0.427 \, \text{kW}$$

Rate of heat loss through ventilating air = 0.427 kW

If the energy lost in a given time interval is required it may be obtained by multiplying Q_v by the time interval expressed in seconds.

Heat lost = Q_v × time
$\qquad$ (J) = (W) $\quad$ (s)

Older houses may have ventilation rates of 3 ach or more but an adequate whole house average is considered to be about 0.8 ach. ('Houses in the 80 s', *Architects' Journal*.)

The fabric heat loss rate (Q_u) and the ventilation heat loss rate (Q_v) are combined to give the total heat loss rate (Q) from an enclosure, when steady temperatures are assumed both inside and outside.

$$Q = Q_u + Q_v$$
$$Q = \Sigma A U(t_i - t_o) + \tfrac{1}{3}Nv(t_{ai} - t_{ao}) \qquad\qquad [1.11]$$

Total$\quad$ Fabric $\qquad\qquad$ Ventilation
(W)$\quad\quad$ (W) $\qquad\qquad$ (W)

Recommended and statutory thermal transmittance values

The Building Regulations (1990 Edition, Part L, Conservation of fuel and power) require that reasonable provision shall be made for the conservation of fuel and power in buildings. This requirement applies to buildings or parts of buildings intended for occupation as separate dwellings (houses, flats or maisonettes) and other buildings when the floor area exceeds 30 m^2.

The loss of heat through the building fabric may be limited by an *Elemental Approach* or by *Calculation*. In the *Elemental Approach* maximum U values are specified for relevant elements of construction (Table 1.11). Where cold bridges (lintels, beams, columns etc.) occur, it may be necessary to add extra insulation to bring that part of the element up to the required standard. Lintels, jambs and sills associated with windows and rooflights may be counted as part of the window or rooflight area but their U values should not exceed 1.2 W/m^{2}K.

These maximum *U* values do not control the total rate of heat loss through the building fabric unless the ratio of opaque structure to glazed area is specified in some way. The average *U* value of perimeter walls depends upon the *U* value of the external wall, the proportion of glazing in the external wall and whether it is single or double glazing and the ratio of partition wall to perimeter wall. If the total wall area is A_w and has a *U* value U_w and the total glazed area is A_g with a *U* value U_g, then the average *U* value for this wall and its glazing is given by

$$\frac{A_w U_w + A_g U_g}{A_w + A_g}.$$

When more components are to be considered,

average U value $= \Sigma A U / \Sigma A$.

When calculating the average U value certain values are to be assumed:

5.7 W/m²K   for single glazing
2.8 W/m²K   for double glazing
2.0 W/m²K   for double glazing with low emissivity coating
2.0 W/m²K   for triple glazing

Example 1.22   Calculate the average U value for the perimeter wall of a modern semi-detached house with the following construction elements:

76 m² of external wall with a U value of 0.96 W/m² K
40 m² of party wall for which a U value of 0.5 W/m² K

is to be assumed.

Single glazed window area of 10 m², $U = 5.7$ W/m² K (assumed)

Double glazed window area of 6 m², $U = 2.8$ W/m² K (assumed)

Average U value $= \Sigma A U / \Sigma A$, where A represents the area of each window or wall comprising the perimeter wall, and U equals the U value of each window or wall comprising the perimeter wall.

Construction element	Area (m²)	U value (W/m² K)	AU (W/K)
External wall	76	0.96	73.0
Party wall	40	0.5	20.0
Window (single glazed)	10	5.7	57.0
Window (double glazed)	6	2.8	16.8
	$\Sigma A = 132$		$\Sigma A U = 166.8$

Average U value $= \dfrac{\Sigma A U}{\Sigma A} = \dfrac{166.8}{132} = 1.26$ W/m² K

The average U value for the perimeter wall $= 1.26$ W/m² K.

In dwellings the maximum permitted area of single glazed windows and rooflights together is 15 per cent of the total floor area. In other building types the maximum single glazed rooflight area is 20 per cent of the roof area. Maximum single glazed window areas, specified as a percentage of the exposed wall area, are 25 per cent in other residential (including hotels and institutional) buildings, 35 per cent in places of assembly, offices and shops and 15 per cent in industrial and storage buildings.

Calculation Procedure 1 allows variations in the level of insulation of individual elements (see table 1.11 for limiting U values) and areas of windows and rooflights. Calculation should show that the rate of heat loss through the envelope of the proposed building is not greater than that through a notional building of the same size and shape which complies with the *Elemental* requirements of the Regulations. With *Calculation Procedure 2* any valid energy conservation measure and useful heat gains (solar and internal) may be allowed for.

The Regulations (1990 Edition, Part L) also require measures to control as appropriate the output of the space heating and hot water systems and to limit the heat loss from hot water storage vessels, pipes and ducts. In order to reduce the condensation risk in roofs and roof spaces, the size of ventilation openings is specified (1990 Edition, Part F2, Condensation).

Good insulation coupled with controlled ventilation can reduce energy consumption considerably and provide improved thermal

Table 1.11 Elemental and limiting U values

Building element	Maximum U value (W/m^2K)	
	Dwellings	All other buildings
Elemental Approach		
1. Roofs	0.25*	0.45
2. Exposed walls	0.45	0.45
Exposed floors		
Ground floors		
3. Semi-exposed walls and floors	0.6	0.6
Calculation Procedure		
4. Roofs	0.35	0.6
5. Exposed walls	0.6	0.6

Notes
Exposed element means an element exposed to the outside air.
* For loft conversions in existing dwellings it would be reasonable to have a roof U value of 0.35 W/m^2K.
As a simple alternative to the levels of insulation shown for dwellings (*Elemental Approach*) all windows may be double glazed, exposed walls at 0.6 W/m^2K, the roof at 0.35 W/m^2K and the floor uninsulated.

comfort with inside surface temperatures much closer to the internal air temperature.

Steady heat loss from an enclosed structure

The total steady heat loss rate (Q) from an enclosed structure is the sum of the fabric heat loss rate (Q_u) and the ventilation heat loss rate (Q_v).

$$Q = Q_u + Q_v$$

$$Q = \Sigma AU (t_i - t_o) + \tfrac{1}{3}NV (t_{ai} - t_{ao}) \qquad [1.11]$$

Total heat = fabric heat + ventilation heat
loss rate = loss rate + loss rate
(W)   (W)   (W)

The heat loss per unit time for unit temperature difference provides a convenient measure of the thermal effectiveness of an enclosure.

$$Q/K = \Sigma AU + \frac{Nv}{3} \qquad [1.12]$$

Total heat = fabric heat + ventilation
loss/second K = loss/second K + loss/second K
(W/K)   (W/K)   (W/K)

Example 1.23 Calculate the fabric heat loss rate per K in a semi-detached house with the levels of thermal insulation indicated in Table 1.12. Assume a ventilation rate of 1 ach.

It is usual to assume that adjoining dwellings are at the same temperature and therefore that no net heat transfer occurs across the party wall. The ground floor, external walls and roof form the boundaries of the enclosure from which heat is lost.

The U values for the 1930s house represent solid brick wall (220 mm), uninsulated pitched roof, suspended timber floor, and single glazed doors and windows.

Table 1.12

Building element	Area (m²)	(1930s) U value (W/m² K)	(1960s) U value (W/m² K)	(1976) U value (W/m² K)	(Early 1980s) U value (W/m² K)
Ground floor	40	0.65	0.7 ·	0.7	0.7
External wall	90	2.1	1.5	1.0	0.6
Windows + doors	21	5.7	5.7	5.7	5.7
Roof	40	1.5	1.5	0.6	0.4

1930s house

Building element	Area (A) (m^2)	U value (U) $(W/m^2 K)$	Heat loss rate/K (AU) (W/K)
Floor	40	0.65	26
External wall	90	2.1	189
Windows + doors	21	5.7	119.7
Roof	40	1.5	60
			$\Sigma AU = 394.7$

Fabric heat loss rate per degree = 394.7 W/K

The 1960s house construction replaces the solid wall by a cavity wall and the timber floor by a solid ground floor.

1960s house

Building element	Area (A) (m^2)	U value (U) $(W/m^2 K)$	Heat loss rate/K (AU) (W/K)
Floor	40	0.7	28
External wall	90	1.5	135
Windows + doors	21	5.7	119.7
Roof	40	1.5	60
			$\Sigma AU = 342.7$

Fabric heat loss rate per degree = 342.7 W/K

Improvement in the U values for wall and roof may be obtained by the replacement of the inner leaf of brickwork in the traditional cavity wall by lightweight concrete block and by the addition of about 50 mm of insulation to the roof space as in a typical 1976 house.

1976 house

Building element	Area (A) (m^2)	U value (U) $(W/m^2 K)$	Heat loss rate/K (AU) (W/K)
Floor	40	0.7	28
External wall	90	1.0	90
Windows + doors	21	5.7	119.7
Roof	40	0.6	24
			$\Sigma AU = 261.7$

Fabric heat loss rate per degree = 261.7 W/K

The wall U value of 0.6 W/m^2 K, selected for the early 1980s house, can be achieved by insulation of the traditional brick and blockwork wall or by timber framed construction methods. 100 mm of insulation in the roof space will provide a U value slightly better (lower) than 0.4 W/m^2 K. The glazed area has not been reduced.

Early 1980s house

Building element	Area (A) (m^2)	U value (U) (W/m^{2}K)	Heat loss rate/K (AU) (W/K)
Floor	40	0.7	28
External wall	90	0.6	54
Windows + doors	21	5.7	119.7
Roof	40	0.4	16
			$\Sigma AU = 217.7$

Fabric heat loss rate per degree = 217.7 W/K

In this early 1980s house the percentage of fabric heat loss through the glazed areas is 55 per cent

$$\frac{\text{fabric heat loss rate through windows + doors}}{\text{total fabric heat loss rate}} = \frac{119.7}{217.7} \times 100$$

$$= 55 \text{ per cent.}$$

This excessive heat loss may be reduced by restricting the glazed areas and where appropriate introducing double glazing.

Example 1.24 Use the fabric heat loss rate per degree as calculated in Example 1.23 to obtain (a) the fabric heat loss rate and (b) the fabric heat loss from the houses in that example during a 12 hour period. Assume that the internal environmental temperature, $t_{ei} = 18\,°C$ and the external air temperature, $t_{ao} = 5\,°C$.

(a) $\qquad Q_u = \Sigma AU \times (t_{ei} - t_{ao})$

$$\underset{\text{(W)}}{\underset{\text{loss rate}}{\text{fabric heat}}} = \underset{\text{(W/K)}}{\underset{\text{loss rate/K}}{\text{fabric heat}}} \times \underset{\text{(K)}}{\text{temperature difference}}$$

temperature difference = $(t_{ei} - t_{ao}) = 18 - 5 = 13\,°C$

(b) Fabric heat loss = fabric heat loss rate $\times$ time

$\quad$ (J) $\qquad\qquad\qquad$ (W) $\qquad\qquad\qquad$ (s)

time = 12 hours = $12 \times 60 \times 60 = 43\,200$ s

(1) House type	(2) Heat loss rate/K (AU) (W/K)	(3) Heat loss rate (Q_u) $AU \times (t_{ei} - t_{ao})$ (kW)	(4) Heat loss $Q_u \times$ time (MJ)
1930s	394.7	5.131	221.6
1960s	342.7	4.455	192.5
1976	261.7	3.402	147.0
early 1980s	217.7*	2.830*	122.3*

*Further reductions to be expected by restriction of glazed areas.
Columns (1) and (2) are obtained from the previous example.
Column (3) is obtained by multiplying column (2) by the temperature difference and dividing by 1000 to convert watts to kilowatts. Column (4) is column (3) multiplied by the time (expressed in seconds) during which the heat loss is required. $1\,kW \times 1\,s = 1\,kJ$. The numbers are large so the heat loss in kilojoules (kJ) is divided by 1000 to convert it to megajoules (MJ). $1\,MJ = 1 \times 10^6\,J$.

(a) The fabric heat loss rate (kW) may be read from column (3).
(b) The fabric heat loss (MJ) may be read from column (4).

The ventilation heat loss rate (Q_v) and the ventilation heat loss rate for unit temperature difference (Q_v/K) may be calculated for the house in the previous examples.

Example 1.25 Calculate (i) the ventilation heat loss rate per degree, (ii) the ventilation heat loss rate, (iii) the ventilation loss, during a 12 hour period, from the house in the previous examples. Assume house volume $= 264\,m^3$, internal air temperature $= 18\,°C$, external air temperature $= 5\,°C$, number of air changes per hour $= 1$ ach.

$$Q_v = \frac{Nv}{3}(t_{ai} - t_{ao})$$

ventilation heat $= \frac{1}{3} \times$ number of $\times$ volume of $\times$ temperature
loss rate   air changes   enclosure   difference
    (W)   (ach) (m^3) (K)

For unit temperature difference between internal air and external air,

$$Q_v/K = \frac{Nv}{3}$$

ventilation heat $\frac{1}{3} \times$ number of $\times$ volume of
loss rate/K   =   air changes   enclosure
  (W/K)   (ach) (m^3)

ventilation   ventilation heat $\times$ time
heat loss   =   loss rate
  (J)   (W)   (s)

$N = 1$ ach, $t_{ai} = 18\,°C$, $t_{ao} = 5\,°C$, $v = 264\,m^3$, time = 12 hours.

(i)
$$Q_v/K = \frac{Nv}{3} = \frac{1 \times 264}{3} = 88\,W/K$$

ventilation heat loss rate per degree = $88\,W/K$

(ii) $Q_v = \dfrac{Nv}{3}(t_{ai} - t_{ao}) = \dfrac{1 \times 264}{3}(18 - 5) = 1144\,W$

ventilation heat loss rate = $1.144\,kW$

(iii) ventilation heat loss = $Q_v \times$ time = $1.144 \times 12 \times 60 \times 60 =$
49 421 KJ = $49.4\,MJ$
ventilation heat loss in 12 hours = $49.4\,MJ$

In Example 1.25 a ventilation rate of 1 ach (air change per hour) is assumed. Draught proofing may be necessary, but not to the extent of producing an airtight house, in order to maintain ventilation losses at 1 ach. In kitchens and bathrooms a higher ventilation rate is often required, perhaps 3 ach, but over the remainder of the house it may be as low as 0.5 ach, giving an overall ventilation rate of slightly less than 1 ach. A house with obvious draughts will have at least 3 ach. Ventilation is especially required in kitchens, bathrooms and roof spaces in order to reduce the risk of condensation incidence.

The total losses from this house may be obtained by combining fabric and ventilation losses. It is apparent from the values of the total heat loss rate/K in Table 1.13 that as the house is improved thermally the ventilation heat loss rate/K becomes a greater proportion of the total heat loss rate/K.

The total heat loss rate from an enclosure may be obtained by the addition of the fabric heat loss rate and the ventilation heat loss rate. These two quantities have been calculated previously (Examples 1.24 and 1.25) and are included in Table 1.14 along with the total heat loss rate.

Table 1.13 Heat loss rate/K

House type	Fabric heat loss rate/K Q_u/K (W/K)	Ventilation heat loss rate/K Q_v/K (W/K)	Total heat loss rate/K Q/K (W/K)
1930s	394.7	88	482.7
1960s	342.7	88	430.7
1976	261.7	88	349.7
Early 1980s	217.7*	88	305.7

*Further reduction to be expected by restriction of glazed area.

Table 1.14 Total heat loss rate

House type	Fabric heat loss rate Q_u (kW)	Ventilation heat loss rate Q_v (kW)	Total heat loss rate $Q = Q_u + Q_v$ (kW)
1930s	5.131	1.144	6.275
1960s	4.455	1.144	5.599
1976	3.402	1.144	4.546
Early 1980s	2.830	1.144	3.974

The total heat loss from an enclosure over a given period of time may be calculated by multiplying the total heat loss rate (Q) by the time period. For completeness the values previously calculated for the fabric heat loss and the ventilation heat loss are collected and added together in Table 1.15 to give the total heat loss over the 12 hours.

Table 1.15 Total heat loss from an enclosure

House type	Fabric heat loss (MJ)	Ventilation heat loss (MJ)	Total heat loss (MJ)
1930s	221.6	49.4	271
1960s	192.5	49.4	241.9
1976	147.0	49.4	196.4
Early 1980s	122.3*	49.4	171.7

*The fabric heat loss may be further reduced by the restriction of the area of single glazing or by double glazing. However this is only one of the several factors which should be considered in the determination of the area of glazing. The high fabric heat loss through glass may be more than offset by the solar energy transmitted into the building (R. R. Wilberforce). Since maximum solar gain is obtained through a south facing window it would appear advantageous to utilise this fact in building design in order to lessen energy consumption.

Double glazing in the house insulated to the early 1980s standard reduces the fabric heat loss until it is only twice the ventilation heat loss. The total heat loss is approximately half that in an uninsulated house with solid walls (1930s house type) and ventilation controlled to 1 ach.

Early 1980s house with double glazing, $t_{ei} = 18°C$, $t_{ai} = 18°C$, $t_{ao} = 5°C$,
$N = 1$ ach, $v = 264 \, m^3$.

Building element	Area (A) (m^2)	U-value (U) (W/m^2 K)	AU (W/K)	$AU(t_{ei} - t_{ao})$ (W)
Floor	40	0.7	28	364
External wall	90	0.6	54	702
Windows	21	2.8	58.8	764
Roof	40	0.4	16	208
			$\Sigma AU =$ 156.8 W/K (157 W/K)	$\Sigma AU(t_{ei} - t_{ao}) =$ 2038 W (2.0 kW)

Fabric heat loss rate/K, $Q_u/K = 157 \, W/K$

Fabric heat loss rate, $Q_u = 2.0 \, kW$

Ventilation loss rate/K, $Q_v/K = \dfrac{Nv}{3}$

$$= \frac{264}{3} = 88 \, W/K$$

Ventilation loss rate, $Q_v = \dfrac{Nv}{3} (t_{ai} - t_{ao})$

$$= \frac{264 \times 13}{3} = 1144 \, W = 1.1 \, kW$$

Total heat loss rate, $Q = Q_u + Q_v$
$$= 2.0 + 1.1 = 3.1 \, kW$$

(Compare with 1930s house type heat loss rate of 6.3 kW.)

Much greater thermal insulation, such as a U value of 0.35 W/m^2 K or less for walls and double glazing, controlled ventilation, the use of heat recovery systems and heat pumps and the best utilisation of the available solar energy by orientation and design of buildings are some of the methods by which low energy consumption in buildings may be achieved in the future.

Heat gain

Solar heat gain

Solar radiation incident upon a surface is reflected, absorbed and transmitted in a manner which is characteristic both of the wavelength of the incident radiation and the physical properties of the surface.

The absorbed radiation appears as heat in the body of which the surface is a boundary. This heat is partly dissipated by convection to the surrounding air and by reradiation. The wavelength at which this reradiation occurs is determined by the surface temperature and the lower this temperature the longer the wavelength. In contrast to the shortwave radiation from a high temperature surface such as the sun, the radiation emitted by building surfaces and the ground is low temperature, longwave radiation.

How much solar energy is available depends upon latitude, season and the orientation and inclination of the surface. In summer greatest heat gain occurs on a horizontal surface, whereas on a vertical surface peak values occur on east and west facing surfaces in June and on south faces in March and September (Fig. 1.18). The *solar radiation intensity* (I) measures the quantity of solar energy incident upon unit area ($1 \, \text{m}^2$) in unit time ($1 \, \text{s}$) in W/m^2.

Solar heat gain in a building takes place through the opaque elements, the wall and roof and through non-opaque elements such as glass and translucent plastic material. The external surface of a building is exposed to both shortwave radiation and longwave radiation. The total solar radiation (shortwave radiation) which reaches the building surface is a combination of direct radiation from the sun and diffuse radiation reflected from the sky and the surrounding surfaces. The longwave radiation is received from nearby surfaces which have been warmed by the incidence of solar radiation.

Solar radiation absorbed at an opaque external surface is partly transmitted to the interior of the building and results in an increased internal temperature. In the absence of solar radiation a similar increased internal temperature could be expected from a higher external air temperature. This higher external air temperature which produces the same increased internal temperature as is obtained from the combination of solar radiation and the actual external air temperature is known as the sol – air temperature (t_{eo}). The sol – air temperature may be calculated from the expression

$$t_{eo} = t_{ao} + R_{so}(\alpha I_t - \epsilon I_L) \qquad [1.13]$$

where t_{ao} = external air temperature (°C)

R_{so} = external surface resistance ($\text{m}^2 \, \text{K}/\text{W}$)

α = solar absorptivity of the surface (typically $\alpha = 0.5$ for light coloured surfaces and 0.9 for dark surfaces)

I_t = intensity of solar radiation (direct + diffuse) (W/m^2)

ϵ = emissivity of the surface ($\epsilon = 0.9$ for masonry materials)

I_L = net longwave radiation exchange between a black body at external air temperature and the external environment.

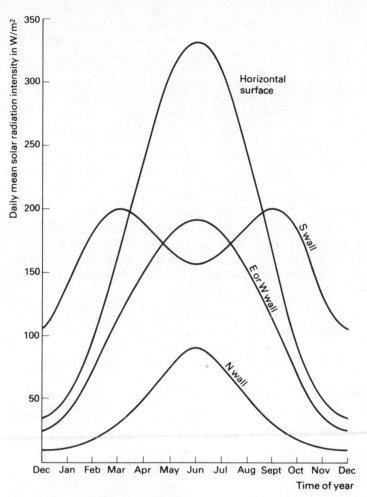

Fig. 1.18 Total solar radiation intensity (daily mean value) received throughout the year on vertical surfaces (for different orientations) and on a horizontal surface. (Curves plotted from data given in the *IHVE Guide* for latitude 51.7 °N.)

For cloudless sky conditions a value of $100\,\text{W/m}^2$ is assumed for a horizontal roof. For a vertical wall I_L is zero since it is assumed that the longwave radiation gained from the ground balances the longwave radiation lost from the wall to the sky.

Air temperature, solar radiation intensity and sol–air temperature may be obtained from tables (*IHVE Guide*, Section A6). The sol–air temperature alters with time of day because of the variation in external air temperature and solar radiation intensity used in its calculation. The external air temperature reaches a

68

maximum in the afternoon but does not depend upon orientation. Solar radiation intensity changes with orientation of surface, time and season. The peak solar radiation intensity on a vertical wall occurs in the morning on the east face, at noon on the south face and in the afternoon on the west face. The occurrence of the maximum value of sol – air temperature on a particular surface orientation may not coincide with the maximum external air temperature.

The heat flow rate through the fabric of a building (Q_u) under steady state conditions may be obtained from equation [1.8], $Q_u = \Sigma A U(t_i - t_o)$, where t_i is the temperature of the internal environment and t_o that of the external environment. In the presence of solar radiation it would be appropriate to calculate the heat transfer rate through the fabric by substituting the sol–air temperature (t_{eo}) for t_o, if steady state conditions existed. However the changing sol – air temperature means that variable and not steady state conditions exist. When the building is not in thermal equilibrium with its environment the thermal capacity of its elements influences the heat flow rate. There is a time lag, dependent upon the thickness and density of the building element, between the instant when the sol – air temperature changes due to incident solar radiation and the transmission of the temperature change through the building fabric. The amplitude or swing about the mean heat flow rate is reduced or damped. This moderation in diurnal swing in heat flow rate and hence in internal temperature also depends upon the thickness of the building element and is termed the decrement or decrement factor (Fig. 1.19).

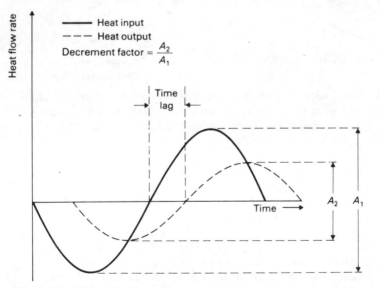

Fig. 1.19   Time lag and decrement factor

In all but the most lightweight of building elements there is a time lag and a moderation of temperature variation. A heavyweight building is less likely to suffer from solar overheating than a lightweight one. Solar heat gain through the walls is usually small compared with solar gain through windows, but solar heat gain through poorly insulated roofs may cause overheating in factories and other single storey buildings. The solar heat gain in a building is determined chiefly by the proportion of glass.

The solar radiation falling upon a clear glass surface is reflected, absorbed and transmitted in proportions similar to those indicated in Fig. 1.20. These quantities depend upon the angle of incidence (i) and the proportion of direct and diffuse radiation. The angle of incidence (i) is the angle measured between the incident light beam

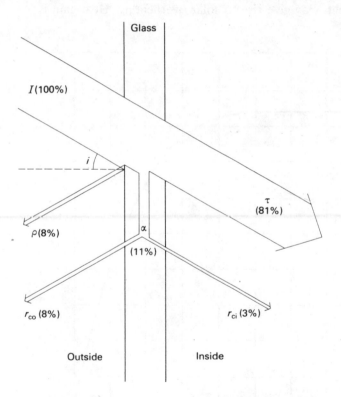

I = incident solar radiation, τ = directly transmitted solar radiation,
ρ = reflected radiation, α = radiation absorbed by glass,
r_{co} = heat loss by radiation (longwave), convection and conduction to outside,
r_{ci} = heat loss by retransmission of radiation (longwave), convection
and conduction to inside i = angle of incidence

Fig. 1.20 Typical proportions of incident solar radiation, reflected, absorbed, transmitted and retransmitted by glass

and the normal to the plane of the glass. When the angle of incidence approaches about 45° the transmitted solar energy decreases while reflection at the outer surface increases (Fig. 1.21).

The absorbed radiation heats the glass and part of this heat reaches the room surfaces by convection and radiation from the inside surface of the glass. The solar heat gain is obtained by adding this inwards released heat to the directly transmitted component of the incident solar radiation. Absorption of this solar heat gain by the internal surfaces raises their temperature. These heated surfaces behave as low temperature, longwave radiators. Since glass transmits shortwave radiation in the range 0.3 to 2.8 μm but is opaque to longwave radiation from low temperature surfaces, the solar heat gained is trapped within the enclosure causing an internal temperature rise. This phenomenon, frequently referred to as the greenhouse effect, may give rise to solar overheating. Heat gain is

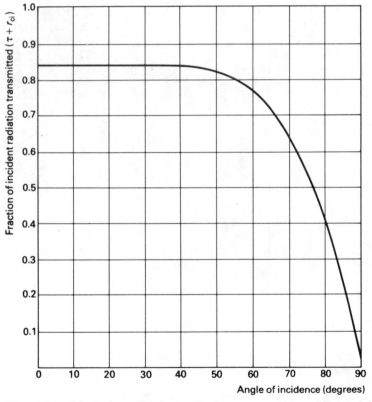

Fig. 1.21 Variation of solar radiation transmission (direct (τ) and retransmitted (r_{ci})) with angle of incidence. (Curve plotted from data given in the *IHVE Guide* for 4 mm clear glass)

directly proportional to the area of glass exposed to solar radiation and therefore large glazed areas will permit a large and rapid heat gain.

Control of solar heat gain

Solar radiation is a source of energy constantly moving from surface to surface of the building, but never quite in the same way from one day to the next. In summer the sun rises north of east and sets north of west (Fig. 1.22). The longer daily path of the sun accounts for the increased daylight hours. From June to December the daily arc of the sun across the sky gradually shortens and moves southward. The winter sun rises and sets towards the south (Fig. 1.22). The sun at noon is 46° lower in the sky in late December than it is in June.

When the sun is low in the sky (small solar altitude angle), the incidence of radiation is nearly normal (small angle of incidence, i) on a vertical surface. East and west facing vertical surfaces in the United Kingdom receive solar radiation at small angles of incidence. In northern latitudes the angle of incidence of the sun's rays on a south facing wall is small in winter because of the low solar altitude, but large in summer because of the high solar altitude. Figure 1.21 illustrates the effect of angle of incidence upon the transmission of solar radiation through clear glass.

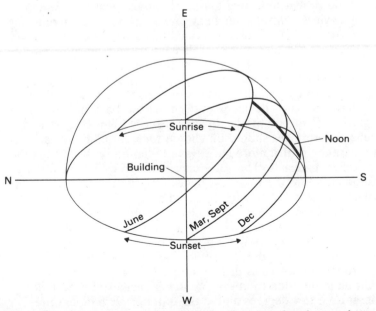

Fig. 1.22 Daily movement of the sun across the sky as observed from the building during June, March and September, and December

Solar radiation entering a building through a glazed surface contributes to the heating requirements in winter but may lead to excessive temperatures during the summer months. Restriction of large glazed apertures on east, south and west walls is necessary since excessive solar heat gain may create uncomfortable internal conditions even when outdoor air temperatures are low. Once the window size has been established the most effective method of reducing solar heat gain is to prevent the transmission of shortwave radiation through the glass by external shading.

Balconies and horizontal projections are useful external shading devices, especially on south facing walls. The overhang is designed so that the window beneath it is shaded from the sun in summer (high solar altitude), but transmits solar radiation from the low solar altitude in winter (Fig. 1.23). In the past small windows with deep reveals prevented many of today's solar overheating problems.

Adjustable shading devices such as awnings, venetian blinds, roller blinds and vertical slatted blinds may be installed outside the window, in order to reduce the amount of solar radiation transmitted through the window glass into the building. While external blinds are efficient in the control of solar heat gain and glare there are among other disadvantages those of high initial cost and maintenance.

Solar heat gain may be reduced by using a special solar control glass. Two types of this glass available lessen solar heat gain by increasing either (i) the absorption (heat absorbing glass) or (ii) the reflection (heat reflecting glass), of the incident solar energy.

Shading devices control solar heat gain by preventing or reducing the transmission of solar radiation through glass. Internally fitted shading devices such as venetian or roller blinds and curtains cannot prevent the transmission of the incident solar radiation through the glass but they reduce solar heat gain by reflecting some shortwave radiation back through the glass. Much of the solar energy incident upon an internal shading device is either reflected back through the window glass or absorbed by the shading device. Light colours and special reflective surfaces should be chosen for internal blinds and curtains since they return more shortwave radiation back through the glass than other materials. The solar energy absorbed by the blind will be reradiated into the room as longwave radiation and contribute to the solar heat gain. To be effective in the control of solar heat gain curtains and blinds must be used for shading before solar radiation has entered the room and longwave reradiation from the heated surfaces has caused an excessive temperature rise.

Ventilation should be used to remove as much excess heat as possible. When outdoor air is drawn in to cool the interior of a lightweight building often the best that can be achieved is an indoor air temperature a few degrees above the outdoor air temperature.

The effectiveness of a typical glazing and shading system may be measured in terms of the solar gain factor (S). This is the proportion

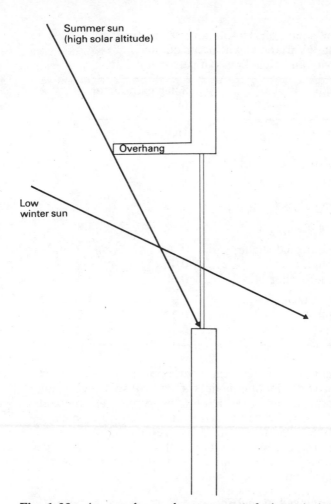

Fig. 1.23 An overhang above a south facing window provides shade from direct solar radiation in summer

of incident solar radiation transmitted by the window and shading device to the interior of an enclosure. The solar gain factor for an unglazed unshaded aperture is unity. This factor decreases as the shading system becomes more effective in reducing solar heat gain.

The transmission of solar radiation through glass and therefore the solar gain factor depend upon the angle of incidence of the radiation (Fig. 1.21) which alters according to latitude, orientation and time of day and season. The average values of solar gain factor calculated for August and latitute 51.7 °N may be applied in the United Kingdom for orientations south of east and west during the summer months (Table 1.16).

Table 1.16 Typical solar gain factors for various types of glazing and shading. (Variations will occur due to changes in reflectivity and cleanliness of surfaces.)

	Solar gain factor (S)
Unglazed aperture	1.00
4 mm clear glass,	
single glazing	0.76
double glazing	0.64
6 mm heat absorbing glass (bronze)	0.59
6 mm heat reflecting glass	0.32 − 0.22
4 mm clear glass + internal shade,	
white venetian blind	0.46
cream holland linen roller blind	0.30
4 mm clear glass + external shade,	
white venetian blind	0.14
canvas roller blind	0.14

The requirements for daylight and ventilation may well conflict with the need to provide shading devices to control solar heat gain, reduce glare and prevent direct radiation falling upon the occupants of an enclosure. The traditional heavyweight building with small windows is unlikely to experience the solar overheating problems which may occur in the excessively glazed lightweight modern office block. Moderate glass areas are desirable to limit both solar heat gains in summer and heat losses in winter.

Heat gain in an enclosed structure

Electric lighting and all other electrical equipment, cookers, office machines, electronic data processing units and industrial machines and processes are some of the items which are normally used for non-heating purposes, but which contribute to the internal heat gains in an enclosed structure. These internal heat gains, frequently referred to as casual or sundry heat gains, also include metabolic heat gains from the occupants and solar radiation and may be a bonus to offset the fabric and ventilation losses in winter or an extra heat load which must be removed from the building if excessive internal temperatures occur during the summer period.

Electric lighting, body heat production and solar radiation are usually the chief sources of casual heat gains within an enclosure.

Almost all the electrical energy consumed by the lighting system contributes to the heat gain because even the light emitted is converted to longwave radiation (heat) by many reflection and absorption processes at the surfaces in the room.

Rate of heat gain = Electrical power consumed
from light source   by light source
(W)   (W)

The daily rate of heat gain from all light sources (Q_L) is calculated over a 24 hour period.

Daily mean rate   = Σ power consumed × number × $\frac{1}{24}$
of heat gain from   sum for   of hours
all light sources   all light   of use
                     sources
(W)   (W)   (h) $\left(\frac{1}{h}\right)$

The total heat gain over the 24 hour period is obtained from:

Daily heat gain = Q_L × time
                  (J)   (W)   (s)

where time = 24 h = 24 × 60 × 60
                  = 86.4×10^3 s

Example 1.26 The power rating and the number of hours of use, during a 24 hour period, of lamps installed in a domestic dwelling are listed in columns (1) and (2) of the table. Estimate for the house (a) the daily mean rate of heat gain and (b) the daily heat gain, from electric lighting.

	(1) **Power** (W)	(2) **Hours of use** (h)	**Power × hours of use** (W × h)
Lamp type			
Fluorescent	65	4	260
Tungsten	100	6	600
Tungsten	100	4	400
Tungsten	60	3	180
Tungsten	60	2	120
			1560 (sum for all lamps)

(a)
$Q_L = \Sigma$ power consumed × hours of use × $\frac{1}{24}$

$= \dfrac{1560}{24} = 65$ W

The daily mean rate of heat gain = 65 W

(b)
Daily heat gain = $Q_L \times$ time
$$= 65 \times 24 \times 60 \times 60 = 5.6 \times 10^6 \, J$$
$$= 5.6 \, MJ$$

The daily heat gain = 5.6 MJ

The heat emission from the people in a building depends upon their activity (Table 1.17).

Table 1.17 Heat emission from the human body. (Adult male, body surface area $2 \, m^2$.)

Degree of activity	Heat emission (q_p) (W)
Seated at rest	115
Light work	140
Medium work	265
Heavy work	440

For women and children multiply these values by 0.85 and 0.75 respectively.

The daily mean rate of heat gain from the occupants (Q_p) may be estimated from

$$Q_p = \Sigma q_p H_p \times \frac{1}{24}$$
$$\text{sum for all occupants}$$

where q_p = heat output rate per occupant

and H_p = number of hours of occupation.

The daily heat gain from the people occupying a building is given by:

Daily heat gain = Q_p × time
(J)   (W)   (s)

where time = 24 h = 24 × 60 × 60
$$= 86.4 \times 10^3 \, s$$

Example 1.27 If 20 people are employed in an office (activity equivalent to light work) for 8 hours during the day estimate (a) the daily mean rate of heat gain and (b) the daily heat gain, from the occupants.

(a)

$$Q_p = \Sigma q_p\, H_p \times \frac{1}{24}$$

$$(W)\,(h) \quad \left(\frac{1}{h}\right)$$

where $q_p = 140\,W$ and $H_p = 8\,h$

$$Q_p = 20 \times 140 \times 8 \times \frac{1}{24}$$

$$= 933\,W$$

Daily mean rate of heat gain from the occupants $= 933\,W$

(b)   Daily heat gain from the occupants $= Q_p \times$ time,
        where time $= 24\,h$
        Daily heat gain $= 933 \times 24 \times 60 \times 60$

$$= 81\,MJ$$

Alternatively the daily heat gain may be calculated directly. Daily heat gain $=$ number of occupants $\times q_p \times$ time, where time $= 8\,h = 8 \times 60 \times 60\,s$ in this example.

Daily heat gain $= 20 \times 140 \times 8 \times 60 \times 60$

$$= 81\,MJ$$

The daily heat gain from occupants $= 81\,MJ$.

The rate of heat gain from solar radiation (Q_s) depends upon the intensity of incident solar radiation (I measured in W/m^2), the area of glazing (A) and the solar gain factor (S).

$$Q_s \;=\; \Sigma \qquad AIS$$
(W) sum over $(m^2)(W/m^2)$
all glazed
areas and
orientations

The daily mean values of the total solar radiation intensities on vertical surfaces may be obtained from tables (*IHVE Guide*, Section A6).

Total solar intensities on vertical surfaces, daily mean values for latitude $51.7\,°N$ are shown in the table

Orientation	21 June (W/m^2)	24 August and 20 April (W/m^2)	22 December (W/m^2)
North	90	50	10
East and West	190	150	25
South	155	185	105

Example 1.28 Calculate (a) the daily mean rate of solar heat gain and (b) the daily solar heat gain for a dwelling in February and October, where the orientations, areas of glazing, solar radiation intensities on vertical surfaces and solar gain factors are as shown in the table.

(a)   Daily mean rate of solar heat gain

Orientation	Area A (m^2)	Solar intensity I (W/m^2)	Solar gain factor S	Solar gain rate $A \times I \times S$ (W)
North	8.2	20	0.76	124.6
South	1.4	180	0.76	191.5
East and West	11.4	70	0.76	606.5
				$\Sigma AIS = 922.6$

$Q_s = \Sigma AIS = 922.6\,\text{W}$

The daily mean rate of solar heat gain in February and October = 922.6 W.

(b)   Since the daily mean values for solar radiation intensities are calculated over a 24 hour period, the daily solar heat gain is obtained from:

Daily solar heat gain = $Q_s \times$ time
  (J)   (W)   (s)
where time = $24 \times 60 \times 60\,\text{s}$

Daily solar heat gain = $922.6 \times 24 \times 60 \times 60$
          $= 79.7\,\text{MJ}$

The daily solar heat gain in February and October = 79.7 MJ.

Casual gains in a typical domestic dwelling amount to an average rate of heat gain of about 1 kW (O'Callaghan, P.W.) with cooking making a major contribution of about 350 W. (The average rate is estimated over an annual period and does not include solar gains.)

An estimate of the total heat gain in an enclosed structure is made by adding the heat gains from lighting, people and solar radiation to any other casual gains which may occur.

Questions

1. Define thermal conductivity
   The surface temperatures of a layer of material 12.5 mm thick are maintained at 15 °C and 20 °C respectively. If the material is:

(a) a good conductor, a metal with thermal conductivity $k = 80$ W/m K; and

(b) a thermal insulator, cellular polystyrene with $k = 0.035$ W/m K;

calculate the rate of heat flow through:
  (i) unit area (1 m^2);
  (ii) 3 m^2
of the layer.

2. Calculate the thermal resistance of:
   (a) brickwork, 105 mm thickness, $k = 0.84$ W/m K;
   (b) aerated concrete blockwork, 100 mm thickness, $k = 0.2$ W/m K;
   (c) urea formaldehyde foam, 50 mm thickness, $k = 0.038$ W/m K.

3. (a) Describe the transmission of heat by convection.
   (b) Evaluate the rate of heat loss by convection from a wall with a surface temperature of 16 °C. Assume that air temperature $= 12$ °C, wall area $= 12 \text{ m}^2$ and the convective heat transfer coefficient $h_c = 3.0$ W/m² K.

4. Define the terms (i) emissivity (ii) absorptivity.
   Estimate the rate of radiative heat transfer from unit area (1 m^2) of a surface at 10 °C to surrounding surfaces at 6 °C. Both the surface and its surroundings are of high emissivity. Assume that the emissivity factor is 0.86 and the radiation coefficient is 5.1 W/m² K.

5. Discuss the factors which influence the resistance of a surface. Calculate the external surface resistance of a wall at which the exposure is considered to be severe (wind speeds of 6 m/s). Assume an emissivity factor of 0.81 since both the wall and its surroundings are of high emissivity ($\epsilon = 0.9$), a mean surface temperature of 0 °C and a radiative heat transfer coefficient of 4.6 W/m² K. For air movement over a surface the convection coefficient (h_c) may be obtained from

$$h_c = 5.8 + 4.1v$$

where v is the wind speed in m/s.

6. Obtain the internal surface resistance (R_{si}) values in Table 1.3. Assume a radiation coefficient (h_r) for surfaces at 20 °C of 5.7 W/m² K and a convection coefficient (h_c) of 3.0 W/m² K for horizontal heat flow, 4.3 W/m² K for upward heat flow and 1.5 W/m² K for downward heat flow.
   As recommended by the *CIBS Guide*, Section A3, use the expression

$$\frac{1}{h_c + \frac{6}{5} E h_r}$$

where E is the emissivity factor to evaluate R_{si}. For a high emissivity surface it is assumed that both the surface and its surroundings have an emissivity of 0.9, so that approximately $\frac{6}{5}E = 6/5 \times 0.9 \times 0.9 = 0.97$. A low emissivity surface is assumed to have an emissivity of 0.05 with surroundings of emissivity 0.9, so that approximately $\frac{6}{5}E = 6/5 \times 0.05 \times 0.9 = 0.05$.

7. Define the term thermal transmittance (U value).
A solid unplastered wall is composed of 105 mm brickwork. The thermal conductivity of brickwork is 0.84 W/m K and the internal and external surface resistances are 0.12 m² K/W and 0.06 m² K/W respectively. What is the thermal transmittance of the wall?

8. A cavity wall is constructed with 105 mm outer and inner leaves of brickwork and plastered internally with 16 mm lightweight plaster. Assume the following thermal conductivity values: brick inner leaf, 0.62 W/m K; brick outer leaf, 0.9 W/m K; and lightweight plaster 0.16 W/m K. If the cavity resistance is 0.18 m² K/W and the internal surface resistance is 0.12 m² K/W, calculate the U value of the wall when the external surface resistance is:
(a) 0.06 m² K/W (normal exposure); and
(b) 0.03 m² K/W (severe exposure).

9. The steady state heat loss through a 220 mm solid brick wall with 16 mm dense plaster on the inside face is 2.2 W/m² K. If the thermal conductivity of plaster is 0.5 W/m K and the inside and outside surface resistances are 0.12 m² K/W and 0.06 m² K/W respectively, calculate the thermal conductivity of the brickwork.

10. (a) Which factors influence the thermal resistance of an air cavity?
(b) What is the thermal transmittance of a wall with 105 mm brick outer skin, 50 mm air cavity, 100 mm lightweight concrete block inner skin which is lined internally with 10 mm of plasterboard on plaster dabs? The air space formed between the plasterboard and the blockwork has a resistance of 0.10 m² K/W. The other thermal resistances are as follows:
outside surface 0.06 m² K/W; 50 mm cavity 0.18 m² K/W; and inside surface 0.12 m² K/W. The thermal conductivities of brick, block and plasterboard are 0.96 W/m K, 0.22 W/m K and 0.16 W/m K respectively.
(c) Obtain the heat flow rate through 20 m² of wall area if the temperature of the internal environment is 20 °C and the external air temperature is 5 °C.

11. (a) Explain how moisture content and density affect the thermal conductivity of insulating materials.
(b) When the 50 mm air cavity of the wall in Question 10 is filled with an insulating material of thermal conductivity

0.04 W/m K, estimate the percentage reduction in heat transmission.

12. (a) Cavity fill is a convenient means of thermally improving many walls. What factors should be considered before this mode of improvement is selected?

(b) The U value of a foam filled cavity wall is 0.6 W/m K. If the internal air temperature is maintained at 18 °C and the external surface resistance is 0.06 m² K/W, to what temperature does the external surface of the wall fall when the external air temperature is −1 °C?

13. (a) Describe two methods of increasing the thermal insulation provided by a building.

(b) What thickness of insulation is required to reduce the heat transmission through a flat roof from 1.5 W/m² K to 0.6 W/m² K?
Assume that the thermal conductivity of the insulation material is 0.1 W/m K.

14. (a) Estimate the thermal transmittance of single glazing for three exposure conditions:
   (i) sheltered, external surface resistance = 0.08 m² K/W;
   (ii) normal, external surface resistance = 0.06 m² K/W;
   (iii) severe, external surface resistance = 0.03 m² K/W.
   The internal surface resistance is 0.12 m²K/W.

(b) Use the results of your calculations to justify the statement that while it is usually adequate to use the external surface resistance for normal exposure in heat transmission calculations on a well insulated structure, the external thermal resistance appropriate to the degree of exposure must be used in calculations dealing with low thermal resistance elements such as windows.

15. The proposed wall construction in a detached dwelling has a U value of 0.6 W/m² K. What is the maximum percentage of:
(a) single glazing ($U = 5.7$ W/m² K); and
(b) double glazing ($U = 2.8$ W/m² K)
which may be allowed in the design if the average U value of the perimeter wall is not to exceed 1.2 W/m² K?

16. Modern lightweight buildings with large glazed areas experience excessive internal temperatures during periods of sunshine.
Explain why this solar overheating occurs and suggest remedies.

17. (a) Explain the terms:
   (i) fabric heat loss;
   (ii) ventilation heat loss.

(b) The internal environment in a flat roofed building 18 m × 9 m × 3.5 m high is maintained at a temperature of 18 °C when the external air temperature is 3 °C, and the average ventilation rate is 1 air change per hour. The window

and door areas are $36\,m^2$ and $8\,m^2$ respectively and the
building elements have the following thermal properties.

Building element	U value $(W/m^2\,K)$
Floor	0.6
External wall	1.1
Window	5.6
Door	2.8
Roof	0.9

(i) Calculate the total heat loss rate from the building
     enclosure.
(ii) Estimate the percentage reduction in total heat
     transmission if the U value of the roof is improved to
     $0.4\,W/m^2\,K$ by the application of thermal insulation.

18. In the building in Question 17 internal heat gains arise from
    people (1.4 kW), lighting (600 W) and machinery (800 W).
    Calculate the internal temperature at which these casual heat
    gains equal the total heat loss from the building. Assume an
    external air temperature of $-1\,°C$.

Chapter 2

Thermal comfort

The human body is thermally comfortable when the heat constantly produced by bodily processes balances heat losses and gains to and from the environment. The achievement of such a balance depends upon the combined effect of many factors: personal variables such as activity and clothing as well as physical variables. The physical variables: (i) air temperature; (ii) radiant temperature; (iii) humidity; and (iv) air movement are considered to be the four main environmental factors affecting human comfort.

Air temperature

Air temperature (or dry bulb temperature) is an air temperature measurement which is independent of radiant heat from the surroundings and in which air motion relative to the measuring device is not significant. Because the concept of wet bulb temperature is used as a characteristic property of moist air the conventional air temperature is often called the dry bulb temperature.

Many thermometer types are available for air temperature measurement. The traditional mercury in glass thermometer is frequently used in environmental studies, but when a continuous temperature record is required, electrical resistance sensors (such as thermistors and platinum resistance thermometers) or thermocouples · are used with suitably calibrated recording or data logging equipment.

A thermometer suspended within a room reaches thermal equilibrium by convection (the contribution from conduction is very small) to or from the ambient air and by radiation exchange with the surrounding surfaces. A temperature between air temperature (t_a) and mean radiant temperature (t_r) is recorded. In the ideal air temperature measurement the thermometer should be shielded by an open polished metal cylinder, long enough to protect the temperature sensor from thermal radiation exchange with the surrounding surfaces and with a diameter which permits free circulation of air around the sensor.

Mean radiant temperature

Human comfort within an enclosed structure depends upon the radiation exchange between the human body and the surrounding surfaces. The concept of mean radiant temperature (see Appendix) is used to describe this balance.

Mean radiant temperature (t_r), is the temperature of a uniform black enclosure in which a solid body or occupant would exchange

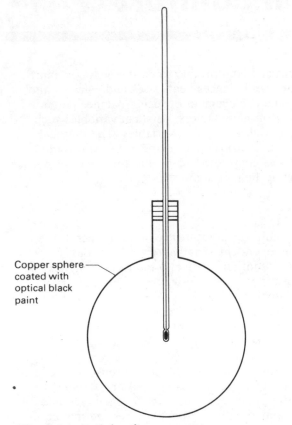

Copper sphere coated with optical black paint

Fig. 2.1 A globe thermometer

the same quantity of radiant heat as in the real non-uniform environment.

The mean radiant temperature within a room may be estimated by the use of a globe thermometer (Fig. 2.1) together with measurements of air temperature and air movement. To record globe temperature (t_g) a mercury in glass thermometer is mounted with its bulb at the centre of a 150 mm diameter hollow copper sphere coated with mat black paint. If the mean radiant temperature is higher than the air temperature, the black globe gains heat by radiation exchange with the surrounding surfaces and loses heat by convection. The temperature recorded by the globe thermometer (t_g) will be above air temperature. Conversely, with the surroundings cooler than the air, the black globe loses heat by radiation exchange with the surrounding surfaces and gains heat by convection. The globe temperature (t_g) will be below air temperature. When a steady globe temperature is reached, globe temperature (t_g), air temperature (t_a) and air velocity (v) are recorded. Air temperature and air velocity must also be measured since convective heat loss from the globe depends upon air temperature and air velocity as well as the globe temperature. Globe temperature lies between air temperature and mean radiant temperature. In still air $(v = 0)$ the globe temperature is equal to the mean radiant temperature.

With these measurements $(t_g, t_a$ and $v)$, mean radiant temperature (t_r) may be obtained from the nomogram supplied with the globe thermometer.

Air movement

Convective heat loss from a surface depends upon the air movement over the surface. Hence air velocity influences the rate of convective heat loss by the human body and is one of the factors which must be considered when assessing human comfort conditions. The low air velocities (between 0.05 m/s and 0.5 m/s) usually encountered in rooms and enclosed spaces are variable in magnitude and direction. The kata thermometer is a suitable instrument for measuring the non-directional magnitude of the air speed, sometimes referred to as the total air speed. Bead thermistor and hot wire anemometers may be used when a more detailed knowledge of the air velocity is required.

The kata thermometer (Fig. 2.2) is an alcohol in glass thermometer with a large silvered bulb (18 mm diameter × 40 mm length) and two temperature graduations such as 54.5 °C and 51.5 °C upon the stem. The thermometer is heated by immersion in hot water until the level of the alcohol is well above the upper marked temperature. The thermometer (bulb and stem) is carefully dried and suspended with stem vertical and still in the region of required air

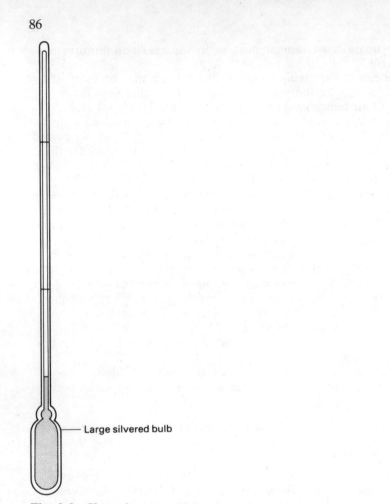

Fig. 2.2   Kata thermometer

velocity measurement. As cooling occurs the time taken by the alcohol level to fall from the upper temperature to the lower temperature is measured with a stopwatch. The process is repeated until agreement is obtained in three consecutive measurements. Air temperature is measured by a mercury in glass thermometer suspended near the kata thermometer.

The kata thermometer bulb is brightly silvered* so that as much incident radiation as possible is reflected. After the hot water heating, the top of the alcohol column should be well above the upper temperature graduation, so that the rate of cooling will be steady during the time measurement. The heat loss from the silvered bulb,

---

* An unsilvered bulb may be used when the difference between mean radiant temperature and air temperature is not large.

which is mainly convective, may be related to the air speed by the equation

$$H = (a + b\sqrt{v})(T - t_a) \qquad [2.1]$$

where a, b and T are constants, H is the mean rate of heat loss per unit area of cooling surface in cooling through the specified temperature range, v is the air speed and t_a is the air temperature. Manufacturers calibrate each instrument and provide a kata factor (F) which is the heat loss from unit area of cooling surface during the cooling period. Hence H is F divided by the measured cooling time. Air speed is either calculated from equation [2.1] or obtained from charts supplied with the instrument.

Air/moisture mixture and its assessment

Evaporation of body moisture from the skin and respiratory tract is the mode of body heat loss most influenced by humidity. Since air is usually composed of a mixture of gases (nitrogen, oxygen, a small quantity of carbon dioxide and traces of inert gases) and water vapour a brief examination of the behaviour of gases and water vapour follows.

Pressure in a gas

In a simple model of an ideal gas, the gas molecules move about at random, making elastic collisions with one another and the walls of the containing vessel. During collisions with the vessel walls the rapidly moving gas molecules impart momentum to the walls. The rate of change of momentum in these collisions exerts the force at the walls which gives rise to the gas pressure.

The pressure exerted by a fixed volume of a gas is proportional to the number of gas molecules present and to the absolute temperature of the gas. If the number of gas molecules within a container is increased the pressure increases because the gas molecules make more collisions with one another and the vessel walls. If the gas is heated within an enclosed container it exerts a greater pressure because the heated gas molecules move faster and experience more impacts with one another and the vessel walls.

Pressure, the force exerted on unit area, is measured in newtons/square metre (N/m^2). The N/m^2 is also known as the pascal (Pa). Standard atmospheric pressure is defined as $101.325 \, kN/m^2$ or $101.325 \, kPa$. Another unit sometimes used to indicate pressure is the bar.

1 bar $= 10^5 \, N/m^2 = 10^5 \, Pa$
1 mbar (millibar) $= 10^2 \, N/m^2 = 10^2 \, Pa$

88

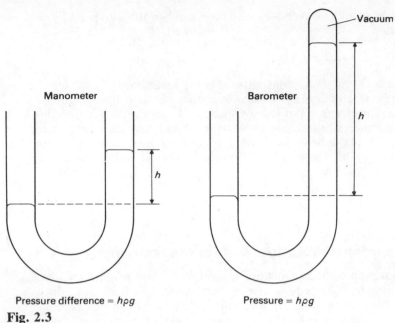

Fig. 2.3

Pressure and pressure differences indicated by barometers and manometers (Fig. 2.3) are often expressed in millimetres (mm) of mercury (or some other liquid). Standard atmospheric pressure corresponds to a barometer reading of 760 mm of mercury. This liquid column height (h) is converted to pressure when multiplied by liquid density (ρ) and acceleration due to gravity (g).

Example 2.1 Use a barometer reading of 760 mm of mercury to express atmospheric pressure in N/m². Assume that the density of mercury, $\rho = 13.6 \times 10^3$ kg/m³ and the acceleration due to gravity, $g = 9.8$ m/s².

$h = 760$ mm $= 0.760$ m

$$\text{Atmospheric pressure} = \underset{(\text{N/m}^2)}{h} \times \underset{(\text{m})}{\rho} \times \underset{(\text{kg/m}^3)}{g} \quad (\text{m/s}^2)$$

$$\text{Atmospheric pressure} = 0.760 \times 13.6 \times 9.8 \times 10^3$$
$$= 101.3 \times 10^3 \, \text{N/m}^2$$
$$= \underline{101.3 \, \text{kN/m}^2}$$

Dalton's law of partial pressures

Each gas in a mixture of gases exerts a pressure known as its partial pressure, which is the same as the pressure it would exert if it were alone and occupying the volume of the mixture. The total pressure exerted by a mixture of gases equals the sum of the partial pressures (Fig. 2.4).

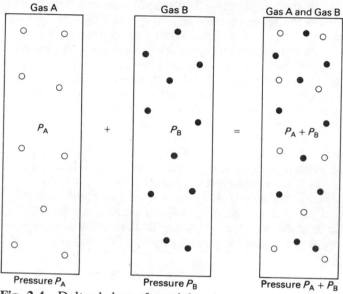

Fig. 2.4 Dalton's law of partial pressure

Saturated and unsaturated vapours

Water vapour evaporates and condenses readily at room temperatures and pressures so it is often convenient to consider this vapour separately, but to treat the mixture of gases in dry air as a single gaseous substance.

When a liquid such as water is placed in a closed container a *saturated vapour* is soon established in the region above the liquid surface. The more energetic liquid molecules tend to leave the region of the liquid surface and become gaseous. Those vapour molecules in the vicinity of the surface of the liquid collide with the surface and some return to the liquid. Initially more molecules leave the liquid than return to it, but when a dynamic equilibrium is established the number of molecules which leave the vapour and return to the liquid (condense) in unit time equals the number of liquid molecules which escape from the liquid surface (evaporate) in unit time. This equilibrium situation depends upon molecular energies (velocities) and hence upon temperature.

No more vapour forms unless the temperature is increased. Raising the temperature increases the molecular velocities and hence the rate of condensation and evaporation at the liquid surface. The number of vapour molecules and hence the vapour pressure in the enclosed region above the liquid is greater. The partial pressure exerted by this saturated vapour, known as the saturation vapour pressure (SVP) or the saturated vapour pressure, increases rapidly with temperature as indicated in Fig. 2.5. The air can hold more water vapour when the temperature is raised. The air can hold less water vapour when the temperature is lowered. The pressure exerted by a saturated vapour is constant for a given temperature of the liquid, even though the volume occupied by the vapour may change. (A saturated vapour does not obey the gas laws.)

A saturated vapour is in equilibrium with its own liquid. If only a small quantity of liquid is placed in a closed container and all the liquid evaporates, the vapour produced is said to be unsaturated. At a given temperature the partial pressure exerted by the *unsaturated vapour*, known as the vapour pressure, is less than the saturation vapour pressure for that temperature. The pressure of a given mass of unsaturated vapour is dependent upon volume as well as temperature. The gas laws are obeyed approximately.

Air is normally composed of a mixture of gases and water vapour. Air entirely without the water vapour present is referred to as dry air. The total air pressure is the sum of the partial pressure of the gas mixture (dry air) and the partial pressure (vapour pressure) of the water vapour.

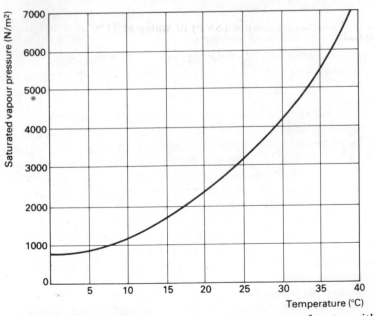

Fig. 2.5 Variation of saturated vapour pressure of water with temperature

Moisture content, a measure of the quantity of water vapour, is the mass of water vapour contained in 1 kg of dry air in an air/water vapour mixture (unit kg/kg or g/kg). Terms such as absolute humidity, mixing ratio, specific humidity and humidity ratio are also used to describe this quantity.

In the absence of condensation or evaporation processes, moisture content is temperature independent, but gives no indication of the closeness of the partial pressure of the water vapour to the saturation vapour pressure of water vapour at the air temperature in question.

Relative humidity

Many practical situations are more dependent upon how near the saturation vapour pressure is than upon the actual quantity of water vapour in the air. *Relative humidity* expresses as a percentage the ratio between the actual water vapour pressure of an air sample and the maximum water vapour pressure (the SVP) it could sustain at the same temperature:

$$\text{relative humidity} = \frac{\text{water vapour pressure}}{\text{SVP}} \times 100 \text{ per cent}$$

where both pressures refer to air at the same temperature.

Example 2.2 Calculate the relative humidity at 21 °C when the water vapour pressure in the air sample is 1.200 kN/m². The saturation vapour pressure (SVP) of water at 21 °C is 2.486 kN/m²:

$$\text{relative humidity} = \frac{\text{water vapour pressure}}{\text{SVP}} \times 100 \text{ per cent}$$

$$= \frac{1.200}{2.486} \times 100$$

$$= 48.3 \text{ per cent at } 21 °C$$

The relative humidity of the air sample at 21 °C is 48.3 per cent.

A high value of relative humidity means that the partial pressure of the water vapour is close to the saturation vapour pressure. At 100 per cent relative humidity the air is said to be saturated, it contains as much water vapour as it can hold at that particular temperature. At a low value of relative humidity the air is far from being saturated. Relative humidity is temperature dependent since saturation vapour pressure varies markedly with temperature.

Percentage saturation or degree of saturation is the ratio expressed as a percentage of the moisture content of air at a given temperature to the moisture content of saturated air at the same temperature.

Percentage saturation and relative humidity are identical only for dry air (relative humidity = 0 per cent) and saturated air (relative humidity = 100 per cent). At ambient air temperatures the difference between percentage saturation and relative humidity is small.

Dewpoint

When unsaturated air is cooled, the dewpoint temperature is reached, the air becomes saturated and cannot continue to hold all its water vapour. Further cooling below this dewpoint temperature results in condensation of some of the vapour. The *dewpoint temperature* may be defined as the temperature of saturated air which has the same vapour pressure as the air under examination.

The equality of the saturation vapour pressure at the dewpoint with the partial pressure of the water vapour in the air under consideration leads to an alternative ratio for relative humidity.

Relative humidity

$$= \frac{\text{SVP of water vapour at the dewpoint temperature}}{\text{SVP of water vapour at the air temperature}} \times 100 \text{ per cent}$$

Example 2.3 Calculate the relative humidity of a sample of air at 21 °C which has a dewpoint temperature of 11 °C. From tables: SVP = 2.486 kN/m² at 21 °C, SVP = 1.308 kN/m² at 11 °C.

$$\text{Relative humidity} = \frac{1.308}{2.486} \times 100$$
$$= 52.6 \text{ per cent at } 21\,°C$$

The relative humidity of the air sample at 21 °C is 52.6 per cent.

Relative humidity measurement

Hygrometers are used to assess air moisture. The most frequently measured quantity is relative humidity since it affects feelings of comfort and warmth especially at high air temperatures. Also certain industrial processes are influenced by the relative humidity. Low humidity may lead to problems with wood shrinkage and static electricity while mould growth and condensation on cool surfaces are often the result of long periods of high humidity (over about 70 per cent). In most situations a relative humidity between about 40 and 70 per cent is acceptable.

Various instruments for relative humidity measurement are available:

(a)   a psychrometer, some type of wet and dry bulb thermometer;
(b)   a dewpoint apparatus;
(c)   others, such as hair or paper hygrometers.

Wet and dry bulb hygrometers

Relative humidity measurements are most frequently made using a hygrometer of the wet and dry bulb type. This method of measurement is based upon the fact that when water evaporates in air, the rate of evaporation depends upon the degree of saturation of the surrounding air. The liquid molecules escape from the water surface at a greater net rate when the surrounding air is relatively dry but at a slower net rate when the air is nearly saturated. The molecules which escape from the liquid surface are usually the most energetic so that a cooling of the liquid (to provide the latent heat of evaporation) is always associated with the evaporation process.

The wet and dry bulb hygrometer (thermometer) consists of two mercury in glass thermometers mounted side by side (Fig. 2.6). One of these with an ordinary dry bulb is used to measure the temperature of the air, which is called the *dry bulb temperature*. The bulb of the second thermometer is kept wet by a sleeve or wick of damp muslin or similar material. This sleeve is continuously moistened by contact with a small reservoir of distilled or clean fresh water. Water evaporates from the sleeve at a rate which depends upon the relative humidity of the surrounding air, the lower the relative humidity the faster the rate of evaporation. The wet bulb is cooled and a constant temperature is reached at which the latent heat absorbed from the wet bulb for the evaporation process is equal to the heat transferred by convection from the air to the cool bulb. The temperature of the cool wet bulb, *the wet bulb temperature*, is recorded. *Wet bulb temperature* is that temperature at which water evaporated into the air brings the air to saturation at the same temperature. If the air is completely saturated with water vapour the temperature recorded by the wet bulb thermometer is the same as the dry bulb temperature.

So that air in the vicinity of the wet bulb remains representative of the surrounding air conditions, water which evaporates from the bulb should be immediately removed to prevent the accumulation of saturated air. The wet bulb temperature depends only upon the relative humidity of the air when heat transfer from the bulb takes place by convection. Both of these conditions may be achieved by increasing the air flow over the two thermometer bulbs to at least 5 m/s. The accumulation of saturated air is prevented and convective heat transfer is increased to such an extent that radiative heat transfer is negligible by comparison. The wet bulb temperature is conveniently independent of environment and air speed and dependent only upon the relative humidity of the air.

The difference between the wet and dry bulb temperatures, known as the wet bulb depression, enables the relative humidity and moisture content of air to be obtained from tables, a slidescale supplied with the psychrometer or from charts.

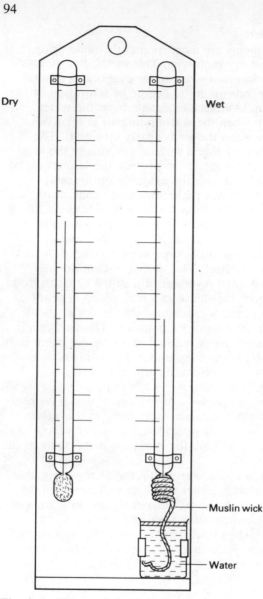

Dry

Wet

Muslin wick

Water

Fig. 2.6 Wet and dry bulb hygrometer

Three types of this hygrometer are: (1) wall mounted (Fig. 2.6); (2) whirling (Fig. 2.7); (3) Assman.

In order to obtain reasonable readings from the wall mounted hygrometer the air around the hygrometer should be disturbed by fanning.

The sling or whirling psychrometer is mounted on a handle

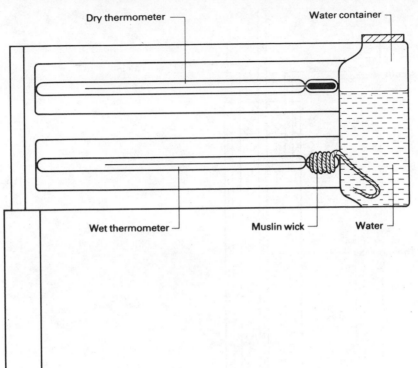

Fig. 2.7   Whirling hygrometer

which enables it to be whirled rapidly (so that air speed over the thermometer bulbs is at least 5 m/s) wherever relative humidity is to be measured. As soon as whirling is stopped the thermometers are read, the wet bulb reading being taken first since it will begin to rise. This procedure should be repeated until three sets of readings obtained in sequence agree closely.

In the Assman or aspirated hygrometer a fan draws air at a controlled rate past the two thermometer bulbs. Wet and dry bulb temperatures are recorded.

A dewpoint apparatus: the Regnault hygrometer

The Regnault hygrometer (Fig. 2.8) determines the dewpoint temperature and hence the relative humidity. A stream of air is blown through the ether in the part-silvered tube, to evaporate the ether and produce cooling. Air in contact with the tube is cooled and eventually condensation may be observed on the silvered surface. The temperature at which this condensation first occurs is recorded. The air stream is stopped and the tube is allowed to warm up. The temperature at which condensation ceases to occur on the silvered

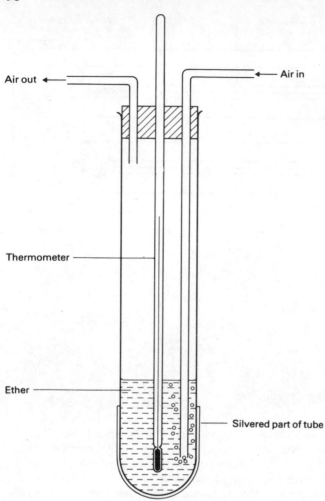

Fig. 2.8 Regnault's hygrometer

surface is recorded. The average value of these two temperatures is taken as the dewpoint temperature. Air temperature is recorded on a thermometer placed in an identical, but empty part-silvered tube.

Since relative humidity

$$= \frac{\text{SVP of water vapour at the dewpoint temperature}}{\text{SVP of water vapour at the air temperature}} \times 100 \text{ per cent}$$

the dewpoint temperature and air temperature enable the relative humidity to be calculated from tables.

Hair and paper hygrometers

In these hygrometers the hair or paper coil (Fig. 2.9) act as the relative humidity sensor. Hair and paper expand because of moisture absorption when the humidity is high. Hair may be used as the humidity sensor in the thermohydrograph, a simple but not very accurate air temperature and relative humidity recording device in which a bimetallic strip measures temperature.

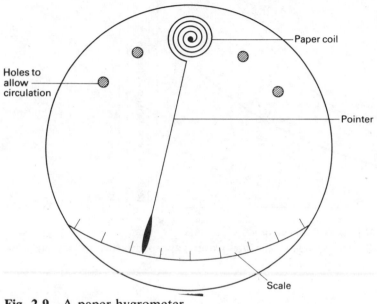

Fig. 2.9 A paper hygrometer

The psychrometric chart

The psychrometric chart represents graphically properties of the air-water vapour mixture related to thermal comfort, condensation and air conditioning calculations.

Each parameter depicted on the psychrometric chart has a separate scale. In the sketch (Fig. 2.10):

1. Dry bulb temperature (°C) increases from left to right on a horizontal scale.
2. Moisture content (mixing ratio) in g/kg (dry air) or kg/kg (dry air) is represented on one of the two vertical scales on the right-hand side of the chart.
3. Vapour pressure in N/m^2, Pa or mbar occupies the second vertical scale on the right-hand side of the chart.

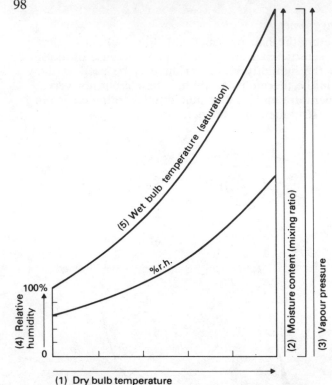

Fig. 2.10 Properties of the air–water vapour mixture as depicted on the psychrometric chart

4. The short vertical scale on the left-hand side of the chart represents relative humidity (per cent r.h.) and refers to the curved lines rising from left to right. On some charts the per cent r.h. value may be written along the curve. The uppermost of these curves represents saturation (100 per cent r.h.).

5. The wet bulb temperature scale (°C) increases from left to right along the saturation curve. Diagonal straight lines moving down from the saturation curve may be used with the wet bulb temperature scale to interpret whirling hygrometer results.

Guidance in the use of the psychrometric chart (Fig. 2.11) is provided in each of the following examples by the psychrometric chart sketches.

Example 2.4 Use the psychrometric chart in Fig. 2.11 to determine the moisture content and vapour pressure of air at 50 per cent relative humidity when the air temperature is (a) 3 °C, (b) 15 °C, (c) 22 °C.

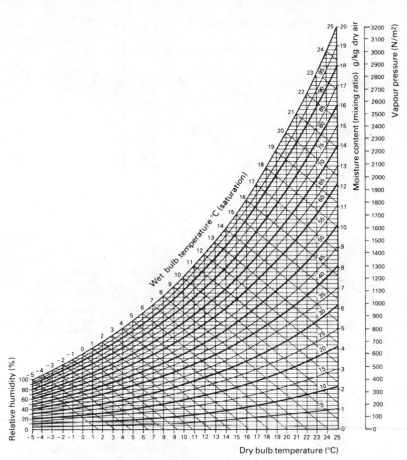

Fig. 2.11 Psychrometric chart

(a) From 3 °C on the dry bulb temperature scale follow the vertical
     line until it intercepts the 50 per cent r.h. curve, point A on the
     sketch in Fig. 2.12. Construct a horizontal line from point A to
     cut the vertical scales for moisture content and vapour pressure.
     Read off the required values from these two scales.

(b), (c) This procedure is repeated for the other temperatures
     (points B and C in the sketch).

50 % r.h.	Temperature (°C)	Moisture content (g/kg)	Vapour pressure (N/m²)
(a)	3	2.4	370
(b)	15	5.4	850
(c)	22	8.3	1330

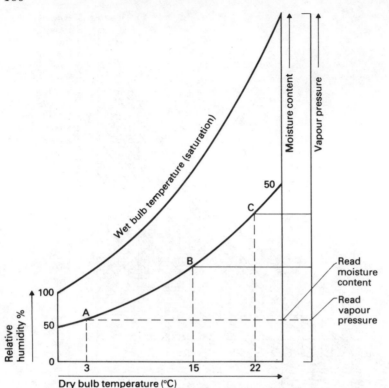

Fig. 2.12 Psychrometric chart sketch (Example 2.4)

Since the relative humidity is constant at 50 per cent all the points A, B, C lie along the 50 per cent r.h. curve.

Example 2.5 The mixing ratio (moisture content) is 9 g/kg (dry air). Estimate the relative humidity at (a) 21 °C, (b) 15 °C, and (c) predict the temperature at which condensation occurs.

In this example the moisture content is constant.

(a) Follow the vertical line from the dry bulb temperature of 21 °C until it intercepts the horizontal line from the 9 g/kg moisture content (point A in Fig. 2.13). Point A lies between the 55 per cent r.h. and 60 per cent r.h. curve. An estimate of the relative humidity at 21 °C is 58 per cent.

(b) Repeat the procedure for the 15 °C dry bulb temperature and the same horizontal moisture content line. Point B (Fig. 2.13) lies between the 85 per cent and 80 per cent r.h. curves. Approximately the relative humidity at 15 °C is 83 per cent.

(c) Condensation occurs when 100 per cent saturation is reached, when the horizontal line from the 9 g/kg moisture content cuts the saturation curve, point C in Fig. 2.13. The dewpoint

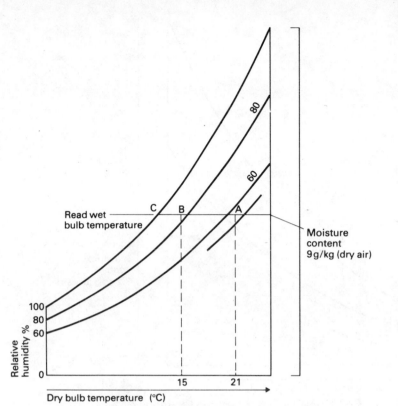

Fig. 2.13 Psychrometric chart sketch (Example 2.5)

temperature, 12.6 °C, is read from the wet bulb temperature scale. (The wet and dry bulb temperatures are equal at saturation (100 per cent r.h.).)

(a) 58 per cent r.h. at 21 °C,   (b) 83 per cent r.h. at 15 °C
(c) 12.6 °C.

Example 2.6   The air in a domestic dwelling is maintained at 18 °C. Initially the moisture content is 3.8 g/kg. Calculate: (a) the relative humidity when an additional 3.4 g/kg moisture content is added to the air by the occupants; (b) how much water vapour must be added to the initial 3.8 g/kg to produce saturated air.

(a)   Total moisture content = 3.8 + 3.4 = 7.2 g/kg. Find the intercept between the vertical line from the dry bulb temperature of 18 °C and the horizontal line from a moisture content of 7.2 g/kg, point A in Fig. 2.14. Point A is just above the 55 per cent r.h. curve. Relative humidity = 56 per cent at 18 °C.

(b)   At saturation the wet bulb temperature is equal to the dry bulb temperature. From point B (Fig. 2.14), at a wet bulb

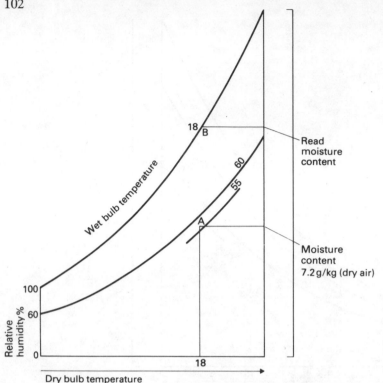

Read
moisture
content

Moisture
content
7.2 g/kg (dry air)

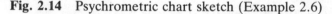

Dry bulb temperature

Fig. 2.14 Psychrometric chart sketch (Example 2.6)

temperature of 18 °C on the saturation curve draw a horizontal
line to intercept the moisture content scale. The moisture
content of saturated air at 18 °C is 12.9 g/kg. The water vapour which must be added to the initial 3.8 g/kg to produce
saturation is $12.9 - 3.8 = 9.1$ g/kg.

(a) 56 per cent r.h. at 18°C   (b) 9.1 g/kg.

Example 2.7   In a determination of relative humidity using a whirling hygrometer, a wet bulb temperature of 12 °C and a dry bulb
temperature of 19 °C are recorded. Use the psychrometric chart
(Fig. 2.11) to estimate the relative humidity.

Follow the vertical line from the dry bulb temperature of 19 °C
until it intercepts the diagonal line from the wet bulb temperature of
12 °C. The relative humidity is estimated from the distance of the
point of intersection (point A in Fig. 2.15) from the adjacent relative
humidity curves.

The relative humidity is estimated as 42 per cent at 19 °C.

Example 2.8   An air temperature of 18 °C and a dewpoint temperature of 10 °C is measured using a Regnault hygrometer. Use the
psychrometric chart in Fig. 2.11 to estimate the relative humidity.

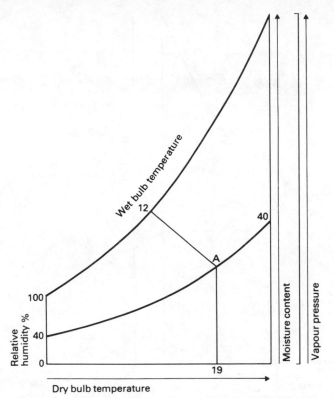

Fig. 2.15 Psychrometric chart sketch (Example 2.7)

$$\text{Relative humidity} = \frac{\text{SVP at the dewpoint temperature}}{\text{SVP at the air temperature}} \times 100 \text{ per cent}$$

The saturation vapour pressure (SVP) at the dewpoint temperature may be obtained by drawing a horizontal line from the wet bulb temperature of $10\,°C$ on the saturation curve (point A in Fig. 2.16) to intercept the vapour pressure scale.

SVP at $10\,°C = 1200\,N/m^2$

The SVP at the air temperature may be obtained by constructing a second horizontal line from the wet bulb temperature of $18\,°C$ on the saturation curve (point B in Fig. 2.16) to intercept the vapour pressure scale.

SVP at $18\,°C = 2050\,N/m^2$

$$\text{Relative humidity} = \frac{1200}{2050} \times 100$$

$$= 58.5 \text{ per cent at } 18\,°C.$$

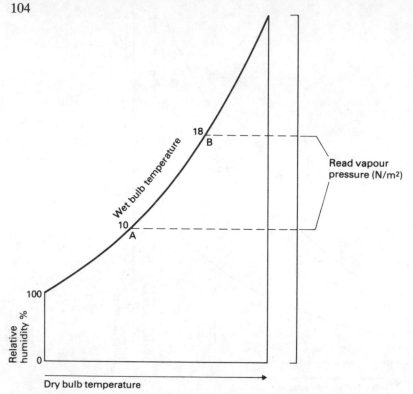

Fig. 2.16 Psychrometric chart sketch (Example 2.8)

Ventilation

Ventilation within an enclosed structure must provide fresh air to:

(a) maintain an adequate supply of oxygen for breathing;
(b) remove carbon dioxide and water vapour (products of respiration);
(c) dilute odours, for example body odours, cooking smells, cigarette smoke, to an acceptable or unnoticeable level;
(d) reduce the risk of airborne infection;
(e) give adequate air movement for a comfortable environment, one which is refreshing but not draughty;
(f) prevent or reduce condensation risk within the enclosure;
(g) remove excess heat.

When minimum ventilation requirements are based upon keeping the carbon dioxide content of the air down to a safe level it is recommended that the carbon dioxide content of air during an 8 hour occupation period should not exceed about 0.5 per cent (*CIBS Guide*, Section A1). Respiration (items (a) and (b)) is adequately catered for. The quantity of fresh air needed to dilute odours is greater, and although more difficult to assess, it is a more suitable criterion for evaluating ventilation requirements. The fresh air supply in an occupied room should be sufficient to remove the effects of tobacco smoke and odours.

Ventilation needs may be expressed in terms of the volume of the fresh air supply per unit time (litres/second, l/s), and are often dependent upon the volume of space per person and the contaminants. The statutory minimum volume per person in factories and offices is $11.5 \, \mathrm{m}^3$. The corresponding minimum outdoor air supply is $4.72 \, \mathrm{l/s}$ per person. In open plan air conditioned offices where some smoking takes place an outdoor air supply of $5 \, \mathrm{l/s}$ per person or $1.3 \, \mathrm{l/s}$ per m^2 of floor area (whichever is greater) is specified. In domestic kitchens and toilets the minimum outdoor air supply is $10 \, \mathrm{l/s}$ per square metre of floor area.

In a domestic dwelling, special ventilation provision may be necessary in the kitchen and bathroom and in overcrowded conditions. Air infiltration normally provides the necessary fresh air supply in the remaining parts of the house.

The effect of varying ventilation rates on human thermal comfort is measured most easily in terms of air velocity rather than air supply (l/s) or air change (ach). Air velocities less than $0.01 \, \mathrm{m/s}$ give rise to stagnant conditions. During cold weather the air in a dwelling or factory which is not overheated is unlikely to be stagnant if the recommended fresh air supply is provided. Greater ventilation rates are usually required during warm weather. Typically, air velocities not exceeding $0.1 \, \mathrm{m/s}$ at air temperature of $16 \, ^\circ\mathrm{C}$ to $20 \, ^\circ\mathrm{C}$ increasing to a maximum velocity of about $0.45 \, \mathrm{m/s}$ at $25 \, ^\circ\mathrm{C}$ provide thermal comfort (dependent upon activity and clothing).

Ventilation rates are expressed in terms of air changes per hour (ach) for use in heating calculations and condensation problems. During the ventilation process fresh air is mixed with the air within the enclosure. One air change per hour (1 ach) implies the diffusion of the room volume of fresh air into the room in 1 hour. Although a similar volume of air is withdrawn from the room the entire initial air content of the room is not exchanged for fresh air. Especially when ventilation rates are rapid, the diffusion process is incomplete and some of the fresh air supplied to the room leaves without carrying off its quota of contaminants.

Human comfort

The thermal environment within an enclosed structure should provide thermal comfort for the highest possible percentage of the occupants.

Thermal comfort or thermal neutrality may be defined as a condition in which a person would prefer neither warmer nor cooler surroundings.

The human body's thermal regulatory system maintains a constant body temperature. A heat balance is achieved where the rate of heat production in the body is equal to the rate of heat loss from it as expressed by the energy balance equation in Fig. 2.17. All the parameters which affect the rate of heat exchange between the body and its environment produce sensations of warmth or coldness.

$$\boxed{M - W} \quad = \quad \boxed{C + E + R} \quad + \quad \boxed{S}$$

total heat production rate   total heat loss rate   heat storage rate

M = metabolic rate
W = rate at which energy is expended in mechanical work
C = rate of heat loss by convection
E = rate of heat loss by evaporation
R = rate of heat loss by radiation
S = rate at which heat is stored within the body
($S = 0$ when the body temperature remains constant)

Fig. 2.17 Energy balance equation

The factors which most influence human thermal comfort within an enclosed structure are:

Personal variables

(a)   activity level;
(b)   thermal resistance of clothing (clo value);

Physical variables

(c)   air temperature;
(d)   mean radiant temperature;
(e)   air velocity;
(f)   humidity (water vapour pressure in ambient air).

The heat flow rate from the human body core to the thermal environment depends upon activity level. When work is performed by the body the metabolic rate increases in order to provide the necessary energy. However, only a small part of this energy is used in the work process, the majority is transferred into heat. Table 2.1 indicates the typical rate of heat emission from the human body for various activity levels (*CIBS Guide*, Section A1).

Clothing

Clothes provide a layer of thermal insulation between the body surface and the environment. The thermal resistance of clothing is

Table 2.1 Typical values of heat emission rate from the human body. (Adult male, body surface area 2 m^2.)

Activity level	Example	Heat emission rate (W)
Sleeping		80
Sitting	Sitting reading or watching television	115
Standing	Serving at a counter (as in a store or a bank), walking slowly	160
Active	Movement of the whole body, light bench work in a factory	235

expressed in terms of a unit known as the clo. 1 clo is equivalent to a thermal resistance of 0.155 m^2 K/W and is approximately the insulation afforded by a man's suit (normal winter indoor wear). Light clothing (indoor summer wear) such as trousers and shirt or skirt and blouse provides an insulation level of about 0.5 clo. (Appropriate underwear is assumed.) The clothing people wear is influenced by the daily mean outdoor temperature appropriate to the time of year (BRE Current Paper CP 53/78).

Activity increases air movement over clothing surfaces and so reduces thermal resistance.

Thermal comfort is obtained when a balance is achieved between the personal variables, activity and clothing level and the physical variables within the enclosure.

Thermal indices

The thermal environment within a room may be assessed from four measurements; air temperature, mean radiant temperature, air velocity and wet bulb temperature (for humidity and water vapour pressure in ambient air).

Many thermal indices have been devised in an attempt to represent thermal comfort conditions by a single temperature. The index temperature for comfort, which the CIBS recommend for use in the UK, is the dry resultant temperature (t_{res}). *Dry resultant temperature* is the temperature (in °C) measured at the centre of a blackened globe 100 mm in diameter (*CIBS Guide*, Section A1). Recent work suggests that the temperature recorded by a smaller globe (approx. 40 mm diameter) is to be preferred as an index of subjective warmth for normal indoor conditions (BRE Current Paper CP 9/78).

108

Dry resultant temperature is given by

$$t_{res} = \frac{t_r + t_{ai} \sqrt{10v}}{1 + \sqrt{10v}} \qquad [2.2]$$

where t_{ai} = internal air temperature

   t_r = mean radiant temperature

   v = air velocity

Indoor air speeds are frequently low. Substituting $v = 0.1$ m/s (typical of still air) into equation [2.2] simplifies the expression for t_{res}

$$t_{res} = \tfrac{1}{2}t_r + \tfrac{1}{2}t_{ai}$$

In still air conditions the dry resultant temperature or globe temperature (globe diameter 100 mm or less) combines the effect of

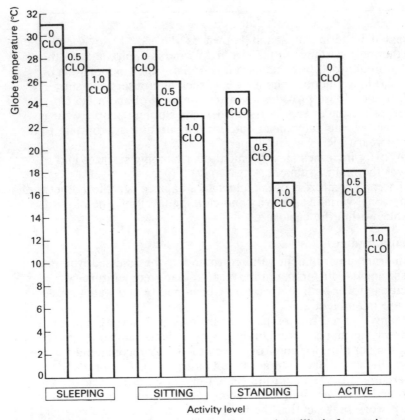

Fig. 2.18 Comfortable room temperatures in still air for various activity and clothing levels. (Plotted from data available in *BRE Digest*, 226.)

surface temperatures and air temperatures. Figure 2.18 indicates thermal comfort temperatures (measured by globe thermometer) for various activity and clothing levels in still air.

Examples of comfort temperatures (t_{res}), selected from the many recommended design values listed in the *CIBS Guide*, Section A1 are given in Table 2.2.

Table 2.2 Recommended values for internal dry resultant temperatures (t_{res})

Indoor environment	t_{res} (°C)
Dwellings	
Living room	21
Bedroom	18
Offices	20
Factories	
Light work	16
Heavy work	13

At an air speed of 0.2 m/s or greater an increase in dry resultant temperature is required to compensate for the cooling effects of air movement. The temperature corrections needed to maintain comfort conditions are indicated in Fig. 2.19. Normally air speeds in excess of about 0.3 m/s are only acceptable during warm weather or in too hot an environment.

Humidity

Humidity, the fourth physical variable in the thermal environment is the one which most influences the rate of heat loss from the body by evaporation of moisture from the skin and respiratory tract.

The effect of humidity on thermal comfort is thought to be very small when the resultant temperature is close to the recommended comfort value (appropriate to activity and clothing level). At higher temperatures (above about 25 °C) and at higher rates of activity (especially in warm conditions) sweating is an important mechanism for bodily heat loss. At temperatures above about 29 °C almost all the heat loss from the body is by evaporation. In these cases a reduction in relative humidity improves comfort but a high water vapour pressure aggravates discomfort.

While it is not possible to provide a thermal environment which will be satisfactory for everyone, the majority of people will prefer neither warmer nor cooler surroundings in rooms where the dry

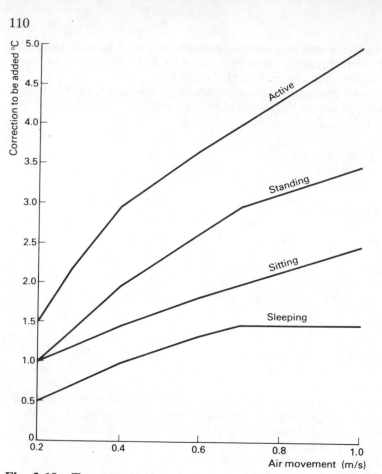

Fig. 2.19 Temperature corrections needed to maintain comfort conditions at air speeds above 0.2 m/s. (Plotted from data available in *BRE Digest*, 226.)

resultant temperature is between 19 °C and 23 °C, the air is nominally still and the relative humidity lies between 40 and 70 per cent.

Air conditioning

Air conditioning may be required to provide:

(a) a comfortable environment for human occupation;
(b) the special environment necessary in many industrial processes; when these conditions cannot be achieved by other means.

An air conditioning system:

(i) cleans and purifies the air so that it is free from dust, odours and other impurities when it enters the conditioned space;

(ii)   adjusts the moisture content of the air by humidification or dehumidification;

(iii) controls the air temperature by heating or cooling;

(iv)   supplies sufficient outdoor air for the ventilation needs of the occupants or the process.

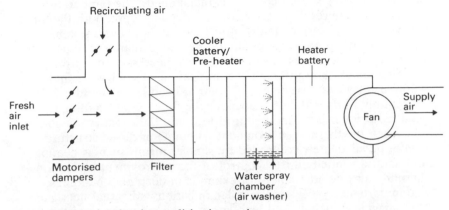

Fig. 2.20 A simple air conditioning unit

In a simple air conditioning unit (Fig. 2.20) motorised dampers control the volume of air drawn through air filters by the fan. This is usually a mixture of outdoor air and recirculated room air. One hundred per cent outside air is supplied to the conditioned space in circumstances where the air may be contaminated, for example in operating theatres, radioactive areas and process work. Filters reduce or eliminate particles (such as dust, smoke), vapours, gases, bacteria and other contaminants from the air. Many types of filter are available. The filter type is selected according to the nature of the impurities in the air.

Dehumidification occurs when air is cooled below its dewpoint temperature because some water vapour condenses out and the moisture content of the air is reduced. The outdoor and recirculated air mixture is cooled to a specified temperature by the cooler battery before it enters the water spray chamber (air washer). In this chamber the air is saturated with water vapour and cleaned (hence the alternative term air washer). Humidification occurs when the cooler battery/water spray temperature is above the dewpoint temperature of the mixed air stream. The moisture content of the air is increased until the air becomes saturated.

Since the moisture content of saturated air depends upon its temperature, adjustment of air temperature by the preheater (in winter), the cooler battery and the water spray temperature provide control of the moisture content of air entering the heater battery.

When heat is supplied by the heater battery the air temperature rises and the relative humidity decreases until the supply air conditions are reached.

The supply fan forces this conditioned air along ductwork and through diffusers into the conditioned spaces within the building. The speed at which the air enters the conditioned space should be uniform but not excessive so that comfort is maintained. The temperature for comfort, the resultant temperature, depends upon both the air temperature and surface temperatures. Ideally the average temperature of the room surfaces should be higher than air temperature. If surface temperatures are low it may be necessary to increase the air temperature although a combination of cold walls and warm air may cause stuffiness. The air at head level should not be perceptibly warmer than the air near the floor. When cooling is required cold air is normally introduced into the room at a high level.

When the thermal environment within an enclosed structure cannot be adequately controlled by simpler means such as heating and natural ventilation, full control of the environment by air conditioning may be advisable. For example in deep plan buildings, and in particular office blocks, solar radiation and casual heat gains from machines, people and high lighting levels may be used to offset the heating loads in winter. However, in summer these casual gains combined with the solar heat gains give rise to high internal temperatures and a considerable cooling load. Adequate ventilation rates may be achieved by natural means in low rise buildings with normal room sizes. In a deep plan office window opening does not provide sufficient ventilation and allows the ingress of dust, fumes, smoke, traffic and aircraft noise and draughts. Sealed windows and conditioned air correctly distributed within the occupied space, lead to a quieter, cleaner and draught-free environment. When the humidity is high dehumidification removes excess moisture from the air to improve comfort and reduce condensation risk. Humidification increases the moisture content of dry air so that discomfort to occupants, shrinkage of furniture and problems with static electricity are prevented.

In many industrial processes air conditioning is necessary to maintain cleanliness and values of temperature, humidity and ventilation rate within well defined limits for all external conditions.

Questions

1. Thermal comfort conditions within an enclosed structure may be assessed from measurements of four physical variables: dry bulb temperature; wet bulb temperature; globe temperature; and air speed. For each variable, describe the instrument and the method of measurement. Illustrate your answer with sketches

and mention any precautions which should be taken to minimise errors.

2. (a) Which factors influence the body's feeling of thermal comfort?
   (b) Describe the effect of these factors upon the rate of heat transfer from the body.

3. Define
   (a) vapour pressure
   (b) relative humidity
   Describe how a whirling (or sling) psychrometer and a psychrometric chart may be used to determine the relative humidity of the atmospheric air.

4. (a) Distinguish between a saturated and an unsaturated vapour.
   (b) Use the psychrometric chart (Fig. 2.11) to estimate the relative humidity and the percentage saturation of moist air at atmospheric pressure. Measurements using a sling psychrometer indicate that the wet and dry bulb temperatures are 14 °C and 21 °C respectively.

5. (a) The air temperature within a five person dwelling is maintained at 20 °C.
     (i) What is the relative humidity when the air moisture content (mixing ratio) is 6.8 g/kg (dry air)?
     (ii) The moisture content is increased to 10.8 g/kg (dry air). Estimate the relative humidity and the dewpoint temperature.
     (Use the psychrometric chart in Fig. 2.11.)
   (b) Explain how the dewpoint temperature can be determined in practice.

6. (a) Which environmental factors are controlled by the installation of air conditioning?
   (b) Discuss the need for air conditioning in industrial and commercial buildings.

Chapter 3

Condensation

When outside air enters a building its moisture content is increased by the addition of water vapour produced by the occupants and their activities and in the case of newly constructed buildings by the addition of residual moisture.

The dewpoint, the temperature at which the air becomes saturated with water vapour, depends upon the amount of water vapour present in the air. Warm air can hold more moisture than cold air. Whenever the temperature is low enough (at or below the dewpoint temperature), water is deposited as:

(a) surface condensation, when water vapour condenses upon a surface;
(b) interstitial condensation, when water vapour condenses within a structural element.

Condensation may lead to mould growth, deterioration of materials, reduction in the effectiveness of thermal insulation and even serious structural damage.

Control of condensation

Thermal insulation, a supply of heat, ventilation, and vapour resistance are important factors in the control of condensation in buildings.

The correct use of thermal insulation will not make a cold room warm but it should eliminate cold bridges and provide a building envelope with sufficient thermal resistance to enable the occupants to afford adequate heating. Warm internal surfaces lessen the risk of surface condensation and warm air has an increased water vapour carrying capacity.

Means of ventilation should be provided especially in areas where activities yield large quantities of water vapour, for example, cooking, washing, clothes drying, the combustion of oil or gas and certain industrial processes. Water vapour produced within the building should be carried outside by the ventilating air from a point as near the source of moisture as possible. Condensation risk is enhanced if moisture laden air is allowed to diffuse through the building fabric (interstitial condensation), or on to cold surfaces in unheated rooms remote from the source of moisture (surface condensation).

A vapour barrier or a vapour check may be incorporated within a structural element to increase its vapour resistance and reduce the likelihood of interstitial condensation.

Surface condensation

Since water vapour condenses upon a surface whenever the surface temperature is at or below the dewpoint temperature, surface condensation risk may be predicted by a comparison of the surface temperature and the dewpoint temperature.

The temperatures and moisture conditions recommended in BS 5250:1975 as a basis for design to control surface condensation and to prevent mould growth (Table 3.1) will be used in the following steady state examples.

Table 3.1

Environment	Air temperature (°C)
Internally	
Occupied living rooms	20
Minimum room temperature	10
Externally	
Roof	−5
Wall	0
Wall, particularly affected by radiant heat loss to a cold night sky	−3

In a domestic dwelling (except in kitchens and bathrooms) an internal air moisture content of 6.8 g/kg (of dry air) should be assumed for normal occupancy conditions provided ventilation is adequate, a minimum of 1.0 air change per hour in occupied living rooms and 0.5 air change per hour in bedrooms and unoccupied living rooms. Assume that the relative humidity of the external air is 90 per cent.

Example 3.1   In a domestic dwelling assume that the internal moisture content is $6.8\,\text{g/kg}$ and the internal air temperatures are

(a) $20\,°C$ in living rooms,
(b) $10\,°C$ in unheated rooms,

when the external air temperature is $0\,°C$ (assumptions recommended in BS 5250:1975).

Predict whether condensation will occur on the internal surfaces of:

(i)   single glazed windows;
(ii)   double glazed windows (cavity resistance, $R_a = 0.18\,\text{m}^2\,\text{K/W}$);
(iii) an external solid wall
       (thermal transmittance, $U = 2.1\,\text{W/m}^2\,\text{K}$).
       The internal surface resistance, $R_{si} = 0.12\,\text{m}^2\,\text{K/W}$ and the external surface resistance, $R_{so} = 0.06\,\text{m}^2\,\text{K/W}$.

On the psychrometric chart, Fig. 2.11, construct a horizontal line from a moisture content of $6.8\,\text{g/kg}$ to intercept the saturation curve and read the dewpoint temperature from the wet bulb temperature scale. A moisture content of $6.8\,\text{g/kg}$ corresponds to a dewpoint temperature of $8.5\,°C$.

Window calculations are based on the surface resistances $(R_{si} + R_{so})$ and the insulating effect of introducing an air cavity (R_a) for double glazing.

(i)   Single glazed window (Fig. 3.1(a))
As in Chapter 1 the temperature drop across the surface resistance of the window may be calculated from

$$\frac{\text{temperature difference (room air to window surface)}}{\text{temperature difference (air to air)}}$$

$$= \frac{\text{surface resistance}}{\text{total resistance}}$$

$$\frac{t_i - t_s}{t_i - t_o} = \frac{R_{si}}{R_{si} + R_{so}}$$

where t_s is the internal surface temperature, and t_i and t_o are the temperatures of the inside and the outside air respectively.

$$t_i - t_s = \frac{(t_i - t_o)}{(R_{si} + R_{so})} \times R_{si}$$

$t_o = 0\,°C$, $R_{si} = 0.12\,\text{m}^2\,\text{K/W}$, $R_{so} = 0.06\,\text{m}^2\,\text{K/W}$

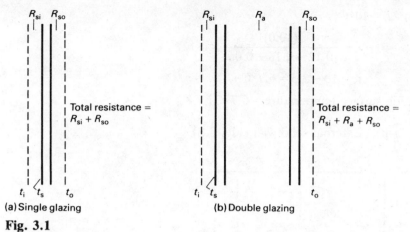

(a) Single glazing   (b) Double glazing

Fig. 3.1

(a) $t_i = 20\,°C$

$$20 - t_s = \frac{(20 - 0)}{(0.12 + 0.06)} \times 0.12$$

$$t_s = 20 - 13.3 = 6.7\,°C$$

Inside surface temperature = $6.7\,°C$

(b) $t_i = 10\,°C$

$$10 - t_s = \frac{(10 - 0)}{(0.12 + 0.06)} \times 0.12$$

$$t_s = 10 - 6.7 = 3.3\,°C$$

Inside surface temperature = $3.3\,°C$

(ii)   Double glazed window (Fig. 3.1(b))

$$t_i - t_s = \frac{(t_i - t_o)}{(R_{si} + R_a + R_{so})} \times R_{si}$$

$t_o = 0\,°C$, $R_{si} = 0.12\,m^2\,K/W$, $R_a = 0.18\,m^2\,K/W$, R_{so}
$= 0.06\,m^2\,K/W$

(a) $t_i = 20\,°C$

$$20 - t_s = \frac{(20 - 0)}{(0.12 + 0.18 + 0.06)} \times 0.12$$

$$t_s = 20 - 6.7 = 13.3\,°C$$

Inside surface temperature = $13.3\,°C$

(b) $t_i = 10\,°C$

$$10 - t_s = \frac{(10.0)}{(0.12 + 0.18 + 0.06)} \times 0.12$$

$$t_s = 10 - 3.3 = 6.7\,°C$$

Inside surface temperature $= 6.7\,°C$

(iii)   External solid wall (Fig. 3.2)

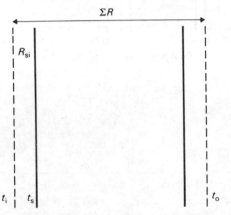

Fig. 3.2   External solid wall in Example 3.1

The temperature drop across the surface resistance of the wall

$$= \frac{\text{temperature difference (air to air)}}{\text{total thermal resistance of wall}} \times \text{surface resistance}$$

The total thermal resistance of the wall, $\Sigma R = 1/U$, where

$U = 2.1\,W/m^2\,K$

$$\Sigma R = \frac{1}{2.1} = 0.48\,m^2\,K/W$$

$t_o = 0\,°C,\ R_{si} = 0.12\,m^2\,K/W,\ \Sigma R = 0.48\,m^2\,K/W$

(a) $t_i = 20\,°C$

$$20 - t_s = \frac{(20 - 0)}{0.48} \times 0.12$$

$$t_s = 20 - 5 = 15\,°C$$

Inside surface temperature $= 15\,°C$

(b) $t_i = 10\,°C$

$$10 - t_s = \frac{(10-0)}{0.48} \times 0.12$$

$$t_s = 10 - 2.5 = 7.5\,°C$$

Inside surface temperature $= 7.5\,°C$

Surface condensation incidence is predicted whenever the inside surface temperature (t_s) is less than the dewpoint temperature of $8.5\,°C$.

		Surface condensation incidence	
		(a) $t_i = 20\,°C$	(b) $t_i = 10\,°C$
(i)	Single glazed window	Yes	Yes
(ii)	Double glazed window	No	Yes
(iii)	External solid wall	No	Yes

Example 3.2 Surface condensation risk (at an internal air temperature of $10\,°C$) may be prevented in an external wall ($U = 2.1\,W/m^2\,K$) by the application of insulation. By using Fig. 3.3, to what value must the thermal transmittance of this wall be improved to prevent surface condensation when the dewpoint temperature is $8.5\,°C$ and the external air temperature is $0\,°C$? The internal surface resistance, $R_{si} = 0.12\,m^2\,K/W$.

Estimate the thermal transmittance which maintains the internal surface temperature (t_s) at the dewpoint temperature ($8.5\,°C$), when the internal air temperature, $(t_i) = 10\,°C$ and the external air

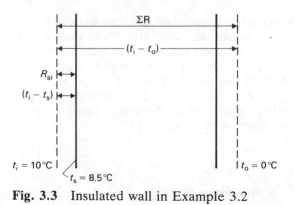

Fig. 3.3 Insulated wall in Example 3.2

temperature, $(t_o) = 0\,°C$. Any U value better (lower) than this will prevent condensation incidence.

$t_i = 10\,°C$, $t_s = 8.5\,°C$, $t_o = 0\,°C$, $R_{si} = 0.12\,m^2\,K/W$

The temperature drop across surface resistance of the wall

$$= \frac{\text{temperature difference (air to air)}}{\text{total thermal resistance of the wall}} \times \text{surface resistance}$$

$$t_i - t_s = \frac{t_i - t_o}{\Sigma R} \times R_{si}$$

where $\Sigma R =$ thermal resistance necessary to keep the inside surface temperature (t_s) at the dewpoint temperature $(8.5\,°C)$.

$$10 - 8.5 = \frac{10 - 0}{\Sigma R} \times 0.12$$

$$\Sigma R = \frac{10 \times 0.12}{1.5} = 0.8\,m^2\,K/W$$

$$U = \frac{1}{\Sigma R} = \frac{1}{0.8} = 1.25\,W/m^2\,K$$

A thermal transmittance value lower than $1.25\,W/m^2\,K$ should remove the surface condensation risk in these environmental conditions.

The improved thermal requirements of Part F of the Building Regulations (England & Wales) 1976 were introduced to help reduce the incidence of condensation and mould growth on surfaces. Scottish Regulations also require construction to be designed so as to avoid damage to the building fabric as a result of moisture transfer from the interior.

Interstitial condensation

Interstitial condensation risk may be estimated from a knowledge of the thermal and vapour characteristics of the building fabric and the environment.

For most building materials it is useful to draw an analogy between heat flow and water vapour diffusion. Heat flows from a high temperature region to a region of lower temperature, similarly water vapour will move from a place of high vapour pressure to a place of lower vapour pressure. As with rate of heat flow and temperature difference (equation [1.1]), the rate of water vapour transfer through a slab of material is directly proportional to the water vapour pressure difference on opposite sides of the material [3.1]

$$G = \frac{d_v}{L} \times \Delta p \qquad\qquad [3.1]$$

$$\frac{\text{vapour diffusion}}{\text{rate}} = \frac{\text{vapour diffusivity}}{\text{thickness}} \times \text{vapour pressure difference}$$

where,

G = mass of water vapour (in g or kg) transferred through unit area ($1\,m^2$) of the material in unit time ($1\,s$)

L = thickness in metres

Δ_p = water vapour pressure difference on opposite sides of the material (in MN/m^2 or N/m^2)

d_v = water vapour diffusivity (in g m/MN s or kg m/N s)

In equation [3.1], d_v, known either as the water vapour diffusivity or the permeability of the material, measures the ability of water vapour to diffuse through the pores of the material.

When vapour diffusivity (d_v) is expressed in the units g m/MN s (gram metre per meganewton second as in *BRE Digest*, 110 and BS 5250:1975) it may be defined as the rate of water vapour transfer in grams per second through unit area ($1\,m^2$) of a material of 1 metre thickness under a vapour pressure difference of 1 meganewton per square metre ($1\,MN/m^2$).

It is also convenient to express moisture properties in terms of vapour resistivity (r_v) and vapour resistance (R_v). Vapour resistivity (r_v) is the reciprocal of vapour diffusivity.

$$r_v = \frac{1}{d_v} \quad (MN\,s/g\,m)$$

Vapour diffusivity (or permeability) and resistivity are properties of the material, terms analogous to thermal conductivity and thermal resistivity in heat transfer.

The vapour resistance (R_v) of a given thickness of material measures its effectiveness in hindering the diffusion of water vapour through it. The vapour resistance (R_v) of a layer of material of thickness L may be calculated from

$$R_v = \frac{L}{d_v} \quad = r_v L$$

$$\frac{\text{vapour}}{\text{resistance}} = \frac{\text{thickness}}{\substack{\text{vapour} \\ \text{diffusivity}}} = \frac{\text{vapour}}{\text{resistivity}} \times \text{thickness}$$

$(MN\,s/g)$ $\qquad\qquad (MN\,s/g\,m \times m)$

While vapour resistivity values are available for many building materials, vapour resistance values are quoted for thin sheet materials and coatings such as paint films (Table 3.2).

There is some resistance to the transfer of vapour between a surface and the adjacent air, but it is small enough to be neglected in calculations in comparison to the vapour resistance of most solids.

A measure of the overall resistance of a composite partition to

Table 3.2 Typical values of (a) vapour resistance and (b) vapour resistivity

(a) Membrane	Vapour resistance (R_v) (MN s/g)
Bitumen felt (1 layer) laid in hot bitumen	Impermeable
Aluminium foil	4000
Polythene sheet (60 μm)	110 to 120
Gloss paint	7.5 to 40

(b) Material	Vapour resistivity (r_v) (MN s/g m)
Brickwork	25 – 100
Concrete	30 – 100
Plaster	60
Plasterboard	45 – 60
Timber	45 – 75
Expanded polystyrene	100 – 600
Mineral wool	5

vapour diffusion may be obtained by summing the vapour resistances of all the constituent parts.

Since condensation occurs whenever the structural temperature falls below the dewpoint temperature, condensation risk may be predicted from a comparison of these temperatures. Under steady state conditions the temperature distribution through a construction element may be obtained as in Chapter 1. The dewpoint temperature distribution may be calculated in an analogous way.

To estimate condensation risk consider the following:

1. The temperature of the internal and external environment.
2. Thermal resistance.
3. The temperature drop across each element.
4. The moisture conditions of the internal and external environment.
5. Vapour resistance.
6. The vapour pressure drop across each element.
7. The dewpoint temperature at the boundary of each element.
8. A plot of structural and dewpoint temperatures.

Example 3.3   Indicate any condensation zone within a cavity wall when the internal air temperature is (a) 20 °C, (b) 10 °C. Assume that internally there is a moisture content addition of 3.4 g/kg due to normal activities. The temperature and relative humidity of the external air are 0 °C and 90 per cent respectively. The thermal and vapour properties of the construction elements are indicated in Table 3.3.

Table 3.3

Building element	Thickness (m)	Thermal conductivity (W/m K)	Vapour resistivity (MN s/g m)
Plaster	0.013	0.16	60
Concrete block	0.100	0.19	40
Brick	0.105	0.84	25
		Thermal resistance (m² K/W)	Vapour resistance (MN s/g)
Internal surface (R_{si})		0.12	Negligible
External surface (R_{so})		0.06	Negligible
Cavity	0.050	0.18	Negligible

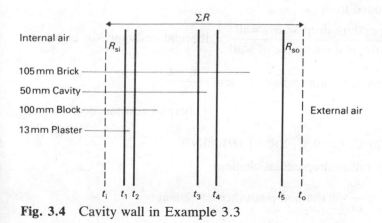

Fig. 3.4   Cavity wall in Example 3.3

In the wall example (Fig. 3.4) it is convenient to work through items (1) to (8) in turn.

1. *Temperature of internal and external environment*
    The internal temperatures in this example correspond to:
    (a) $t_i = 20\,°C$, a suitable internal design temperature for occupied living rooms; and
    (b) $t_i = 10\,°C$, a suitable internal temperature for unoccupied/unheated rooms.
    The external air temperature, $t_o = 0\,°C$

2. *Thermal resistance*

Building element	Thickness (m)	Thermal conductivity (W/m K)	Thermal resistance (m² K/W)
R_{si}			0.12
Plaster	0.013	0.16	0.08
Block	0.100	0.19	0.526
Cavity	0.050		0.18
Brick	0.105	0.84	0.125
R_{so}			0.06
			$\Sigma R = 1.09$

The total thermal resistance (air to air), $\Sigma R = 1.09\,m^2\,K/W$

3. *The temperature drop across each element*
    The temperature drop across an element of the wall may be calculated from:

    $$\frac{\text{temperature drop across wall}}{\text{total thermal resistance of wall}} \times \text{thermal resistance of element}$$

    Temperature drop across element $= \dfrac{t_i - t_o}{\Sigma R}$

    $\times$ thermal resistance of element

(a) $t_i = 20\,°C$, $t_o = 0\,°C$, $\Sigma R = 1.09\,m^2\,K/W$

    Temperature drop across element

    $$= \frac{20}{1.09} \times \text{thermal resistance of element}$$

    $$= 18.35 \times \text{thermal resistance of element}$$

Building element	Thermal resistance $(m^2\,K/W)$	Temperature drop across element (°C)	Temperature (°C)
			$t_i = 20$
R_{si}	0.12	$18.35 \times 0.12 = 2.20$	$t_1 = 17.8$
Plaster	0.08	$18.35 \times 0.08 = 1.47$	$t_2 = 16.33$
Block	0.526	$18.35 \times 0.526 = 9.64$	$t_3 = 6.69$
Cavity	0.18	$18.35 \times 0.18 = 3.30$	$t_4 = 3.39$
Brick	0.125	$18.35 \times 0.125 = 2.29$	$t_5 = 1.10$
R_{so}	0.06	$18.35 \times 0.06 = 1.10$	$t_o = 0$

The temperature profile is obtained by plotting these temperatures at the appropriate distance through the wall and joining the points by straight lines in Fig. 3.5 (assumes steady state conditions and homogeneous materials).

(b) $t_i = 10\,°C$, $t_o = 0\,°C$, $\Sigma R = 1.09\,m^2\,K/W$

Temperature drop across element

$$= \frac{10}{1.09} \times \text{thermal resistance of element}$$

$$= 9.17 \times \text{thermal resistance of element}$$

The temperature drop across each element and the temperatures at the interfaces between the elements are calculated as in the previous case.

Building element	Temperature drop across element (°C)	Temperature (°C)
		$t_i = 10$
R_{si}	$9.17 \times 0.12 = 1.10$	$t_1 = 10 - 1.10 = 8.90$
Plaster	$9.17 \times 0.08 = 0.73$	$t_2 = 8.90 - 0.73 = 8.17$
Block	$9.17 \times 0.526 = 4.82$	$t_3 = 8.17 - 4.82 = 3.35$
Cavity	$9.17 \times 0.18 = 1.65$	$t_4 = 3.35 - 1.65 = 1.70$
Brick	$9.17 \times 0.125 = 1.15$	$t_5 = 1.70 - 1.15 = 0.55$
R_{so}	$9.17 \times 0.06 = 0.55$	$t_o = 0.55 - 0.55 = 0$

The wall temperatures are transferred to Fig. 3.5.

4. *The moisture conditions of the internal and external environment*
Externally, relative humidity = 90 per cent at 0 °C. The values of moisture content (mixing ratio) and water vapour pressure may be read from the psychrometric chart (Fig. 2.11) as indicated in the sketch (Fig. 3.6).

126

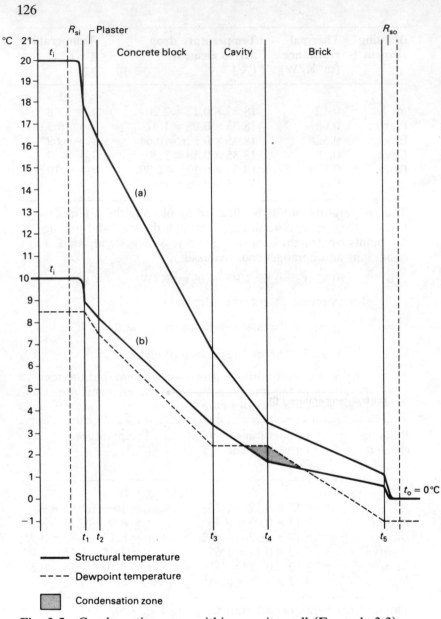

Fig. 3.5 Condensation zone within a cavity wall (Example 3.3)

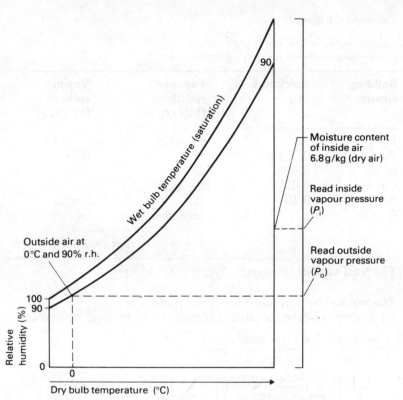

Fig. 3.6 Psychrometric chart sketch (Example 3.3)

Internally, human occupation adds a moisture content of 3.4 g/kg to the moisture content of the external air.

Moisture content = 3.4 + 3.4 = 6.8 g/kg of dry air.

A moisture content of 6.8 g/kg corresponds to a vapour pressure of $1070 \, \text{N/m}^2$ (Fig. 2.11).

Vapour pressure difference between internal environment and external environment = $1070 - 530 = 540 \, \text{N/m}^2$.

The vapour pressure difference is analogous to the temperature difference in a thermal system. Just as heat flows from a high temperature to a lower one, so water vapour diffuses from a high vapour pressure region to a lower one. Water vapour attempts to diffuse from the high water vapour pressure region (usually the internal environment) through the building fabric to the lower water vapour pressure region (usually the external environment).

128

5. *Vapour resistance*

$$R_v = r_v \times L$$

Vapour resistance = vapour resistivity × thickness

Building element	Thickness (m)	Vapour resistivity (MN s/g m)	Vapour resistance (MN s/g)
R_{si}		negligible	—
Plaster	0.013	60	0.78
Block	0.100	40	4.00
Cavity	0.050	negligible	—
Brick	0.105	25	2.625
R_{so}		negligible	—
			$\Sigma R_v = 7.405$

The total vapour resistance, $\Sigma R_v = 7.405$ MN s/g

6. *The vapour pressure drop across each element*
By analogy with temperature difference and thermal resistance

$$\frac{\text{vapour pressure drop across wall}}{\text{total vapour resistance of wall}} = \frac{\text{vapour pressure drop across element}}{\text{vapour resistance of element}}$$

The vapour pressure drop across an element of the wall may be calculated from:

$$\frac{\text{vapour pressure drop across wall}}{\text{total vapour resistance of wall}}$$

$\times$ vapour resistance of element

Vapour pressure drop across element

$$= \frac{P_i - P_o}{\Sigma R_v} \times \text{vapour resistance of element,}$$

where P_i = water vapour pressure inside and P_o = water vapour pressure outside.

$P_i - P_o = 540 \text{ N/m}^2$, $\Sigma R_v = 7.405$ MN s/g

Vapour pressure drop across element

$$= \frac{540}{7.405} \times \text{vapour resistance of element}$$

$$= 72.9 \times \text{vapour resistance of element}$$

Building element	Vapour resistance (MN s/g)	Vapour pressure drop (N/m²)	Vapour pressure (N/m²)
Inside air			1070
Plaster	0.78	57	1013
Block	4.00	292	721
Cavity	—	—	721
Brick	2.625	191	530
Outside air	—	—	530

7. *The dewpoint temperature at the boundary of each element*
   The dewpoint temperature may be read from the psychrometric chart (Fig.2.11). Construct a horizontal line from each vapour pressure value to intercept the saturation curve (100 per cent relative humidity) at the dewpoint temperature (Fig. 3.7). The

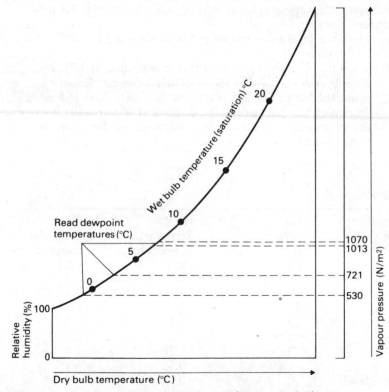

Fig. 3.7 Psychrometric chart sketch (Example 3.3)

dewpoint temperature is estimated from the temperature scale
marked along the saturation curve (Fig. 2.11).

Building element	Vapour pressure (N/m^2)	Dewpoint temperature (°c)
Inside air	1070	8.5
Plaster	1013	7.5
Concrete	721	2.5
Cavity	721	2.5
Brick	530	−1
Outside air	530	−1

8. *A plot of structural and dewpoint temperatures*
    The structural temperatures calculated in (3) and the dewpoint
    temperatures calculated in (7) are collected together in Table 3.4
    and plotted on the scale diagram of the wall. Notice that the
    dewpoint temperature gradient is horizontal across the cavity,
    the internal surface and the external surface, those elements
    with negligible vapour resistance (Fig. 3.5). From Fig. 3.5 it is
    apparent that:
    (a) when the internal environment is maintained at 20 °C,
         condensation does not occur;
    (b) when the internal environment is maintained at 10°C con-
         densation occurs within the structure. The condensation
         zone, shaded in Fig. 3.5 occurs at the inner face and inner
         section of the brickwork (the outer leaf of the wall) because
         the structural temperature has fallen below the dewpoint
         temperature.

Table 3.4

Building element	(a) Structural temperature (°C)	(b) Structural temperature (°C)	Dewpoint temperature (°C)
Inside air	20	10	8.5
R_{si}	17.8	8.9	8.5
Plaster	16.33	8.17	7.5
Block	6.69	3.35	2.5
Cavity	3.39	1.70	2.5
Brick	1.10	0.55	−1
R_{so}/outside air	0	0	−1

Example 3.4   Repeat Example 3.3 when the wall cavity has been filled with mineral wool. Take the thermal conductivity of mineral wool as 0.04 W/m K and its vapour resistivity as 5MN s/g m.

A sketch of the wall is shown in Fig. 3.8. Since the calculation method is the same as in Example 3.3 the result of each calculation stage is summarised in table form.

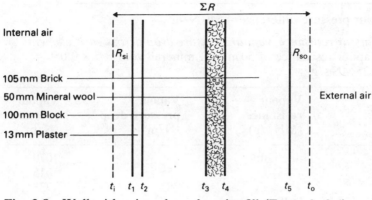

Fig. 3.8   Wall with mineral wool cavity fill (Example 3.4)

1. *Temperature of internal and external environment*
   (a) $t_i = 20\,°C$ (b) $t_i = 10\,°C$
         $t_o = 0\,°C$ $t_o = 0\,°C$

2./3.   *Thermal resistance/temperature drop across each element*
       Thermal resistance of 50 mm of mineral
       wool = $0.050/0.04 = 1.25\ m^2\,K/W$

Building element	Thermal resistance (m^2 K/W)	(a) Temperature drop (°C)	(b) Temperature drop (°C)
R_{si}	0.12	1.11	0.56
Plaster	0.08	0.74	0.37
Block	0.526	4.87	2.44
Mineral wool	1.25	11.56	5.78
Brick	0.125	1.16	0.58
R_{so}	0.06	0.56	0.28
	$\Sigma R = 2.161$		

4. *The moisture conditions of the environment (exactly as Example 3.3)*

Environment	Moisture content (g/kg of dry air)	Vapour pressure (N/m^2)
Internal	6.8	1070
External	3.4	530

Vapour pressure difference $= 540\,N/m^2$.

5./6. *Vapour resistance/vapour pressure drop across each element*
Vapour resistance of 50 mm of mineral wool $= 5 \times 0.050 = 0.25\,MN\,s/g$

Building element	Vapour resistance $(MN\,s/g)$	Vapour pressure drop (N/m^2)	Vapour pressure (N/m^2)
Inside air	negligible	—	1070
Plaster	0.78	55	1015
Block	4.0	282	733
Mineral wool	0.25	18	715
Brick	2.625	185	530
Outside air	negligible	—	530
	$\Sigma R_v = 7.655$		

7. *The dewpoint temperature at the boundary of each element*

Table 3.5

Building element	(a) Structural temperature (°C)	(b) Structural temperature (°C)	Dewpoint temperature (°C)
Inside air	20	10	8.5
R_{si}	18.89	9.44	8.5
Plaster	18.15	9.07	7.5
Block	13.28	6.63	3.0
Mineral wool	1.72	0.85	2.5
Brick	0.56	0.27	−1
R_{so}/outside air	0	0	−1

8. *A plot of structural and dewpoint temperatures*
The temperatures from Table 3.5 are plotted on the scale diagram of the wall in Fig. 3.9. The dewpoint temperature across

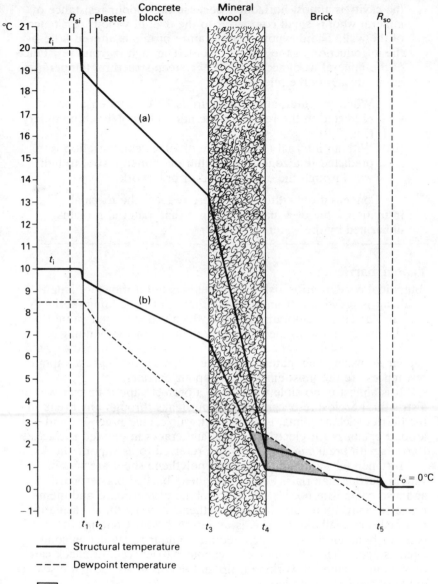

Fig. 3.9 Condensation zone in wall with mineral wool cavity fill (Example 3.4)

the cavity is almost horizontal because the vapour resistance of mineral wool is small compared to the overall vapour resistance of the wall. The dewpoint temperature profile is altered little by the introduction of the mineral wool. The high thermal resistance of the mineral wool accounts for the steep structural temperature gradients across the filled cavity.

(a)   When the internal temperature is 20 °C, condensation is predicted in the region of the mineral wool/brick boundary (Fig. 3.9).

(b)   With an internal temperature of 10 °C, condensation is predicted in a zone from within the mineral wool to half way through the external leaf of brickwork.

Condensation within a structure reduces the thermal insulation value of a material even if that material remains unharmed by the water deposited.

___

Vapour barriers

Interstitial condensation risk may be eliminated if moisture laden inside air is prevented from diffusing through the building fabric. A vapour barrier, a membrane through which water vapour cannot pass, may be incorporated in the structural elements of the building.

Materials used as vapour barriers offer a high resistance to water vapour diffusion. The materials (Table 3.2) with the highest vapour resistances are the most efficient as vapour barriers.

It is almost impossible to provide a perfect vapour barrier within a structural enclosure because of vapour leaks through any break in the barrier such as joints, especially at wall/ceiling junctions and around openings for electrical points and cracks in painted surfaces. Barriers with breaks in continuity are referred to as vapour checks.

Bitumen felt, aluminium foil and polythene sheet are suitable vapour barrier materials. Foil or polythene backed plasterboard and also composite boards, which combine plasterboard and membrane (usually aluminium foil or polythene sheet) with an insulating material, are available for use as an internal skin. Coatings of paint or bitumen may provide adequate vapour resistance in some circumstances. The efficiency of a vapour barrier/vapour check can only be maintained if workmanship and details of jointing and sealing are very carefully attended to.

The thermal resistance of a vapour barrier membrane is negligible but the vapour resistance of such a membrane produces a significant change in the dewpoint temperature profile within the structural element. The dewpoint temperature falls sharply at a total vapour barrier and remains constant on the outside of the barrier, since no outward diffusion of water vapour takes place. A vapour check allows some water vapour transfer, so it is essential that the remainder of

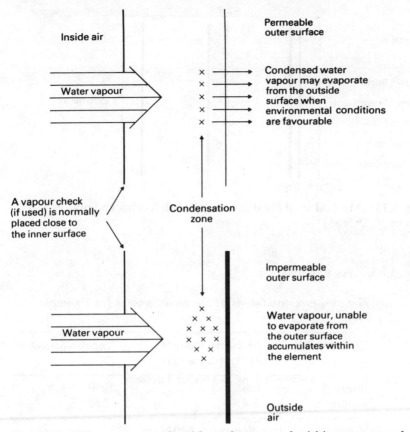

Fig. 3.10 Water vapour should not be trapped within a structural element

the structural element is permeable in order to allow this vapour to diffuse to the outside air (Fig. 3.10).

Example 3.5 What is the effect of: (a) a highly efficient vapour barrier material such as aluminium foil with a vapour resistance of 4000 MN s/g; and (b) gloss paint which provides a more modest vapour resistance of 10 MN s/g, upon the cavity wall of Example 3.4? Does condensation still occur?

Since the thermal resistances of the aluminium foil and the gloss paint are negligible, the structural temperatures are unchanged and it is only necessary to recalculate the dewpoint temperature profile through the wall.

(a) The aluminium foil may be applied as a lining to plasterboard (Fig. 3.11) which has the same thermal and vapour properties as plaster.

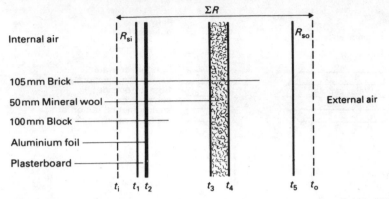

Fig. 3.11 Mineral wool filled cavity wall with foil backed plasterboard (Example 3.5(a))

1.–3.   As Example 3.4.

4.   *The moisture conditions of the environment* (as Example 3.4).

Environment	Moisture content (g/kg (dry air))	Vapour pressure (N/m^2)
Internal	6.8	1070
External	3.4	530

Vapour pressure difference = 540 N/m^2

5.   *Vapour resistance*

Building element	Thickness (m)	Vapour resistivity (MN s/g m)	Vapour resistance (MN s/g)
R_{si}			—
Plasterboard	0.013	60	0.78
Aluminium foil			4000
Block	0.100	40	4.0
Mineral wool	0.050	5	0.250
Brick	0.105	25	2.625
R_{so}			—
			$\Sigma R_v = 4007.655$

The total vapour resistance $\Sigma R_v = 4007.655$ MN s/g

6. *The vapour pressure drop across each element*

Building element	Vapour resistance (MN s/g)	Vapour pressure drop (N/m^2)	Vapour pressure (N/m^2)
Inside air	—		1070
Plasterboard	0.78	0.11	1069.89
Aluminium foil	4000	538.97	530.92
Block	4.0	0.54	530.38
Mineral wool	0.250	0.03	530.35
Brick	2.625	0.35	530
Outside air	—	—	530

7. *The dewpoint temperature of each element*
   The dewpoint temperature may be read from the psychro-metric chart (Fig. 2.11).

Table 3.6

Building element	Vapour pressure (N/m^2)	Dewpoint temperature (°C)	Structural temperature (°C)
Inside air	1070	8.5	10
R_{si}	1070	8.5	9.44
Plasterboard	1070	8.5	9.07
Aluminium foil	531	8.5	9.07
Block	530	−1	6.63
Mineral wool	530	−1	0.85
Brick	530	−1	0.27
R_{so}/outside air	530	−1	0

The dewpoint temperatures together with the structural temperature previously calculated in Example 3.4 are collected together in Table 3.6. The vapour resistances in (5) and the vapour pressures in (6) were calculated exactly for completeness. However, when an efficient vapour barrier material such as aluminium foil is used, its vapour resistance is so high that the vapour resistances of the remaining elements are negligible by comparison. Almost the entire 540 N/m^2 pressure difference occurs across the aluminium foil.

8. *A plot of structural and dewpoint temperatures*
   The temperatures from Table 3.6 are plotted on the scale diagram of the wall in Fig. 3.12. The very high dewpoint

138

temperature gradient across the aluminium foil is represented by a vertical line. The insulation provided by the mineral wool filled cavity is not reduced since condensation does not occur within the wall.

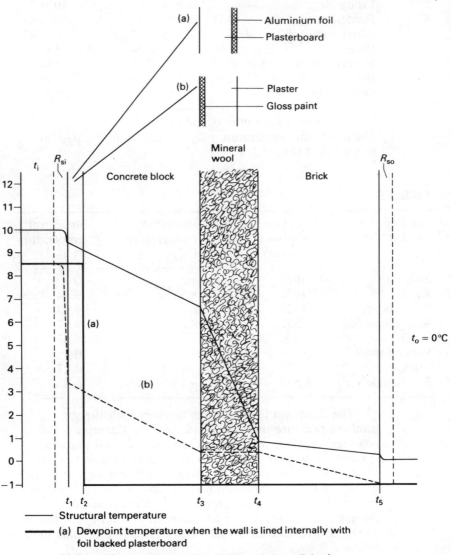

Fig. 3.12 The vapour barrier eliminates condensation risk in this case (Example 3.5)

(b) 1.–4. As in (a) above.
   5. *Vapour resistance*

Building element	Thickness (m)	Vapour resistivity (MN s/g m)	Vapour resistance (MN s/g)
R_{si}			—
Gloss paint			10
Plaster	0.013	60	0.78
Block	0.100	40	4.0
Mineral wool	0.050	5	0.250
Brick	0.105	25	2.625
R_{so}			—
			$\Sigma R_v = 17.655$

The total vapour resistance $\Sigma R_v = 17.655$ MN s/g

   6. *The vapour pressure drop across each element*

Building element	Vapour resistance (MN s/g)	Vapour pressure drop (N/m²)	Vapour pressure (N/m²)
Inside air	—		1070
Gloss paint	10.0	306	764
Plaster	0.78	24	740
Block	4.0	122	618
Mineral wool	0.250	8	610
Brick	2.625	80	530
Outside air	—	—	530

   7. *The dewpoint temperature at the boundary of each element*
       The dewpoint temperature may be read from the
       psychrometric chart (Fig. 2.11).

   8. *A plot of structural and dewpoint temperatures*
       The dewpoint temperatures from Table 3.7 are plotted
       with the previous ones on the scale diagram of the wall
       in Fig. 3.12. While gloss paint does not provide the very
       high vapour resistance of aluminium foil, it affords some
       resistance to moisture transfer from the interior. Condensation
       does not occur within the wall and the full insulation value
       of the mineral wool is retained.

   In regions of continuously high humidity, a vapour barrier should
not be the only means of combating interstitial condensation. If the

Table 3.7

Building element	Vapour pressure (N/m^2)	Dewpoint temperature (°C)	Structural temperature (°C)
Inside air	1070	8.5	10
R_{si}	1070	8.5	9.44
Gloss paint	764	3.3	9.44
Plaster	740	3.0	9.07
Block	618	0.3	6.63
Mineral wool	610	0.3	0.85
Brick	530	−1	0.27
R_{so}/outside air	530	−1	0

water vapour which is prevented from diffusing through a structural element is not removed from within the enclosure, surface condensation risk is increased. Vapour barriers/vapour checks, combined with the correct use of ventilation, heat and thermal insulation, can be expected to alleviate interstitial condensation problems.

The positioning of vapour barriers

In permeable heavyweight structural elements such as brick or block walls a vapour barrier is not usually necessary because fabric deterioration is unlikely even if interstitial condensation does occur.

In lightweight walls, tiled and slated roofs, most flat roofs, and where it is necessary to improve the thermal performance of heavyweight walls by the addition of internal insulating linings, a vapour barrier or vapour check is advisable. A vapour barrier reduces interstitial condensation risk by preventing the diffusion of water vapour from the internal environment to a colder region of the structure where the temperature may fall to the dewpoint temperature. Therefore the vapour barrier should be placed where the structural temperature remains above the dewpoint temperature. A vapour barrier or vapour check is installed on the warm side of the insulation, and in many wall systems, close to the internal surface of the wall, so that there is no risk of condensation occurring on the inside surface of the barrier or check (Fig. 3.13).

Since it is very difficult to guarantee the completeness of a vapour barrier any water vapour transmitted through the barrier should not be trapped within the structural element, but should either be able to diffuse through the remainder of the element and any external cladding, or be removed by sufficient ventilation of cavities and air spaces (Fig. 3.14).

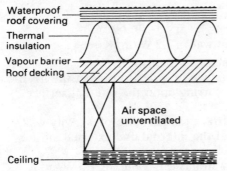

Waterproof roof covering

Thermal insulation

Vapour barrier

Roof decking

Air space unventilated

Ceiling

The vapour barrier is placed on the warm side of the thermal insulation. Condensation incidence is prevented if the level of insulation is sufficient to ensure that the vapour barrier is maintained at a temperature above the dewpoint.

Fig. 3.13 Lightweight flat roof (a warm roof)

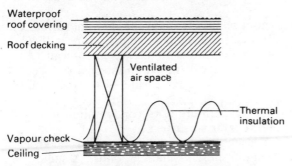

Waterproof roof covering

Roof decking

Ventilated air space

Thermal insulation

Vapour check

Ceiling

The quantity of water vapour entering the roof structure is reduced by a vapour check (aluminium foil backed plaster-board or polythene sheet) at ceiling level. The air space above the insulation must be adequately ventilated so that any water vapour transmitted through the vapour check is removed from the roof structure.

Fig. 3.14 Lightweight flat roof (a cold roof)

Questions

1. (a) Explain what happens when air in an unsaturated condition is cooled to a temperature below its dewpoint upon contact with surfaces within the building enclosure.

   (b) Calculate the internal and external surface temperatures of a solid brick wall when the internal and external air temperatures are 20 °C and 0 °C respectively. Use the following values:

Internal surface resistance $= 0.12 \, m^2 \, K/W$

Thermal transmittance of wall $= 2.2 \, W/m^2 \, K$

External surface resistance $= 0.06 \, m^2 \, K/W$

2. (a) Which factors must be known in order to predict the likelihood of condensation occurring upon the internal surface of a composite partition?

(b) Calculate the internal surface temperature of a cavity wall when the temperatures of the internal and external air are 10 °C and 0 °C respectively.
Assume that the cavity wall has a U value of $1.5 \, W/m^2 \, K$ and an internal surface resistance of $0.12 \, m^2 \, K/W$.

(c) Does surface condensation occur when the wet bulb temperature of the inside air is 9 °C?
Use the psychrometric chart in Fig. 2.11.

3. (a) Why does surface condensation occur?

(b) What problems are associated with surface condensation?

(c) Discuss ways in which such condensation may be avoided or minimised.

4. (a) Verify that condensation occurs upon the internal surface of a solid brick wall when inside, the air temperature is 16 °C and the dewpoint temperature is 13 °C. Assume a wall thermal transmittance of $2.1 \, W/m^2 \, K$, an internal surface resistance of $0.12 \, m^2 \, K/W$ and an outside air temperature of 0 °C.

(b) At what outside air temperature does condensation just begin to occur on the internal surface?

(c) Does surface condensation still occur if the wall is lined internally with 25 mm of fibre insulating board? Assume that the thermal conductivity of fibre insulating board is 0.05 W/m K and the dewpoint and air temperatures are as in (a) of this question.

5. A cavity wall consists of a 105 mm thick brick outer leaf, a 50 mm cavity filled with urea formaldehyde foam, a 100 mm thick aerated concrete inner leaf and 16 mm of internal plastering. The temperature and relative humidity of the external air are 0 °C and 90 per cent respectively. Internally the air temperature is 20 °C and human occupation adds a moisture content of 3.4 g/kg (dry air) to the moisture content of the outside air. Estimate the structural and dewpoint temperatures through the wall. Plot these temperatures at the appropriate position on a diagram of the wall. Indicate any region within the wall where there is a likelihood of condensation incidence.

Use the psychrometric chart (Fig. 2.11) and the following data:

Internal surface resistance $0.12 \, m^2 \, K/W$

External surface resistance $0.06 \, m^2 \, K/W$

Material	Thermal conductivity (W/m K)	Vapour resistivity (MN s/g m)
Plaster	0.46	60
Aerated concrete	0.19	40
Urea formaldehyde foam	0.036	20
Brick	0.84	25

6. A composite board of a 10 mm thickness of plasterboard backed by a 15 mm thickness of expanded polystyrene is used as an internal lining to a 220 mm thick brick wall. The temperature and relative humidity of the outside air are 0 °C and 90 per cent respectively. The temperature and mixing ratio of the inside air are 20 °C and 6.8 g/kg (dry air) respectively.
   (a) Estimate the structural and dewpoint temperatures through the wall. Plot these temperatures at the appropriate position on a diagram of the wall. Indicate any region within the wall where there is a likelihood of condensation incidence.
   (b) A vapour check, in the form of a polythene film, is placed between the plasterboard and the expanded polystyrene. Recalculate and plot the dewpoint temperatures on the wall diagram. Comment upon the probability of condensation incidence.
   Use the psychrometric chart (Fig. 2.11) and the following values:
   Internal surface resistance 0.12 m² K/W
   External surface resistance 0.06 m² K/W
   Vapour resistance of polythene film 50 MN s/g

Material	Thermal conductivity (W/m K)	Vapour resistivity (MN s/g m)
Plasterboard	0.16	60
Expanded polystyrene	0.036	300
Brick	0.84	25

7. (a) Describe the conditions under which condensation may occur within the fabric of a building.
   (b) Discuss the methods employed to prevent or reduce such condensation.
8. Discuss the reasons for increased condensation incidence in domestic buildings and the ways in which such condensation may be controlled.

Section II

Sound

Chapter 4

Sound and its measurement

Sound is an aural sensation caused by pressure variations in the air which are always produced by some source of vibration. They may be from a solid object or from turbulence in a liquid or gas. These pressure fluctuations may take place very slowly, such as those caused by atmospheric changes, or very rapidly and be in the ultrasonic frequency range. The velocity of sound is independent of the rate at which these pressure changes take place and depends solely on the properties of the material in which the sound wave is travelling.

The nature and propagation of sound waves

Sound waves in liquids and gases

In air or any other fluid there is only one possible type of wave. This is a longitudinal one, in which the molecular vibrations are in the same direction as that of the sound transmission. Imagine a line of billiard balls, each a few millimetres from the next. When the end one is tapped it hits the second and so on down the line. The middle balls will vibrate along the line of balls and not across them. When a regular series of impulses is given to a line of molecules a regular series of compressions and rarefactions will follow travelling outwards at the velocity unique to that particular material (Fig. 4.1).

In practice, of course, it is not just one line of molecules which is affected. The pressure fluctuations are travelling outwards from the source in three dimensions rather like a pulsating balloon.

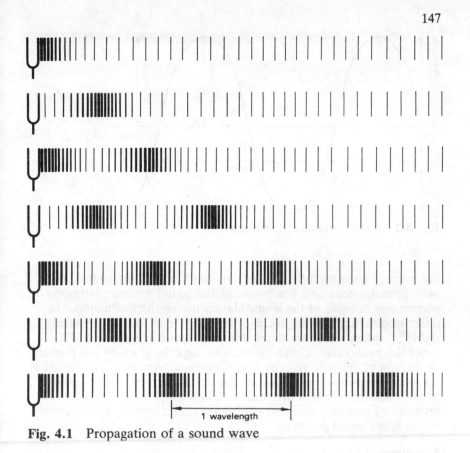

Fig. 4.1 Propagation of a sound wave

The magnitude of these pressure fluctuations is very small indeed compared to the static pressures. In air the normal atmospheric pressure is $10^5 \, \text{N/m}^2$ while the magnitudes of sound pressure affecting the ear vary from $2 \times 10^{-5} \, \text{N/m}^2$ at the threshold, up to $200 \, \text{N/m}^2$ in the region of instantaneous damage. These pressure fluctuations are superimposed upon the normal pressure in the fluid (Fig. 4.2). It can be seen that the pressure at any point in the fluid oscillates slightly above and slightly below air pressure.

Sound waves in solids
The same principles apply to sound transmission in solids that apply in fluids. The process of sound transmission requires an elastic medium. Unlike fluids solids can transmit both longitudinal and bending waves because they are able to sustain shear forces.

Mean pressure
It can be seen in Fig. 4.2 that sound pressure varies sinusoidally with time. The average sound pressure would of course be zero. Clearly

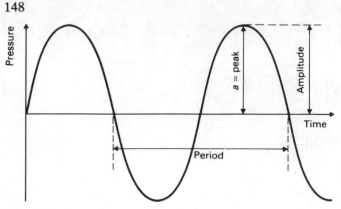

Fig. 4.2   Sound pressure variations

the average gives no indication of the amplitude of the sound. The
peak pressure does give a measure of the sound pressure but not a
correct one because of the sinusoidal nature of the fluctuations. In
order to obtain a realistic average the root mean square pressure
(RMS) is taken. For pure tones the RMS value is equal to 0.707
times the peak value. Other measurable aspects of sound are particle
displacement, particle velocity and particle acceleration. As the ear is
a pressure-sensitive mechanism it is most convenient to use pressure
as the measure of sound magnitude. It will be seen later that sound
intensity or measure of sound energy is related directly to the square
of the RMS sound pressure.

Frequency (f)
This is the number of vibrations or pressure fluctuations per second.
The unit is the hertz (Hz).

Wavelength (λ)
This is the distance travelled by the sound during the period of one
complete vibration.

Amplitude
The amplitude is a measure of the amount of sound – usually the
peak value.

Velocity of sound (V)
The velocity of sound for longitudinal waves in solids is dependent
upon two properties of the material through which it passes. These
are the density of the material and its modulus of elasticity.

Hence $V = \sqrt{\dfrac{E}{d}}$

where $V =$ velocity in m/s

$E =$ modulus of elasticity in newtons/m^2

$d =$ density in kg/m^3

Young's modulus of elasticity is not applicable to fluids as they cannot undergo lateral expansion. The bulk modulus replaces the elastic modulus.

Hence $V = \sqrt{\left(\dfrac{\gamma P}{d}\right)}$

where $P =$ absolute pressure

$\gamma =$ ratio of specific heat

It can be seen from Fig. 4.1 that the wavelength multiplied by the frequency is equal to the distance travelled by the sound in 1 second (s). This is the velocity.
Hence:

velocity = frequency × wavelength

or $V = f\lambda$

where $V =$ velocity of sound in m/s

$f =$ frequency in Hz

$\lambda =$ wavelength in metres (m)

It is sufficiently accurate for the purpose of building acoustics to consider the velocity of sound to be constant at 330 m/s in air. Thus the wavelength of a sound of 20 Hz frequency, at the lowest end of the audible range

$= \dfrac{330}{20}$ m

$= \underline{16.5\,\text{m}}$

Similarly the wavelength of a sound of 20 kHz at the very top end of the audible range even for young children

$= \dfrac{330}{20\,000}$

$= 0.0165\,\text{m}$

$= \underline{16.5\,\text{m}}$

Pure tones and broad band sound

Most of the sounds which we hear are a mixture of frequencies. A pure tone is a sound of one particular frequency alone. The notes on a musical instrument are never pure tones. One particular frequency predominates but with many harmonics present. The different amplitudes of the various harmonics give instruments their different sounds for the same note. For instance if middle C at 261.6 Hz is played then octaves at 523 Hz, 1047 Hz, 2093 Hz, 4186 Hz and 8372 Hz are produced at much lower intensity.

The audible range for human beings extends from approximately 20 Hz to 20 kHz although the upper limit is considerably lower for older people and those who have been exposed to high levels of noise even for fairly short periods.

Many sounds which are produced are broad band in nature containing frequencies across a large part of the audible spectrum. Machines such as circular saws have a fairly broad band spectrum, but with very predominant pure tones resulting from the speed of rotation and cutting. It will be noticed that the pure tone emitted changes slightly according to the load imposed.

The wave form and sound spectra for different sounds can be seen in Figs. 4.3 to 4.14.

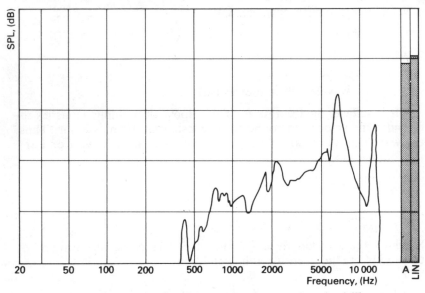

Fig. 4.3 Sound pressure level spectrum for an electric drill

Measurement of sound intensity and loudness

Sound power, sound intensity and sound pressure

The amount of sound produced by a source is measured in watts. This is the rate of doing work or sound power. As one moves away from a source of sound it becomes quieter because the sound is spread over the larger area of the surface of a sphere or hemisphere,

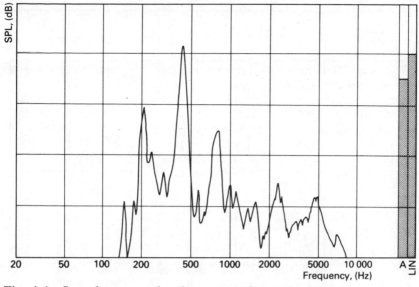

Fig. 4.4 Sound pressure level spectrum for a compressor

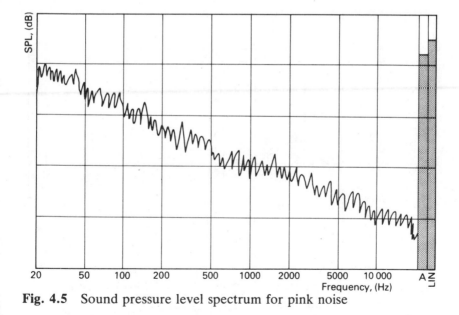

Fig. 4.5 Sound pressure level spectrum for pink noise

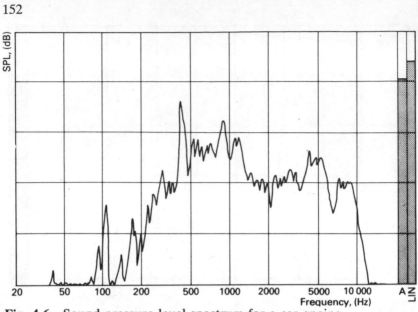

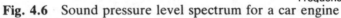

Fig. 4.6 Sound pressure level spectrum for a car engine

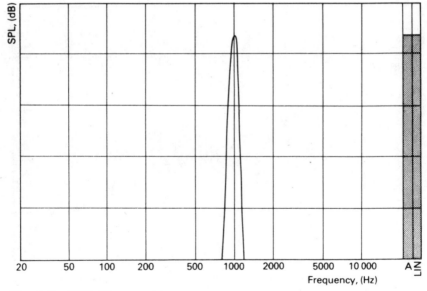

Fig. 4.7 1000 Hz pure tone

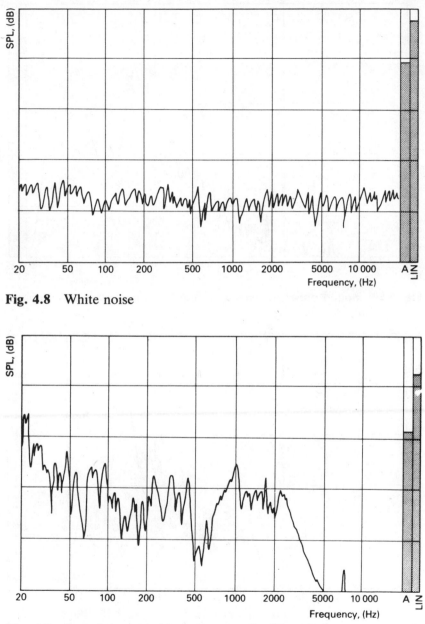

Fig. 4.8 White noise

Fig. 4.9 Sound pressure level spectrum for fan noise

154

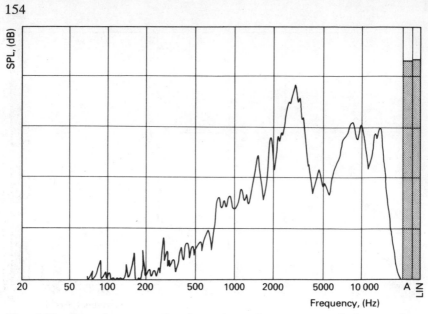

Fig. 4.10   Sound pressure level spectrum for compressed air escaping

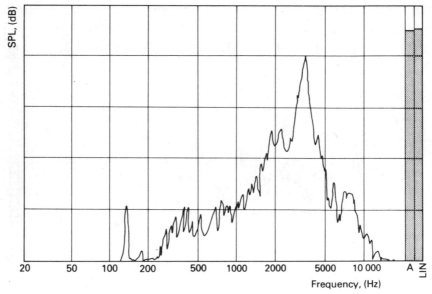

Fig. 4.11   Sound pressure level spectrum for a circular saw not cutting

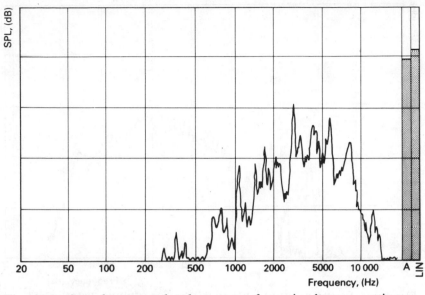

Fig. 4.12 Sound pressure level spectrum for a circular saw cutting

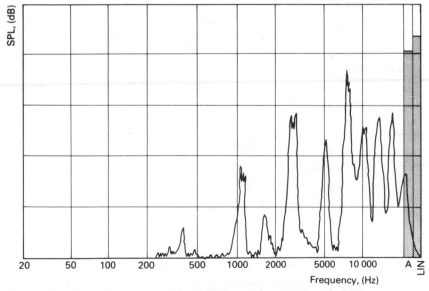

Fig. 4.13 Sound pressure level spectrum for an electric bell

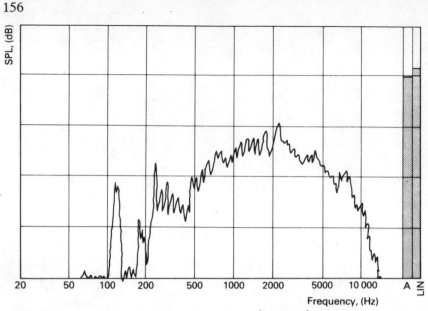

Fig. 4.14 Sound pressure level spectrum for running water

etc. It can be seen from Fig. 4.15 that the intensity at a point is the power divided by surface area or watts per square metre (W/m^2).

Sound intensity is proportional to the square of the sound pressure or

$$I = \frac{P^2}{\rho c}$$

where
  I = intensity in W/m^2
  P = sound pressure in N/m^2
  ρ = density
  c = velocity of sound
  ρc = 410 rayles in air

$$\therefore \quad I = \frac{P^2}{410} \, W/m^2$$

It can be thus seen that the threshold sound intensity which corresponds to a threshold sound pressure of approximately $2 \times 10^{-5} \, N/m^2$ is $10^{-12} \, W/m^2$.

Loudness

The loudness of a sound is a subjective effect which is a function of the ear and the brain as well as amplitude and frequency of vibration.

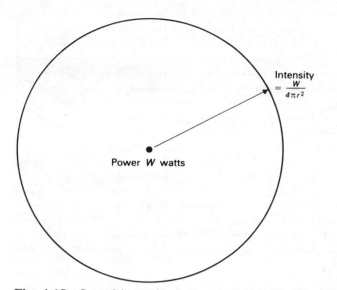

Fig. 4.15 Sound intensity is the power divided by the surface area

In practice it is usual to consider people with normal hearing and correlate only amplitude and frequency with loudness.

Threshold levels vary with frequency as is shown in Fig. 4.16. Equal loudness contours are shown in Fig. 4.17. Response to stimuli

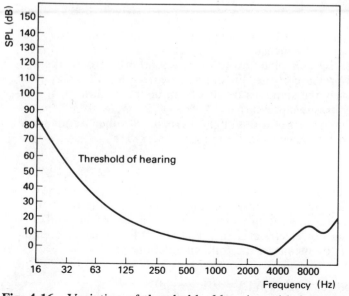

Fig. 4.16 Variation of threshold of hearing with frequency

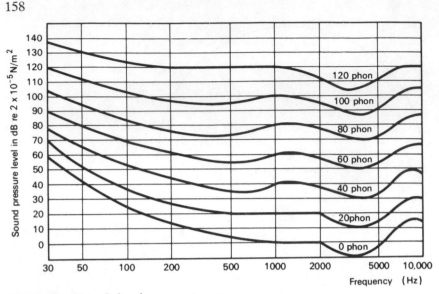

Fig. 4.17 Equal–loudness contours

above the threshold is not linear but proportional to the logarithm of the stimulus.

Response ∝ log of sound pressure.

Threshold levels

Because of the nature of human response to sound pressure it is logical to devise a scale of measurement on a logarithmic scale. As is the case with other measurements levels need to be compared to a base level. In the case of sound the base sound pressure is taken as the sound pressure at the threshold of hearing for a frequency of 1000 Hz. Clearly the standard derives from people with normal hearing. The threshold pressure at 1 kHz is taken as $2 \times 10^{-5} \, \text{N/m}^2$. In practice human sensitivity is slightly greater at other frequencies from 1 kHz to 5 kHz with maximum acuity at 3 kHz.

Decibels

The logarithmic scale for the measurement of sound pressure level, sound intensity level and sound power level is called the decibel scale. The decibel scale is fundamentally a ratio of powers and is frequently used for electrical measurements. It is applied to sound measurements by means of the following definitions:

1. Sound power level (PWL) $= 10 \log \dfrac{W}{W_0} \, \text{dB}$

where $W =$ the power in watts

$W_0 =$ the reference power level

$= 10^{-12}\,\text{W}$

2. Sound intensity level (SIL) $= 10\log\dfrac{I}{I_0}\,\text{dB}$

where $I =$ intensity in W/m^2

$I_0 =$ reference intensity

$= 10^{-12}\,\text{W/m}^2$

3. Sound pressure level (SPL) $= 10\log\left(\dfrac{P}{P_0}\right)^2\,\text{dB}$

$$= 20\log\left(\dfrac{P}{P_0}\right)\,\text{dB}$$

where $P =$ pressure level in N/m^2

$P_0 =$ reference pressure level

$= 2\times 10^{-5}\,\text{N/m}^2$

Addition of sound levels

Due to the fact that the decibel scale is logarithmic it is not possible to add decibels using normal rules of arithmetic. Either pressures or intensities must be added.

Example 4.1   What is the increase in sound pressure level (in dB) if two equal sounds are added?

Increase in SPL $= 10\log\left(\dfrac{2P^2}{P_0^2}\right) - 10\log\left(\dfrac{P^2}{P_0^2}\right)$

$$= 10\log 2 + 10\log\left(\dfrac{P^2}{P_0^2}\right) - 10\log\left(\dfrac{P^2}{P_0^2}\right)$$

$$= 10\log 2$$

$$= 10\times 0.3$$

$$= \underline{3\,\text{dB}}$$

Alternatively if intensities are considered:

Increase in SPL $= 10\log\left(2\dfrac{I}{I_0}\right) - 10\log\left(\dfrac{I}{I_0}\right)$

$$= 10\log 2$$

$$= \underline{3\,\text{dB}}$$

Example 4.2 If three identical sounds are added what is the increase in sound pressure level in decibels?

$$\text{Increase in SPL} = 10\log 3\frac{I}{I_0} - 10\log\frac{I}{I_0}$$
$$= 10\log_{10} 3$$
$$= 10 \times 0.4771$$
$$= \underline{5\,dB}$$

Example 4.3 What is the increase in sound pressure level if the sound pressure is doubled?

$$\text{Increase in SPL} = 20\log_{10}\frac{2P}{P_0} - 20\log_{10}\frac{P}{P_0}$$
$$= 20\log 2$$
$$= 20 \times 0.3$$
$$= \underline{6\,dB}$$

Example 4.4 Two sources of sound individually produce sound pressure levels of 77 dB and 80 dB at a certain point. What is the resultant sound pressure level of the two together?

$$80 = 10\log\frac{I_1}{I_0}$$

$$\therefore \quad \frac{I_1}{I_0} = \text{antilog } 8$$
$$= \underline{1 \times 10^8}$$

$$77 = 10\log\frac{I_2}{I_0}$$

$$\therefore \quad \frac{I_2}{I_0} = \text{antilog } 7.7$$
$$= \underline{5.0119 \times 10^7}$$

$$\therefore \quad \text{SPL} = 10\log\left(\frac{I_1 + I_2}{I_0}\right)$$
$$= 10\log(1 \times 10^8 + 5.0119 \times 10^7)$$
$$= \underline{81.76\,dB}$$

Example 4.5 A certain noise was analysed into octave bands. The sound pressure levels in each were measured as shown below. What is the total level?

Octave band centre frequency (Hz)	125	250	500	1000	2000	4000
SPL (dB)	80	82.5	77.5	70	65	60

Total intensity $I = I_1 + I_2 + I_3 + \cdots I_6$

$$\text{But } SPL = 10 \log \frac{I_1}{I_0}$$

Frequency (Hz)	SPL (dB)	Antilog	$I/I_0 \times 10^8$
125	80	8	1.0000
250	82.5	8.25	1.7782
500	77.5	7.75	0.5623
1000	70	7	0.1000
2000	65	6.5	0.0316
4000	60	6	0.0100
			3.4821

$$\therefore \quad SPL = 10 \log_{10} 3.4819 \times 10^8$$
$$= 10 \times 8.5418$$
$$\simeq \underline{85.5 \, dB}$$

Example 4.6 If 10 identical sources of sound produce a sound pressure level of 80 dB what would each produce alone?

$$80 = 10 \log_{10} 10 \, \frac{I}{I_0}$$

$$\therefore \quad 10 \, \frac{I}{I_0} = \text{antilog } 8$$

$$\therefore \quad \frac{I}{I_0} = \frac{\text{antilog } 8}{10}$$

$$= \frac{1 \times 10^8}{10}$$

$$= 1 \times 10^7$$

$$\therefore \quad 10 \log \frac{I}{I_0} = \underline{70 \, dB}$$

Each machine produces 70 dB at the position concerned.

Example 4.7 Two machines produce 81 dB at certain point. If one would be 77 dB what sound pressure level would the other produce?

$$81 = 10 \log \frac{I}{I_0}$$

$$\therefore \quad \frac{I}{I_0} = \text{antilog } 8.1$$

$$= \underline{1.2589 \times 10^8}$$

$$77 = 10 \log \frac{I_1}{I_0}$$

$$\therefore \quad \frac{I_1}{I_0} = \text{antilog } 7.7$$

$$= \underline{5.0118 \times 10^7}$$

$$\therefore \quad \text{SPL of other machine} = 10 \log (1.2589 \times 10^8 - 5.0118 \times 10^7)$$
$$= 10 \log_{10}(1.2589 - 0.50118) \times 10^8$$
$$= 10 \log (0.7577 \times 10^8)$$
$$= 78.795$$
$$\simeq \underline{\underline{79 \, \text{dB}}}$$

The method shown in the preceding examples can be very quick and convenient if a calculator is available, particularly if it is programmable. If a calculator is not available it is very convenient to use a specially prepared chart such as the one shown in Fig. 4.18.

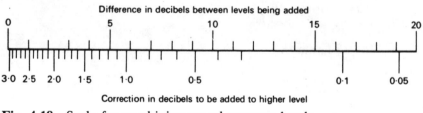

Fig. 4.18 Scale for combining sound pressure levels

To use the chart simply calculate the difference in the two sound pressure levels being added and find the addition needed to the higher level for the sum. Should more than two levels need to be added they can be summed in pairs.

Example 4.8 A noise which had been analysed into octave bands had the sound pressure levels shown. What is the total level?

Centre frequency (Hz)	125	250	500	1000	2000	4000
SPL (dB)		80	82.5	77.5	70	65	60

Frequency (Hz)	SPL (dB)	Difference (dB)	Add dB	Result	Add dB	Result	Result
125	80	2.5	1.95	84.5	0.95	85.5	
250	82.5						
500	77.5	7.5	0.75	78.25			85.5
1000	70						
2000	65	5.0	1.2	67.2		67.2	
4000	60						

It will be seen throughout these calculations that the final results are rounded to the nearest 0.5 dB. It is not possible in practice to give results more accurate than that and in many cases this would be too optimistic.

Sound pressure level and sound power level

It is often convenient to express the total sound output from a source in terms of sound power level. In order to obtain this, measurements of the sound pressure level at given distances must be made.

For the situation of a sound source of power output W radiating uniformly in space then:

$$\text{Sound power level, PWL} = 10\log_{10}\frac{W}{W_0}$$

$$= 10\log_{10}\frac{W}{10^{-12}}$$

$$= 10\log_{10} W - 10\log_{10} 10^{-12}$$

$$= 10\log_{10} W + 120$$

$$\text{Sound pressure level, SPL} = 10\log_{10}\frac{W}{4\pi r^2} + 120$$

$$= 10\log_{10} W - 10\log 4\pi r^2 + 120$$

$$= (10\log_{10} W + 120) - 10\log 4\pi r^2$$

$$= \text{PWL} - 20\log_{10} r - 10\log 4\pi$$

$$= \text{PWL} - 20\log_{10} r - 11$$

If the more common situation applied where the sound is radiated over non-absorbent ground then the relation would become:

$$SPL = PWL - 20\log_{10} r - 8$$

Example 4.9 A compressor with a sound power level of 109 dB is radiating uniformly over a flat non-absorbent surface. Calculate the sound level at a distance of:
(a) 10 m;
(b) 45 m.

(a) At a distance of 10 m

$$\begin{aligned} SPL &= 109 - 20\log_{10} r - 8 \\ &= 109 - 20\log_{10} 10 - 8 \\ &= 109 - 20 - 8 = \underline{81\,dB} \end{aligned}$$

(b) At a distance of 45 m

$$\begin{aligned} SPL &= 109 - 20\log_{10} 45 - 8 \\ &= 109 - 20 \times 1.6532 - 8 \\ &= 109 - 33 - 8 \\ &= \underline{68\,dB} \end{aligned}$$

Weighting scales

It can be seen from Fig. 4.16 that the threshold of hearing varies with frequency. Hearing sensitivity also varies with frequency above the threshold although slightly differently from the variations at threshold. In order that direct readings of noise may be made which correspond with the subjective assessments weighting networks have been devised.

The three weighting scales 'A', 'B' and 'C' correspond to different equal loudness contours (Fig. 4.17). The 'A' weighting scale corresponds approximately to the shape of the 40 dB equal loudness contour, the 'B' scale to the 70 dB contour and the 'C' scale is flat from 200 to 1250 Hz with some attenuation above and below (Fig. 4.19).

In practice it has been shown that the dB(A) scale is the most conveniently accurate one to use where subjective response to noise is involved. Measurements of sound level in dB(A) are called sound levels whereas the linear measurements are called sound pressure levels.

The 'D' weighting is a specialised characteristic being the proposed standard for aircraft noise measurements.

The 'A', 'B', 'C' and 'D' weightings are shown numerically in the following table.

Frequency (Hz)	Curve A (dB)	Curve B (dB)	Curve C (dB)	Curve D (dB)
10	−70.4	−38.2	−14.3	−26.5
12.5	−63.4	−33.2	−11.2	−24.5
16	−56.7	−28.5	− 8.5	−22.5
20	−50.5	−24.2	− 6.2	−20.5
25	−44.7	−20.4	− 4.4	−18.5
31.5	−39.4	−17.1	− 3.0	−16.5
40	−34.6	−14.2	− 2.0	−14.5
50	−30.2	−11.6	− 1.3	−12.5
63	−26.2	− 9.3	− 0.8	−11.0
80	−22.5	− 7.4	− 0.5	− 9.0
100	−19.1	− 5.6	− 0.3	− 7.5
125	−16.1	− 4.2	− 0.2	− 6.0
160	−13.4	− 3.0	− 0.1	− 4.5
200	−10.9	− 2.0	0	− 3.0
250	− 8.6	− 1.3	0	− 2.0
315	− 6.6	− 0.8	0	− 1.0
400	− 4.8	− 0.5	0	− 0.5
500	− 3.2	− 0.3	0	0
630	− 1.9	− 0.1	0	0
800	− 0.8	0	0	0
1 000	0	0	0	0
1 250	0.6	0	0	2.0
1 600	1.0	0	− 0.1	5.5
2 000	1.2	− 0.1	− 0.2	8.0
2 500	1.3	− 0.2	− 0.3	10.0
3 150	1.2	− 0.4	− 0.5	11.0
4 000	1.0	− 0.7	− 0.8	11.0
5 000	0.5	− 1.2	− 1.3	10.0
6 300	− 0.1	− 1.9	− 2.0	8.5
8 000	− 1.1	− 2.9	− 3.0	6.0
10 000	− 2.5	− 4.3	− 4.4	3.0
12 500	− 4.3	− 6.1	− 6.2	0
16 000	− 6.6	− 8.4	− 8.5	− 4.0
20 000	− 9.3	−11.1	−11.2	− 7.5

Sound level meters

A sound level meter is an instrument which responds to sound in approximately the same way as the human ear but it gives objective measurements of sound level.

The sound signal is converted into an identical electrical signal by means of a microphone (Fig. 4.20) This transducer must be of high quality to ensure that there is no distortion of the signal. Normally

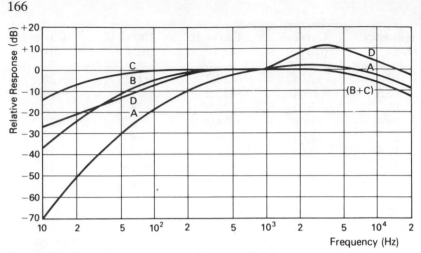

Fig. 4.19 Frequency response of the weighting networks

condenser microphones are essential for precision grade instruments while piezo-electric microphones are commonly used for industrial grade meters. The latter are less accurate but more robust. Since the signal is quite small it is necessary to amplify it. The amplification is carried out in two stages – one part before the weighting networks and the second afterwards. Some meters have 'A', 'B', 'C' and 'D' weighting networks, while others may be more limited. After the second amplifier the signal is large enough to drive the meter. However because sound levels entail RMS measurements a root mean square rectifier is inserted between the second amplifier and the DC meter.

Unfortunately no meter could satisfactorily give direct reading across the range from say 30 dB to 120 dB and attenuators are needed. Frequently these are in 10 dB switched steps and the actual reading is the attenuation setting plus the meter reading.

All instruments for sound measurement consist of the same basic instrumentation but may have additional facilities included. A common facility is an output socket for tape recorder or other instrumentation and this is connected between the rectifier and meter. In addition many meters have facility for connection of octave or third-octave filters.

Calibration of sound level meters

It is important that each time a sound level meter is used that its calibration and battery operation are checked. The battery check is a simple switch on the meter to measure battery voltage. The calibration check is carried out using an acoustic calibrator (Fig. 4.21). This is a standard source of sound to which level the meter can be adjusted.

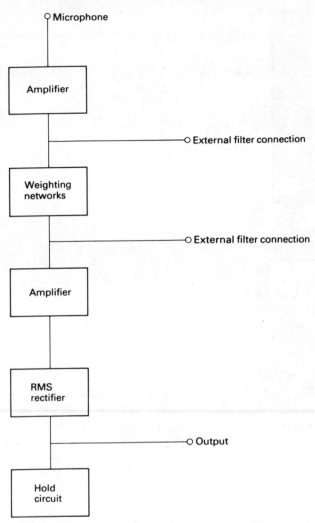

Fig. 4.20 A block diagram of the sound level meter

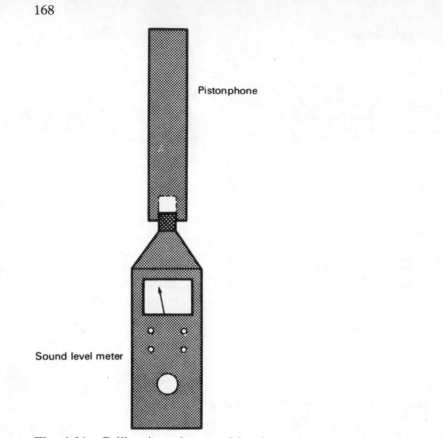

Fig. 4.21 Calibration of a sound level meter using a pistonphone

Experiments

Experiment 4.1   Determination of the velocity of sound using a resonance tube

Apparatus
Glass resonance tube 1.5 m long, approximately 30 mm diameter, slightly larger diameter tube of similar length with a rubber bung and filled with water, set of tuning forks covering the octave range 256 to 512 Hz, metre rule, retort stand, thermometer.

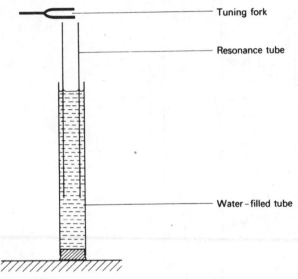

Fig. 4.22

Method
The glass tube containing water is placed with its base on the floor adjacent to a bench where it is held by the retort stand. The resonance tube is then inserted well down into the water. With a tuning fork held above the mouth, the tube is raised until resonance occurs. The length (l_1) is measured. This is repeated several times and the mean value obtained.

The tube is now raised further out of the water and a second resonant length (l_2) obtained in similar manner.

The experiment is repeated with each tuning fork in turn. The mean value for the velocity of sound at the room temperature is then calculated.

Theory

At resonance the air must be at rest at the water surface, which is therefore a node. Maximum vibration occurs at the top end which is an antinode. In practice, the antinode is a short distance (x) above the top.

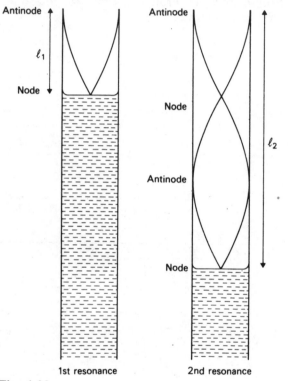

Fig. 4.23

$$\therefore \quad l_1 + x = \frac{\lambda}{4} \qquad \qquad [4.1]$$

Where l_1 = 1st resonant length of tube in metres
  λ = wavelength of note in metres

and $l_2 + x = \frac{3\lambda}{4} \qquad \qquad [4.2]$

Where l_2 = 2nd resonant length of tube in metres

By subtracting [4.1] from [4.2]

$$l_2 - l_1 = \frac{\lambda}{2} \qquad\qquad [4.3]$$

If V is the velocity of sound and f the frequency of the note,

$$V = f\lambda \qquad\qquad [4.4]$$

or $\quad \lambda = \dfrac{V}{f}$

$\therefore \quad l_2 - l_1 = \dfrac{V}{2f}$

$\quad \therefore \quad V = 2f(l_2 - l_1) \qquad\qquad [4.5]$

Results

Frequency, f (Hz)	l_1 (m)	Mean	l_2 (m)	Mean	$l_2 - l_1$	$V = 2 \times f \times (l_2 - l_1)$

Average = _____ m/s

Temperature = _____ °C

$\therefore$ Velocity of sound in air at _____ °C = _____ m/s

Experiment 4.2 Determination of the velocity of sound in and modulus of elasticity for brass using Kundt's tube

Apparatus

Glass tube 1 m long, approximately 30 mm diameter, with a movable plunger at one end and the loose-fit piston at the other attached to a brass rod about 0.8 m long. Clamp, lycopodium powder, resined cloth, metre rule.

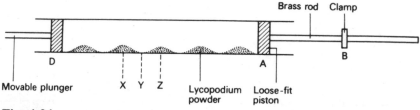

Fig. 4.24

Method

The glass tube is carefully dried to ensure that the powder does not stick. Lycopodium powder is then lightly sprinkled inside and the tube placed horizontally as shown in Fig. 4.24. The brass rod is clamped firmly at its midpoint. The rod is then stroked lengthways, at end C, plunger D being adjusted until resonance of the air column occurs. This is seen by the vibration of the powder which settles down into heaps at the nodal positions. At the sharp resonance position, the lengths XZ and AC are measured, being half wavelengths of the frequency in air and brass. The velocity of sound in brass and the modulus of elasticity are calculated.

Theory

If V and v are the velocities of sound in brass and air respectively then:

$$\frac{V}{v} = \frac{AC}{XZ}$$

But $\quad v = 330 \, \text{m/s}$

$$\therefore \quad V = \frac{AC}{XZ} \times 330 \, \text{m/s}$$

Now $\quad V = \sqrt{\frac{E}{\rho}}$

Where $\quad E =$ Young's modulus of elasticity for brass

$\qquad \rho =$ density of brass

$$\therefore \quad E = V^2 \rho$$

Results

Density of brass   = _____ kg/m^3

Length of brass, AC   = _____ m

Length XZ   = _____ m

Velocity of sound in air = 330   m/s

$\therefore \quad V =$ _____ m/s

$E =$ _____ kN/m^2

Experiment 4.3 Determination of Young's modulus of elasticity for concrete by an electrodynamic method

Apparatus

Variable-frequency oscillator, vibrator, 400 × 100 × 100 mm concrete beam, clamps, piezo-electric (crystal) pick-up, cathode ray oscilloscope.

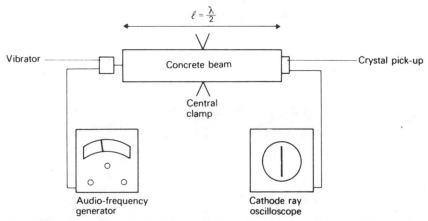

Fig. 4.25 Measurement of the velocity of sound in solids, and Young's modulus of elasticity by the electrodynamic method

Method

The concrete beam is clamped at its centre while it is still wet from the curing tank. The vibrator connected to the oscillator is placed touching one end of the beam. At the other end the pick-up is attached to the beam by means of vaseline. This is connected across the Y plates of the cathode ray oscilloscope with the time base switched off so that a vertical line appears.

The oscillator is switched on to set the beam into forced longitudinal vibration. The frequency is increased slowly until the line on the oscilloscope increases greatly in length, indicating resonance. This may be checked by finding further resonances at multiples of the fundamental frequency. The fundamental frequency is noted.

The density of the wet concrete is found by weighing and noting the dimensions.

Theory

$$V = \sqrt{\frac{E}{d}}$$

where V = velocity of sound in the concrete
         E = Young's modulus of elasticity in the concrete
         d = density of the concrete

But $V = f\lambda$

where f = frequency at resonance
         λ = wavelength of sound at resonance
         $= 2l$

where l = length
    $\therefore$ $V = f \times 2l$

and $V^2 = 4f^2l^2$

Now $V^2 = \dfrac{E}{d}$

    $\therefore$ $E = V^2 d$
         $= 4f^2l^2 d$

This is correct for longitudinal vibration. Other maxima may be found due to bending waves.

Results

Mass of beam   = _____ kg

Length of beam, l = _____ m

Height of beam   = _____ m

Width of beam   = _____ m

$\therefore$ volume of beam = _____ m^3

Density, $d = \dfrac{\text{mass}}{\text{volume}}$

        = _____ kg/m^3

Frequency at resonance, $f =$ _____ s^{-1}

        $E = 4f^2l^2 d$ (kg/m s^2)

         = _____ N/m^2

Note: The value of Young's modulus of elasticity for concrete ranges from approximately $10\,\text{GN/m}^2$ at early ages to $50\,\text{GN/m}^2$ for high-quality concrete at greater ages.
    Materials other than concrete may be used, but if the bars are thin compared with their length, transverse vibration may be large.

Experiment 4.4 Determination of Young's modulus of elasticity for concrete using ultrasonic pulse-velocity apparatus

Apparatus

Concrete, ultrasonic pulse-velocity apparatus.

Two different types of readout are available. One uses a cathode ray tube and the times must be calculated from the pulses; the other has a direct readout of time. Variations in method will be needed according to which of these is available.

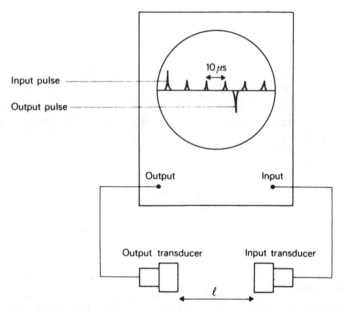

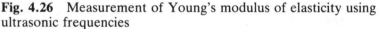

Fig. 4.26 Measurement of Young's modulus of elasticity using ultrasonic frequencies

Method

The apparatus is connected up as shown in Fig. 4.26 with the two transducers placed either side of the concrete, a little Vaseline ensuring good contact –50 Hz pulses of ultrasonic frequency (approximately 150 kHz) are passed through the concrete sample. The time taken for the pulse to travel through the concrete is measured from the input and output pulses on the screen. The density of the concrete must be found from the mass of a known volume or assumed. The distance (l) between the transducers is measured.

Theory

$$V = \sqrt{\frac{E}{d}}$$

where V = velocity of sound in concrete

E = Young's modulus of elasticity in the concrete

d = density of the concrete

But $V = \dfrac{\text{thickness of specimen}}{\text{time } (t)}$

$$= \frac{l}{t}$$

$\therefore$ $E = V^2 d$

$$= \frac{l^2}{t^2} d$$

Results

Density of concrete, $d =$ _____ kg/m^2

Distance between transducers, $l =$ _____ m

Pulse velocity time, $t =$ _____ s

$$\therefore \quad E = \frac{l^2 d}{t^2} \left(\frac{m^2}{s^2} \frac{kg}{m^3} \right)$$

$$= \underline{\hspace{3cm}} N/m^2$$

Experiment 4.5   To verify that the frequency of a stretched string is inversely proportional to the length

Apparatus
Sonometer with steel wire, hanger and weights, set of tuning forks, metre rule.

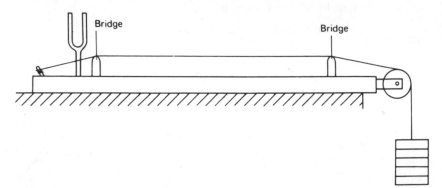

Fig. 4.27   The sonometer

Method
The lowest frequency tuning fork is chosen and made to vibrate by gently tapping it on rubber or other soft material. Weights are adjusted on the sonometer so that the longest length of wire reasonably possible resonates to the tuning fork. This is achieved by adjusting the bridges until the loudest sound is heard with the base of the tuning fork on the sonometer box as shown. For those with musical training this is relatively easy. For those who find adjustment by ear more difficult a small V-shaped piece of paper inverted on the middle of the wire may help. The length at resonance and frequency from the tuning fork are noted.

The experiment is then repeated for the other tuning forks, keeping the tension constant, and a graph of frequency against the reciprocal of length is plotted. A straight line should result.

Theory
For a vibrating wire,

$$f = \frac{1}{l}\sqrt{\frac{T}{m}}$$

where f = frequency in hertz (Hz)
        l = length in m (or mm)
        T = tension in N
        m = mass per unit length of wire

Thus for a particular wire where m is a constant if the tension is fixed then frequency depends upon length.

i.e. $f \propto \dfrac{1}{l}$

Hence a graph of natural frequency against the reciprocal of length will give a straight line.

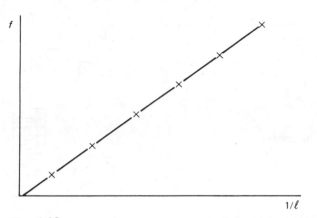

Fig. 4.28

The principle of the sonometer is used in the vibrating-wire type of strain gauge. In that application, the length is kept constant (as is m) and the tension changes with strain. The resonant frequency is found, from which strain is determined. This type of strain gauge has applications within concrete such as roads and bridges.

Results

Frequency (Hz)	Length (l) (mm)	$\dfrac{1}{l}$

Experiment 4.6   To verify that the frequency of a stretched wire is proportional to the square root of the tension

Apparatus
Sonometer with steel wire, hanger and weights, set of tuning forks.

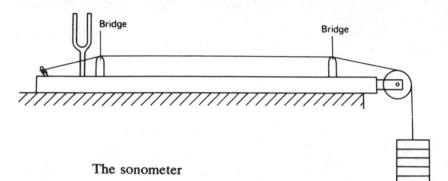

The sonometer

Fig. 4.29

Method
The lowest frequency tuning fork is made to vibrate by gently tapping it on rubber or other soft material. The weights are adjusted so that the longest length of wire reasonably possible resonates to this tuning fork. This may be done by ear or using a tiny inverted 'V' of paper as described in Experiment 4.5.

It is important that the length of wire is not changed during the experiment. The next lowest frequency tuning fork is taken and the weight for resonance determined. This is repeated for all the tuning forks, noting frequency and tension each time. A graph is then drawn of frequency against square root of tension.

Theory
For a vibrating wire,

$$f \propto \frac{1}{l}\sqrt{\frac{T}{m}}$$

where f = frequency in hertz (Hz)
   l = length in m (or mm)
   T = tension in N
   m = mass per unit length of wire

Thus for a particular wire (where m is constant) if the length is fixed,

$$f \propto \sqrt{T}$$

Hence a graph of resonant frequency against square root of tension will give a straight line.

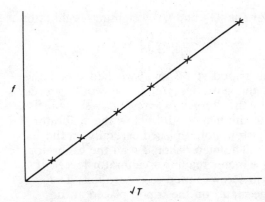

Fig. 4.30

This principle is used in the vibrating-wire strain gauge as explained in the previous experiment.

Results

Frequency (Hz)	Tension (T) (N)	$\sqrt{T}$

Experiment 4.7   Use of a sound level meter to measure sound pressure level and sound level

Apparatus

Sound level meter with linear (or 'C') and 'A' weightings, calibrator, tripod stand.

Method

The meter is set up on the tripod stand and switched on to the 'battery check' position. If the batteries are in full working order the switch is then turned to the 'lin' or 'C' response setting. The attenuator is then adjusted to the nearest '10' below the calibrator level and the calibrator (or pistonphone) placed carefully on the microphone and switched on. The meter should read the calibrator level (i.e. attenuator setting + meter reading = calibrator level). If necessary, the fine-adjustment screw is used to ensure this. The calibrator is removed and the meter on the tripod placed at the position where it is desired to measure a steady noise.

The attenuator is adjusted to obtain a reading between '0' and '10' on the meter. The sound pressure level (SPL) is then the attenuator setting + meter reading. With some meters the attenuator setting is displayed on the meter scale and with digital meters the SPL is read directly. The SPL is noted.

The response is now changed to 'A' weighting and the sound level similarly found.

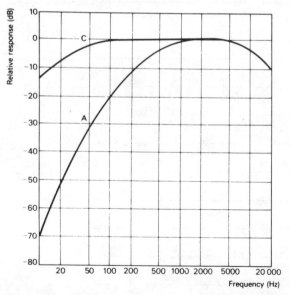

Fig. 4.31   Relative response of A and C weighting scales

This should be repeated for different sounds such as an electric drill, stationary diesel engine or motor car, band saw, etc. Any noticeable difference between the types of noise should be observed such as low pitch or high pitch. This may be compared with the differences between dB and dB(A).

Theory

The sound pressure level (SPL) of a sound, in decibels, is equal to 20 times the logarithm to the base 10 of the ratio of the RMS sound pressure to the reference sound pressure ($2 \times 10^{-5} \, \text{N/m}^2$ in air).

Sound level is the 'A' weighted value of the sound pressure level, and is written as dB(A).

The dB(A) scale is an attempt to obtain a reading with a sound level meter which correlates with loudness. The weighting is shown in Fig. 4.31.

Results

Noise source	Measuring distance	Sound pressure level (dB)	Sound level (dB(A))

Experiment 4.8   Measurement of the L_{10} levels of traffic noise using a sound level meter

Apparatus
Sound level meter (ideally this should be of precision grade but this is not important for the purpose of this experiment), wind shield, tripod stand and acoustic calibrator.

Method
The sound level meter is set up on the tripod stand at the appropriate position. For traffic noise the microphone should be 1.2 m above the ground. The meter is then switched on to the 'battery check' position to ensure that the battery is functioning correctly. The calibrator is then placed on the microphone, the meter attenuator being set at the appropriate position for the calibration. Thus if the calibrator produces 94 dB, the attenuator is set at 90 dB. The meter is now switched to the 'slow' linear position, or if the calibrator is of 1 kHz frequency, to slow 'A' position. Hence the meter response is checked and, if necessary, adjustment is made using the fine-adjustment screw.

The meter should now be set on 'dB(A) slow' with a suitable attenuation to ensure that the needle does not go beyond the top of the scale for the highest levels of noise to be measured. This attenuation setting is entered in the centre box of the results chart. The wind shield must be used for all external measurements.

The meter reading is taken by glancing at the meter and entering a tick on the chart at the appropriate level (the 'five-bar gate' method is convenient). This is repeated for 15 minutes ensuring that a reading is taken at least once every 4 seconds. The total counts at each level are obtained and the grand total (as shown in Table 4.1), and thus the level in dB(A) exceeded for 10 per cent of the time (counts) is obtained.

Theory
The L_{10} level is the level of noise in dB(A) exceeded for 10 per cent of the measurement time. For traffic noise this is normally measured 1 m from the facade of dwellings at 1.2 m from the ground for ground floors, each hour for the 18 hours from 06.00 to 24.00 hours. The 18 measurements are then arithmetically averaged to determine the 18 hour L_{10} value.

The L_{10} value is easily obtained by counting 10 per cent of the total counts from the top and reading the level in dB(A) to the nearest 0.5 dB(A). It will be realised that the levels shown are actually a range of levels, e.g. 72 is a range from 71.5 dB(A) to 72.5 dB(A). Thus the L_{10} shown is $(72.5 - 1/21)$ dB(A) = 72.5 dB(A) to the nearest 0.5 dB(A). The L_{50} or L_{90} values may be found in a similar manner, where L_{50} = level in dB(A) exceeded for 50 per cent of the measurement time; L_{90} = level in dB(A) exceeded for 90 per cent of the measurement time.

Table 4.1

60	Counts		Total counts
1			
2			
3	1		1
4	111		3
5																												25
6																																									1		41
7																		15
8																																									1		41
9																																																			111		53
70																																				11		37
1																																						34
2																					1		21
3																		15
4	||||| 111	30 counts	8
5	111		3
6	11		2
7	1		1
8			
9			
80			
			Total 300

Results

dB(A)		Number of counts	Total counts
	0		
−9	1		
−8	2		
−7	3		
−6	4		
−5	5		
−4	6		
−3	7		
−2	8		
−1	9		
	0		
	1		
	2		
	3		
	4		
	5		
	6		
	7		
	8		
	9		
	0		

$L_{10} = $ _____ dB(A)   Total _____

Experiment 4.9 Determination of the sound power level of a machine in a reverberation room

Apparatus
Reverberation room, electric drill, tape measure, microphone, extension cable, audio-frequency spectrometer, level recorder, sine random generator, amplifier (75 to 100 W), two loudspeakers (50 W each).

Method
The drill whose sound power level is to be measured is mounted on a retort or tripod stand at least 1.5 m from any wall. The sound level is measured in dB at four different positions around the drill at distances of 0.5 m and the average value obtained. This is repeated at a distance of 1 m and any other distances desired. For all these measurements it is desirable that the microphone is set up in the reverberation room and the audio-frequency spectrometer is remote from it. It is important that the apparatus is calibrated initially.

The reverberation time of the room is measured as described in Experiment 5.1.

The measurements are repeated in dB(A) and in octave bands with centre frequencies at 250 Hz, 500 Hz, 1 kHz and 2 kHz.

Theory
If a sound is produced in a reverberation room, the sound pressure level increases until equilibrium is reached where the rate of production of sound energy equals the rate of absorption by the room surfaces. Then:

Sound power level, (PWL) = sound pressure level (SPL)
$$+ \; 10 \log V - 10 \log t - 14 \, \text{dB}$$

where
V = volume of room in m^3

t = reverberation time in s

It can be seen that the equation is independent of distance, and the measurement positions do not matter provided they are not too close to the machine or walls. The method described, while not conforming to the international standard, is capable of reasonably accurate results with the minimum of equipment. If a reverberation room is not available, any fairly reverberant room may be used.

The average is obtained by adding together the four sound pressure levels in the usual way and subtracting 6.

(*References*: ISO 3741, 3742, 3743, 3744, 3745, 3756.)

Results

1. *dB* (linear)

Position	SPL (dB)	Total
1	_____	
2	_____	
3	_____	
4	_____	_____

$\therefore$ average sound pressure level = Total − 6

$$= \underline{\hspace{2cm}} - 6$$

$$= \underline{\hspace{2cm}} \text{ dB}$$

Volume of room, V $\qquad = \underline{\hspace{2cm}} \text{ m}^3$

Reverberation time, t $\qquad = \underline{\hspace{2cm}} \text{ s}$

Now: PWL = SPL $\qquad + 10\log_{10} V \qquad - 10\log_{10} t \qquad - 14$

$\qquad = \underline{\hspace{2cm}} + 10\log_{10} \underline{\hspace{1cm}} - 10\log_{10} \underline{\hspace{1cm}} - 14$

$\qquad = \underline{\hspace{2cm}} + \underline{\hspace{2cm}} - \underline{\hspace{2cm}} - 14$

$\qquad = \underline{\hspace{2cm}} \text{ dB}$

2. *'A' weighted sound power level*

Position	SPL (dB)	Total
1	_____	
2	_____	
3	_____	
4	_____	_____

$\therefore$ average sound pressure level = Total − 6

$$= \underline{\hspace{2cm}} - 6$$

$$= \underline{\hspace{2cm}} \text{ dB}$$

Volume of room, V $\qquad = \underline{\hspace{2cm}} \text{ m}^3$

Reverberation time, t $\qquad = \underline{\hspace{2cm}} \text{ s}$

Now: PWL = SPL $\qquad + 10\log_{10} V \qquad - 10\log_{10} t \qquad - 14$

$\qquad = \underline{\hspace{2cm}} + 10\log_{10} \underline{\hspace{1cm}} - 10\log_{10} \underline{\hspace{1cm}} - 14$

$\qquad = \underline{\hspace{2cm}} + \underline{\hspace{2cm}} - \underline{\hspace{2cm}} - 14$

$\qquad = \underline{\hspace{2cm}} \text{ dB}$

3.   250 *Hz*

Position	SPL (dB)	Total
1	_____	
2	_____	
3	_____	
4	_____	_____

∴   average sound pressure level = Total − 6

$$= \underline{\hspace{2cm}} - 6$$
$$= \underline{\hspace{2cm}} \text{dB}$$

Volume of room, V $= \underline{\hspace{2cm}} \text{m}^3$

Reverberation time, t $= \underline{\hspace{2cm}} \text{s}$

Now: $PWL = SPL \quad + 10\log_{10} V \quad - 10\log_{10} t \quad - 14$

$$= \underline{\hspace{2cm}} + 10\log_{10} \underline{\hspace{1cm}} - 10\log_{10} \underline{\hspace{1cm}} - 14$$
$$= \underline{\hspace{2cm}} + \underline{\hspace{2cm}} - \underline{\hspace{2cm}} - 14$$
$$= \underline{\hspace{2cm}} \text{dB}$$

4.   500 *Hz*

Position	SPL (dB)	Total
1	_____	
2	_____	
3	_____	
4	_____	_____

∴   average sound pressure level = Total   − 6

$$= \underline{\hspace{2cm}} - 6$$
$$= \underline{\hspace{2cm}} \text{dB}$$

Volume of room, V $= \underline{\hspace{2cm}} \text{m}^3$

Reverberation time, t $= \underline{\hspace{2cm}} \text{s}$

Now: $PWL = SPL \quad + 10\log_{10} V \quad - 10\log_{10} t \quad - 14$

$$= \underline{\hspace{2cm}} + 10\log_{10} \underline{\hspace{1cm}} - 10\log_{10} \underline{\hspace{1cm}} - 14$$
$$= \underline{\hspace{2cm}} + \underline{\hspace{2cm}} - \underline{\hspace{2cm}} - 14$$
$$= \underline{\hspace{2cm}} \text{dB}$$

5. 1 kHz

Position	SPL (dB)	Total
1	_____	
2	_____	
3	_____	
4	_____	_____

$\therefore$ average sound pressure level = Total − 6

$$= \underline{\hspace{2cm}} - 6$$

$$= \underline{\hspace{2cm}} \text{dB}$$

Volume of room, V $\quad\quad = \underline{\hspace{2cm}} \text{m}^3$

Reverberation time, t $\quad\quad = \underline{\hspace{2cm}} \text{s}$

Now: PWL = SPL $\quad\quad + 10 \log_{10} V \quad\quad - 10 \log_{10} t \quad\quad - 14$

$$= \underline{\hspace{2cm}} + 10 \log_{10} \underline{\hspace{1cm}} - 10 \log_{10} \underline{\hspace{1cm}} - 14$$

$$= \underline{\hspace{2cm}} + \underline{\hspace{2cm}} - \underline{\hspace{2cm}} - 14$$

$$= \underline{\hspace{2cm}} \text{dB}$$

6. 2 kHz

Position	SPL (dB)	Total
1	_____	
2	_____	
3	_____	
4	_____	_____

$\therefore$ average sound pressure level = Total $\quad\quad$ − 6

$$= \underline{\hspace{2cm}} - 6$$

$$= \underline{\hspace{2cm}} \text{dB}$$

Volume of room, V $\quad\quad = \underline{\hspace{2cm}} \text{m}^3$

Reverberation time, t $\quad\quad = \underline{\hspace{2cm}} \text{s}$

Now: PWL = SPL $\quad\quad + 10 \log_{10} V \quad\quad - 10 \log_{10} t \quad\quad - 14$

$$= \underline{\hspace{2cm}} + 10 \log_{10} \underline{\hspace{1cm}} - 10 \log_{10} \underline{\hspace{1cm}} - 14$$

$$= \underline{\hspace{2cm}} + \underline{\hspace{2cm}} - \underline{\hspace{2cm}} - 14$$

$$= \underline{\hspace{2cm}} \text{dB}$$

Experiment 4.10   Determination of the sound power level of a machine under free field conditions outdoors

Apparatus
Sound level meter with octave filters, tape measure, quiet field (empty football pitch or car park) with no reflecting surfaces.

Method
The background level of sound should be measured first at several positions in the area being used and the mean obtained. It should be at least 15 dB quieter than the lowest level to be measured from the machine.

The sound pressure level is measured at four positions at 1 m distance from the machine. The average is obtained. Measurements are repeated at 2 m, 3 m, 4 m, etc., up to four times the original distance. A graph is plotted on logarithmic/linear paper of distance against sound pressure level. The sound power level is calculated for each measurement position plotted on the straight-line part of the graph.

This may be repeated in dB(A) and octave bands.

Theory
Sound power level, $\text{PWL} = 10 \log_{10} \dfrac{W}{W_0}$ decibels

where $\qquad\qquad W = $ sound power output of the machine in watts

$\qquad\qquad W_0 = $ reference sound power

$\qquad\qquad\quad = 10^{-12}\,\text{W}$

$\qquad \therefore\quad \text{PWL} = 10 \log_{10} W - 10 \log_{10} 10^{-12}$

$\qquad\qquad\qquad = 10 \log_{10} W + 120$

If the sound pressure level of the source is measured in space, then

intensity $\quad I = \dfrac{W}{4\pi r^2}$

where $\quad$ ‹ $r = $ radius of sphere or distance from source

$\quad \therefore\quad \text{PWL} = 10 \log_{10} 4\pi r^2 I + 120$

$\qquad\qquad\qquad = 10 \log_{10} I + 10 \log_{10} 4\pi r^2 + 120$

But $\quad \text{SPL} = 10 \log_{10} \dfrac{I}{I_0}$

where $\quad I_0 = 10^{-12}\,\text{W/m}^2$

$$\therefore \quad PWL = 10 \log_{10} \frac{I}{I_0} + 10 \log_{10} 4 \pi r^2$$

$$\therefore \quad PWL = SPL + 10 \log_{10} 4\pi + 10 \log_{10} r^2$$
$$= SPL + 10 \log_{10} 12.57 + 20 \log_{10} r$$
$$= SPL + 10.97 + 20 \log_{10} r$$
$$= SPL + 20 \log_{10} r + 11$$

If measurements are taken in the practical situation on a hard reflecting ground, then the sound pressure level would be 3 dB higher and in consequence the equation becomes:

$$PWL = SPL + 20 \log_{10} r + 8$$

If a graph is plotted on log/linear graph paper of distance against SPL, a straight-line graph should result in the 'far' field of the machine where the gradient is 6 dB/doubling of distance.

Results
Linear
Background SPL = _____ dB

1   *Distance* = 1 m

Position	SPL (dB)	Total
1	_____	
2	_____	
3	_____	
4	_____	_____

Mean SPL = Total   − 6

=_____ − 6

=_____ dB

2. *Distance* = 2 m

Position	SPL (dB)	Total
1	_____	
2	_____	
3	_____	
4	_____	_____

Mean SPL = Total   − 6

=_____ − 6

=_____ dB

3. *Distance* = 3 m

Position	SPL (dB)	Total
1	_____	
2	_____	
3	_____	
4	_____	_____

Mean = Total $- 6$

    = _____ $- 6$

    = _____ dB

4. *Distance* = 4 m

Position	SPL (dB)	Total
1	_____	
2	_____	
3	_____	
4	_____	_____

Mean SPL = Total $- 6$

    = _____ $- 6$

    = _____ dB

5. *Distance* = 5 m

Position	SPL (dB)	Total
1	_____	
2	_____	
3	_____	
4	_____	_____

Mean SPL = Total $- 6$

    = _____ $- 6$

    = _____ dB

PWL = Mean SPL $+ 20 \log_{10} r + 8$

    = _____ $+ 20 \log_{10} r + 8$ dB(A)

    = _____ dB(A)

Experiment 4.11   To investigate the way in which loudness varies with frequency

Apparatus

Two audio-frequency generators, two-way switch, amplifier, loudspeaker, sound level meter, tripod stand, a reasonably quiet room.

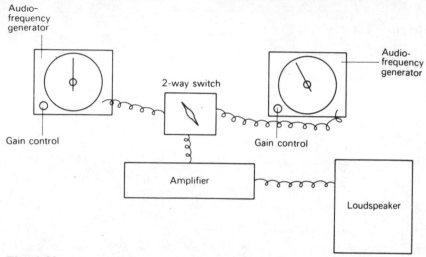

Fig. 4.32

Method

The apparatus is set up as shown in Fig. 4.32 with the subject seated so that he can conveniently reach the switch box and the gain controls on the generators. The sound level meter should be set on its tripod stand to measure as nearly as possible the sound as heard at the subject's ears.

One generator is set to produce a frequency of 1000 Hz and adjusted to produce a sound pressure level of 20 dB at the meter. (If the room is too noisy start at 40 dB.) The second generator is set to 500 Hz and using its gain control the level adjusted until judged equally loud with the 1000 Hz note. This may be compared by quickly switching back and forth between the two generators. The sound pressure level for equal loudness is noted. This is repeated for 250 Hz, 125 Hz, 63 Hz, 2000 Hz, 4000 Hz and 8000 Hz.

The first generator is then adjusted to 40 dB at 1000 Hz and the experiment repeated. The same may be repeated for 1000 Hz levels of 60 dB, 80 dB and possibly 100 dB. (At 100 dB it is advisable that students do not spend more than about half an hour because of potential hearing damage, apart from nuisance to others and temporary threshold shift which will invalidate the results.)

A graph is plotted of equal loudness which may be compared with the equal loudness contours shown in Fig. 4.33.

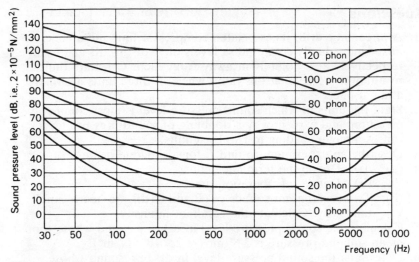

Fig. 4.33 Equal–loudness contours

Theory

Loudness is defined as an observer's auditory impression of the strength of a sound. The loudness of a sound depends upon its amplitude and frequency. Thus for equal loudness at different frequencies the amplitude will have to change. The *phon* is the unit of loudness level for a pure tone of 1000 Hz measured in decibels above 2×10^{-5} N/m^2.

Thus the number of phons for any pure tone is the sound pressure level of a 1000 Hz pure tone which is equally loud.

Results

Sound pressure levels (dB)

1000 Hz	63 Hz	125 Hz	250 Hz	500 Hz	2000 Hz	4000 Hz	8000 Hz
20 dB							
40 dB							
60 dB							
80 dB							
100 dB							

Questions

1. A pneumatic drill, its associated compressor unit and a concrete mixer individually create noise levels of 70 dB, 67 dB and 68 dB respectively at a point 10 m away from each. Calculate the resultant noise level at this point when all three units are operating at once.

2. State the units in which sound pressure level is most commonly measured and explain why a doubling of sound pressure level causes a meter reading to rise by 6 units.

3. Describe an experiment to find E for a concrete beam.

4. (a) (i) Define the term decibel.
    (ii) Define reverberation time.
   (b) Two reverberant sounds of 60 and 65 dB are heard simultaneously. Calculate the resultant intensity.

5. Calculate the sound pressure level in dB of a sound whose root mean squared pressure is $2 \, \text{N/m}^2$ (*re* $2 \times 10^{-5} \, \text{N/m}^2$).

6. Determine the sound pressure level in dB of a sound whose intensity is $0.005 \, \text{W/m}^2$ (*re* $10^{-12} \, \text{W/m}^2$).

7. A noise has the following octave band analysis. Calculate the sound pressure level in dB.

Octave band centre frequency (Hz)	SPL (dB)
125	45
250	53
500	61
1000	58
2000	49

8. Calculate the total intensity in W/m^2 for the sound in Question 7.

9. Calculate the RMS pressure of a sound whose intensity is $0.01 \, \text{W/m}^2$.

10. Define sound power level.
    Calculate the sound pressure level at a distance of 100 m over flat non-absorbent ground of a sound of sound power level 110 dB.

11. Calculate the sound power level of a source which radiates uniformly over flat unobstructed ground and produces a sound pressure level of 62 dB at a distance of 10 m.

Chapter 5

Room acoustics

Requirements for good room acoustics

The aim of the design of any hall for speech or music must be to obtain the optimum acoustic effect on the audience. The normal requirements are as follows:

1. Adequate amount of sound to all parts.
2. An even distribution of the sound.
3. Adequate insulation against outside noise.
4. The rate of decay of sound (reverberation time) should be the optimum for the required use of the ròom.
5. Avoidance of long-delayed echoes.

Absorption of sound

Sound absorption is the damping of a sound wave on passing through a medium or striking a surface. An alternative definition is the property possessed by objects or materials of absorbing sound.

In terms of room acoustics this is not entirely adequate as we are more concerned about reflected sound. An open window for instance actually absorbs no sound but neither would any sound be reflected back into the room. As far as the room is concerned all the sound is absorbed because 100 per cent is transmitted.

Absorption coefficient

This is the fraction of the incident sound which is not reflected by the material. The value varies with frequency and construction as well as material. This can be seen from the examples given in Tables 5.1 and 5.2. It should be realised that sound absorption is quite different from sound insulation. In most cases materials which have high absorption coefficients have poor insulating properties. Absorbent materials are used to reduce the noise of machines within an enclosed space. The reduction is limited to a maximum of about 10 dB even in the best situation. The cost of such reduction is usually very high indeed.

Fibrous absorbers

Porous, fibrous materials are usually the best for general absorption. Mineral fibres are the most commonly used for sound. The mechanism of absorption is to convert the sound energy into thermal energy by vibration of the fibres. It must be appreciated that the amount of sound energy available is very small even for very high noise levels and thus the rise in temperature is also negligible. The efficiency of absorption is dependent upon frequency and thickness of material. The thicker the material the better particularly at low frequencies (Fig. 5.1). An alternative to making the material thicker is to space it away from the hard surface of wall or ceiling. Thus 13 mm mineral fibre acoustic tiles are normally suspended at about 300 mm below the upper floor. This of course has the additional advantage of providing space for services.

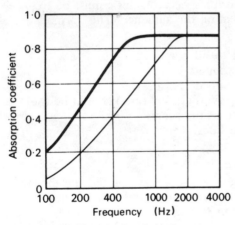

Typical absorption characteristics for porous materials showing

——— Thin material

▬▬▬ Thick material with its increase in absorption at lower frequencies

Fig. 5.1 Typical absorption characteristics for porous materials showing (i) thin material and (ii) thick material with its increase in absorption at lower frequencies

Table 5.1 Absorption coefficients of common building materials

Material and method of fixing	Absorption coefficients			
	Low frequency 125 Hz	Medium frequency 500 Hz	High frequencies	
			2000 Hz	4000 Hz
Boarded roof; underside of pitched slate or tile roof	0.15	0.1	0.1	0.1
Boarding ('match') about 20 mm thick over air space on solid wall	0.3	0.1	0.1	0.1
Brick work – plain or painted	0.02	0.02	0.04	0.05
Clinker ('breeze') concrete unplastered	0.2	0.6	0.5	0.4
Carpet (medium) on solid concrete floor	0.1	0.3	0.5	0.6
Carpet (medium) on joist or board and batten floor	0.2	0.3	0.5	0.6
Concrete, constructional or tooled stone or granolithic finish	0.01	0.02	0.02	0.02
Cork slabs, wood blocks, linoleum or rubber flooring on solid floor (or wall)	0.05	0.05	0.1	0.1
Curtains (medium fabrics) hung straight and close to wall	0.05	0.25	0.3	0.4
Curtains (medium fabrics) hung in folds or spaced away from wall	0.1	0.4	0.5	0.6
Felt, hair, 25 mm thick, covered by perforated membrane (viz. muslin) on solid backing	0.1	0.7	0.8	0.8
Fibreboard (normal soft) 13 mm thick mounted on solid backing	0.05	0.15	0.3	0.3

Table 5.1 (*Cont.*)

Material and method of fixing	Absorption coefficients			
	Low frequency 125 Hz	Medium frequency 500 Hz	High frequencies	
			2000 Hz	4000 Hz
Fibreboard (painted) 13 mm thick mounted on solid backing	0.05	0.1	0.15	0.15
Fibreboard (normal soft) 13 mm thick mounted over air space on solid backing or on joists or studs	0.3	0.3	0.3	0.3
Fibreboard (painted) 13 mm thick mounted over air space on solid backing or on joists or studs	0.3	0.15	0.1	0.1
Floor tiles (hard) or 'composition' flooring	0.03	0.03	0.05	0.05
Glass; windows glazed with up to 3 mm glass	0.2	0.1	0.05	0.02
Glass; 7 mm plate or thicker in large sheets	0.1	0.04	0.02	0.02
Glass used as a wall finish (viz., 'Vitrolite') or glazed tiles or polished marble or polished stone fixed to wall	0.01	0.01	0.01	0.01
Glass wool or mineral wool 25 mm thick on solid backing	0.2	0.7	0.9	0.8
Glass wool or mineral wool 50 mm thick on solid backing	0.3	0.8	0.75	0.9
Glass wool or mineral wool 25 mm thick mounted over air space on solid backing	0.4	0.8	0.9	0.8
Plaster, lime or gypsum on solid backing	0.02	0.02	0.04	0.04

Table 5.1 (*Cont.*)

Material and method of fixing	Absorption coefficients			
	Low frequency 125 Hz	Medium frequency 500 Hz	High frequencies	
			2000 Hz	4000 Hz
Plaster, lime or gypsum on lath, over air space on solid backing, or on joists or studs including decorative fibrous and plaster board	0.3	0.1	0.04	0.04
Plaster, lime gypsum or fibrous, normal suspended ceiling with large air space above	0.2	0.1	0.04	0.04
Plywood mounted solidly	0.05	0.05	0.05	0.05
Plywood panels mounted over air space on solid backing, or mounted on studs, without porous material in air space	0.3	0.15	0.1	0.05
Plywood panels mounted over air space on solid backing, or mounted on studs, with porous material in air space	0.4	0.15	0.1	0.05
Water – as in swimming baths	0.01	0.01	0.01	0.01
Wood boards on joists or battens	0.15	0.1	0.1	0.1
Wood-wool slabs 25 mm thick (unplastered) solidly mounted	0.1	0.4	0.6	0.6
Wood-wool slabs 80 mm thick (unplastered) solidly mounted	0.2	0.8	0.8	0.8
Wood-wool slabs 25 mm thick (unplastered) mounted over air space on solid backing	0.15	0.6	0.6	0.7

Table 5.2 Absorption of special items

	Absorption units m^2			
	Low frequency 125 Hz	Medium frequency 500 Hz	High frequencies	
			2000 Hz	4000 Hz
Air (per m^3)	—	—	0.007	0.020
Audience seated in fully upholstered seats (per person)	0.19	0.47	0.51	0.47
Audience seated in wooden or padded seats (per person)	0.16	0.4	0.43	0.4
Seats (unoccupied) fully upholstered (per seat)	0.12	0.28	0.31	0.37
Seats (unoccupied) wooden or padded or metal and canvas (per seat)	0.07	0.15	0.18	0.19
Theatre proscenium opening with average stage set (per m^2)	0.2	0.3	0.4	0.5

It can be seen that it is not possible to give absorption coefficient figures for fibrous materials unless details of thickness and construction are known. In addition the surface finish may be very important particularly for the higher frequencies. Most acoustic tiles on the market are surface perforated before painting with an acrylic paint. The area of perforations may well be over one-eighth of the total area. It is important on redecoration not to fill these holes with paint which would markedly reduce the effectiveness of the material for sound absorption.

The absorption coefficient for a material will vary with the angle of incidence. In practice sound will be incident at all angles from grazing incidence to normal. Measurements are thus best made under random conditions of incidence.

Membrane absorbers

Fibrous materials are not the only absorbers available although they are the most common. A panel such as a sheet of glass will also act as an absorber. In this case the whole material vibrates, some sound being converted to heat and much being transmitted. As would be expected the absorption properties are quite different from those of fibrous materials. The main difference is that they are far better at absorption of low frequency sound (Fig. 5.2).

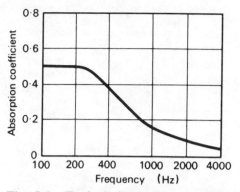

Fig. 5.2 Typical absorption characteristics of a membrane absorber showing increased efficiency at low frequencies

Most of the heavy hard building materials such as bricks and plaster reflect nearly all of the incident sound. Commonly this may be as high as 98 per cent or an absorption coefficient of only 2 per cent (0.02).

It can be seen from what has been said that in any room it becomes important to have the correct sound absorbent finishes. Too little absorption and the room appears to 'echo' and too much makes the room 'dead'. In either case clarity of speech may be low.

Helmholtz or cavity resonator absorbers

These are containers with a small open neck and work by resonance of the air within the cavity. Porous material is introduced into the neck to increase the efficiency of absorption. It can be shown that for a narrow necked resonator of the type shown in Fig. 5.3 the resonant frequency (f) is approximately:

$$f = \frac{Cr}{2\pi} \bigg/ \left(\frac{2\pi}{2(l + \pi r)V} \right)$$

where C = velocity of sound in air

r = radius of neck

l = length of neck

V = volume of cavity

If there is no neck this reduces to:

$$f = \frac{C}{2\pi} \sqrt{\frac{2r}{V}}$$

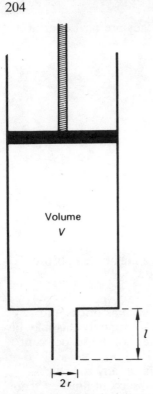

Fig. 5.3 Helmholtz resonator

Efficient absorption is only possible over a very narrow band, as shown in Fig. 5.4 and it is necessary to have many resonators tuned to slightly different frequencies if effective control of reverberation time is to be obtained. Cavity resonators are useful in controlling long reverberation times at isolated frequencies, and are used in a number of concert halls.

Total absorption

The total absorption within a room is the result of the sum of the absorption of all the surfaces and objects in that space. It can be appreciated that the absorption coefficient of a material is also the effective absorption of $1\,\text{m}^2$ of the material. Thus the total absorption by that material in a room may be calculated from the total area of material times the absorption coefficient for the material (i.e. $A \times \alpha$).

Some items within a hall or room cannot conveniently be given by area. Such is the situation for people and seats. In these cases average figures are given. While there will be variations in the contribution which individuals make, depending upon their size and

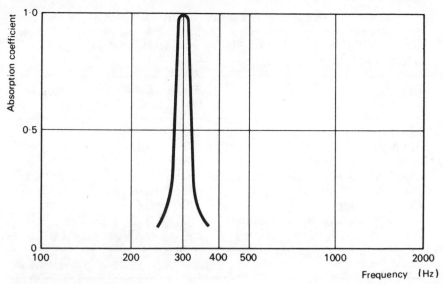

Fig. 5.4 Characteristic very narrow band absorption by Helmholtz or cavity resonator

clothing, on average there will be little variation over a group. Thus for these items their average contribution to sound absorption is given in terms of square metres of absorption.

The total absorption within a room or hall is the sum of the contributions of all materials and items. Allowance may need to be made for shading. Chairs and people may acoustically shade the floor. As may be expected shading is more significant for the higher frequencies. Typical allowance would be a reduction in floor absorption of 40 per cent at 125 to 500 Hz and 60 per cent at 2000 Hz.

Example 5.1 Calculate the total absorption at 500 Hz in a hall containing the following surface finishes:

Plaster on brickwork	$265 \, m^2$
3 mm glass window	$43 \, m^2$
Stage, boards on joist	$70 \, m^2$
25 mm wood-wool slabs	$60 \, m^2$
Plate-glass screen	$96 \, m^2$
Ceiling plaster	$310 \, m^2$
Wood block floor	$300 \, m^2$

Assume that the shading of the floor by the audience effectively reduces its absorption by 40 per cent.

Additional information is needed before the calculation may be performed, that is the absorption coefficients of the materials. These are normally obtained from tables or manufacturers' data. The calculation is best tabulated.

Items	Area (m^2)	Coefficient	(m^2)
Plaster	265	0.02	5.3
3 mm glass	43	0.1	4.3
Stage, boards on joists	70	0.1	7.0
25 mm wood-wool slabs	60	0.4	24.0
Plate glass	96	0.04	3.8
Ceiling plaster	310	0.1	31.0
Wood block floor minus shading	300	0.05–40%	9.0
Audience	250	0.43/person	107.5
Total absorption			$191.9\,m^2$

Reverberation

This is the persistence of sound in an enclosure due to repeated reflections at the boundaries.

Reverberant sound is not the same as an echo. An echo is a single distinct reflection. Reverberant sound is the result of many thousand reflections each second. The result subjectively is a progressive decay in sound level. If a sound decays very fast then there is a strong directional element and clarity can be very high. In situations where the sound persists for a very long time clarity decreases but there is a greater reinforcement of the sound and greater fullness. It will be appreciated that the optimum situation varies with the use of the room or hall, being different for speech from music.

Reverberation time

This is the time taken for a sound to decay by 60 dB. It is dependent upon the volume and amount of absorption in the hall.

Sabine's formula

The actual reverberation time in a hall may be calculated from Sabine's formula:

$$t = \frac{0.16\,V}{A}$$

where t = reverberation time in seconds (s)

V = volume of hall in cubic metres (m^3)

A = area of absorption in square metres (m^2)

Example 5.2   Calculate the reverberation time for Example 5.1 if the volume of the hall is 2500 m^3.

$$t = \frac{0.16 \times V}{A}$$

$$= \frac{0.16 \times 2500}{191.9}$$

$$= 2.1\,\text{s}$$

Optimum reverberation time

The desirable reverberation time is dependent upon the size and use of a room. The optimum is derived from subjective experiments. One empirical formula used is due to Stephens and Bate, is approximate and applicable at 500 Hz,

$$t = r(0.012\sqrt[3]{V} + 0.1070)$$

where t = optimum reverberation time in seconds (s)

V = volume of hall in cubic metres (m^3)

r = 4 for speech

= 5 for orchestral music

= 6 for choral music

It is advisable to increase these figures by 40 per cent at low frequencies (i.e. 125 Hz).

An alternative is to use graphs such as shown in Fig. 5.5.

Volume of a hall

Cost increases with size. At the design stage the number of people for whom the hall is to be built must be decided. From a cost aspect the smaller the volume per person the cheaper will be the building. However the acoustic requirements determine minimum volumes. If the space is to be used solely for speech then a short reverberation time is desirable for clarity and hence the volume can be a minimum. Space rather then acoustic requirements dictate the minimum volume per person. For music longer reverberation times are needed and this decides the minimum volumes possible. In Table 5.3 some suitable figures are suggested for the volumes of different types of hall. Tables 5.4 and 5.5 give acoustical data for some well known concert halls.

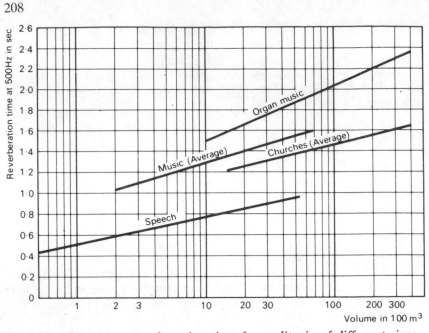

Fig. 5.5 Optimum reverberation time for auditoria of different sizes

Table 5.3 Optimum volume/person for various types of hall

	Minimum	Optimum	Maximum
Concert halls	6.5 m^3	7.1 m^3	9.9 m^3
Italian-type opera houses	4.0 m^3	4.2–5.1 m^3	5.7 m^3
Churches	5.7 m^3	7.1–9.9 m^3	11.9 m^3
Cinemas	—	3.1 m^3	4.2 m^3
Rooms for speech	—	2.8 m^3	4.9 m^3

Shape of hall

Tradition has determined the shape and design of many well known concert halls. Partly as a result three basic plans are in common use for large halls. These are rectangular, fan and horseshoe shaped. Shape is less critical in a hall which seats under 1000 people. As size increases, so also does a preference for the fan shape, so that the audience is seated on average slightly closer to the sound source (Figs. 5.6 and 5.7). Care needs to be taken that the rear of such a hall or ceiling is not concave (Fig. 5.8) as this can cause a focusing of the sound within the hall. A convex shape would be ideal but introduces practical problems. The rear wall must be treated with

Table 5.4 Acoustical data for some well known concert halls

Name	Volume (m³)	Audience capacity	Volume per aud. seat m³	Mid frequency RT for full hall (s)
St Andrew's Hall, Glasgow (built 1877)	16 100	2133	7.6	1.9
Carnegie Hall, New York (1891)	24 250	2760	8.8	1.7
Symphony Hall, Boston (1900)	18 740	2631	7.1	1.8
Tanglewood Music Shed Lennox, Mass. (1938)	42 450	6000	7.1	2.05
Royal Festival Hall (1951)	22 000	3000	7.3	1.47
Liederhalle, Grosser Saal, Stuttgart (1956)	16 000	2000	8.0	1.62
F. R. Mann Concert Hall, Tel Aviv (1957)	21 200	2715	7.8	1.55
Beethovenhalle, Bonn (1959)	15 700	1407	11.2	1.7
Philharmonic Hall, New York (1962)	24 430	2644	9.3	2.0
Philharmonic Hall, Berlin (1963)	26 030	2200	11.8	2.0

absorbent material to avoid strong reflections of sound which can lead to confusion, particularly of speech (Fig. 5.9). Problems can arise from reflections from side walls which may have to be broken up with either large diffusing surfaces or by the use of absorbent material. Opera houses are often horseshoe shaped. The concave surfaces are broken up by tiers of boxes around the walls. The audience provides the absorption of sound. This type of design is excellent for opera where clarity of sound is more important than fullness of tone but would not be considered so good for orchestral music. Covent Garden Opera House is an example of a horseshoe-shaped hall.

Fig. 5.6 Rectangular-shaped hall

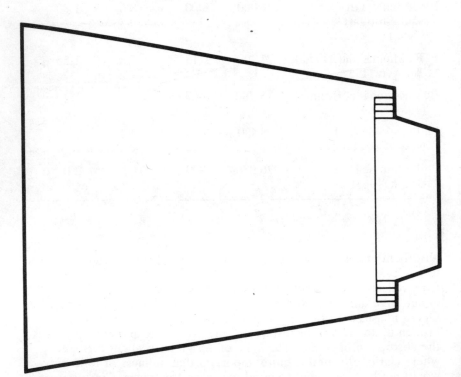

Fig. 5.7 Fan-shaped hall

Table 5.5 Acoustical data for some well known opera houses

Name	Volume (m³)	Audience capacity	Volume per aud. seat (m³)	Mid frequency RT for full hall (s)
Teatro alla Scala, Milan (1778)	11 245	2289	4.91	1.2
Academy of Music, Philadelphia (1857)	15 090	2836	5.32	1.35
Royal Opera House (1858)	12 240	2180	5.6	1.1
Theatre National de L'Opera, Paris (1875)	9960	2131	4.67	1.1
Metropolitan Opera House, New York (1883)	19 520	3639	5.36	1.2

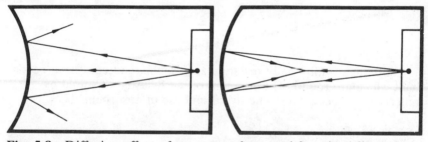

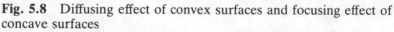

Fig. 5.8 Diffusing effect of convex surfaces and focusing effect of concave surfaces

The traditional rectangular shape has many advantages in construction and as long as reflectors are used over the sound source the difficulty of obtaining sufficient loudness near the back can be overcome. The Royal Festival Hall is an example of a rectangular hall (Fig. 5.6).

Other shapes have been used or suggested. Conventional shapes are probably more popular because the difficulties are less of an unknown quantity. The Royal Albert Hall is one example of an oval construction, but has acoustic problems. The Philharmonic Concert Hall, Berlin, is of irregular shape. The Philharmonic Hall, New York, is a combination of the rectangular and fan shape. Traditional dimensions are typically 2:3:5 for height:width:length.

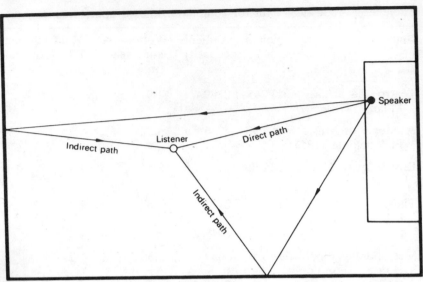

Fig. 5.9 Sound paths differing by 17 m or more can cause confusion of speech

Although traditionally rectangular in shape, churches have been built which are round. The problem of the focusing of sound by reflection has been overcome by careful use of absorbents.

Seating arrangements

Rows of people, particularly at grazing incidence to the sound, represent a most efficient absorber. It is essential in all but the smallest halls (above 200 people) to rake the seating if at all possible. As a simple rule, adequate vision should ensure an adequate sound path. This will mean that the line of sight needs to be raised by 80 to 100 mm for each successive row (Fig. 5.10).

Reflection of sound

Sound is reflected in a similar way to light, the angle of incidence being equal to the angle of reflection. However, it must be remembered that for this to be true, the reflecting object must be at least the same size as the wavelength concerned. It can often be very useful to carry out a limited geometrical analysis. This can prevent

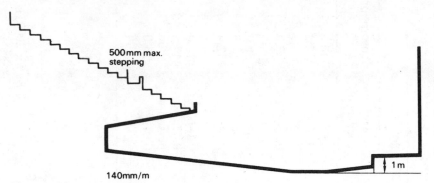

Fig. 5.10 Suitable rake for a hall with a balcony

the problem of long delayed reflections and focusing effects. It is impractical to take a geometrical analysis beyond the first or second reflections but it can prevent gross errors in design.

Selection of absorbent materials

It will be clear that the acoustic design of a hall is complex requiring the correct reverberation time to be obtained for all frequencies from around 100 Hz to 4 kHz. This necessitates the right amount of absorption for each frequency. Thus the choice of absorber is determined by the absorption needed for each frequency. At the very least calculations should be performed for low (125 Hz), medium (500 Hz) and high (2000 Hz) frequencies. The absorption spectra for absorbent materials are important in their selection.

Example 5.3 Calculate the reverberation time at 125 Hz, 500 Hz and 2000 Hz for a hall of volume 2500 m³ having the following surface finishes:

Plaster on brickwork	265 m²
3 mm glass window	43 m²
Stage, boards on joist	70 m²
25 mm wood-wool slabs	60 m²
Plate-glass screen	96 m²
Ceiling plaster	310 m²
Wood block floor	300 m²

Assume the shading of the floor by the audience effectively reduces its absorption by 40 per cent at 125 Hz and 500 Hz and by 60 per cent at 2000 Hz.

Absorption

Item	Area m²	125 Hz		500 Hz		2000 Hz	
		Absorption coefficient	Absorption	Absorption coefficient	Absorption	Absorption coefficient	Absorption
Plaster	265	0.02	5.3	0.02	5.3	0.04	10.6
3 mm glass	43	0.3	12.9	0.1	4.3	0.05	2.2
Stage, boards on joists	70	0.15	10.5	0.1	7.0	0.1	7.0
25 mm wood-wool slabs	60	0.1	6.0	0.4	24.0	0.6	36.0
Plate glass	96	0.1	9.6	0.04	3.8	0.02	1.9
Ceiling plaster	310	0.2	62.0	0.1	31.0	0.04	12.4
Wood block floor minus shading	300	0.05–40%	9.0	0.05–40%	9.0	0.1–60%	12.0
Audience	250	0.17/ person	42.5	0.43/ person	107.5	0.47/ person	117.5
Air	2500 m³					0.01	25.0
Total absorption			157.8		191.9		224.6

Total absorption at 125 Hz = 157.8 m²
Total absorption at 500 Hz = 191.9 m²
Total absorption at 2000 Hz = 224.6 m²

The actual reverberation time by Sabine's formula, $t = (0.16V)/A$,

$$t = \frac{0.16 \times 2500}{157.8} = 2.5 \text{ s for } 125 \text{ Hz} \quad \text{and} \quad t = \frac{0.16 \times 2500}{191.9} = 2.1 \text{ s for } 500 \text{ Hz} \quad \text{and} \quad t = \frac{0.16 \times 2500}{224.6} = 1.8 \text{ s for } 2000 \text{ Hz}$$

Example 5.4 In the previous example calculate the amount of absorption needed at each frequency to achieve the optimum reverberation time for speech.

$$\text{Optimum reverberation time } (t) = 4(0.012 \sqrt[3]{V} + 0.1070)$$
$$= 4(0.012 \sqrt[3]{2500} + 0.1070)$$
$$= 4(0.012 \times 13.57 + 0.1070)$$
$$= \underline{1.08\,\text{s}}$$

Allowing for a 40 per cent increase at 125 Hz the optimum reverberation times will be:

125 Hz   1.5 s
500 Hz   1.1 s
2000 Hz   1.1 s

Thus the amount of absorption needed may be calculated from Sabine's formula:

$$t = \frac{0.16\,V}{A} \quad \text{or} \quad A = \frac{0.16\,V}{t}$$

$$\text{At 125 Hz } A = \frac{0.16 \times 2500}{1.5}$$

$$= \underline{265\,\text{m}^2}$$

$$\text{At 500 Hz } A = \frac{0.16 \times 2500}{1.1}$$

$$= \underline{370\,\text{m}^2}$$

At 2000 Hz $A = \underline{370\,\text{m}^2}$

Thus the extra absorption needed:

At 125 Hz $265 - 158 = 107\,\text{m}^2$
At 500 Hz $370 - 192 = 178\,\text{m}^2$
At 2000 Hz $370 - 225 = 145\,\text{m}^2$

It can be seen that most absorption is needed in the 500 Hz region. It is unlikely that fibrous absorbent will go a long way to solving the problem. No single material is likely to give the perfect solution. It can be seen that if such a material were available 200 m² of a finish giving absorption coefficients of 0.54 at 125 Hz, 0.89 at 500 Hz and 0.73 at 2 kHz would suffice. Allowance would also need to be made for the surface covered. In practice a combination of fibrous absorbers, membrane absorbers and possibly Helmholtz resonator absorbers may need to be used.

Experiments

Experiment 5.1 Measurement of the reverberation time of a room using white noise (Method 1)

Apparatus
White-noise generator or tape recorder with white noise recorded, amplifier, two 50 W loudspeakers, condenser microphone, audio-frequency spectrometer (third octave filters) and level recorder.

Method
The apparatus is set up as shown in Fig. 5.11. The white-noise generator is set to produce a bandwidth greater than one-third of an octave centred around 100 Hz. By adjustment of the amplifier, a sound pressure level of about 100 dB is produced within the room. A suitable microphone position is selected greater than 1 m from any surface and the frequency analyser set at a one-third octave centred around 100 Hz. The level recorder using a 50 dB potentiometer is then adjusted to give full-scale deflection for the sound pressure level produced. With the recording paper running at 30 mm/s (10 mm/s in a room with a very long reverberation time), the noise is switched off instantly. The decay line is recorded on the moving paper, from which the reverberation time may be calculated or determined using protractors.

This procedure should be repeated in one-third octave intervals from 100 to 3150 Hz. The microphone is then moved to other room positions and the whole procedure repeated, six times for frequencies below 500 Hz and three times for frequencies of 500 Hz and above.

Theory
Reverberation time is the time in seconds for a 60 dB decay in the sound pressure level after the source has been switched off.

Standing waves make accurate measurement difficult, particularly for low frequencies, and measurements need to be made at several different room positions. It is for this reason that white noise is used.

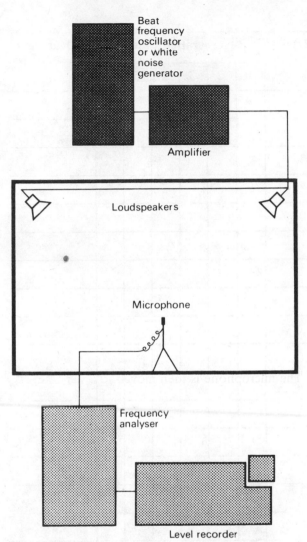

Fig. 5.11 Measurement of reverberation time using a white-noise generator as sound source

Results

Centre frequency (Hz) $\frac{1}{3}$-octave bandwidth	Reverberation time (s)						Mean value
	1	2	3	4	5	6	
100							
125							
160							
200							
250							
315							
400							
500							
630							
800							
1000							
1250							
1600							
2000							
2500							
3150							

Experiment 5.2 Measurement of the reverberation time of a room using a starting pistol (Method 2)

Apparatus
Pistol with blank cartridges (a starting pistol is suitable for small rooms), condenser microphone, audio-frequency spectrometer or third-octave filters and level recorder. (Ear defenders.)

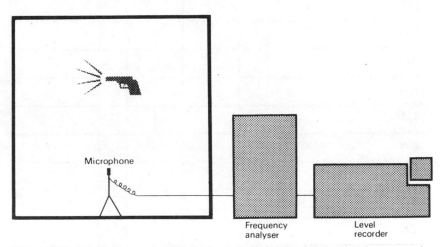

Fig. 5.12 Measurement of reverberation time using a pistol as sound source

Method
The apparatus is set up as shown in Fig. 5.12. A suitable microphone position is selected greater than 1 m from any surface and the frequency analyser set at one-third octave centred around 100 Hz. The level recorder using a 50 dB potentiometer is then adjusted to give full-scale deflection for the sound produced by the pistol. With the recording paper running at 30 mm/s, the pistol is fired. The decay line is recorded on the moving paper from which the reverberation time may be calculated or determined using protractors.

This procedure should be repeated in one-third octave intervals from 100 to 3150 Hz. The microphone is then moved to other room positions and the whole procedure repeated, six times for frequencies below 500 Hz and three times for frequencies of 500 Hz and above.

It is undoubtedly preferable for those concerned to wear ear defenders during these measurements.

Theory
Reverberation time is the time in seconds for a 60 dB decay in the sound pressure level after the source has been switched off.

220

Results

Centre frequency (Hz) $\frac{1}{3}$-octave bandwidth	Reverberation time (s)						Mean value
	1	2	3	4	5	6	
100							
125							
160							
200							
250							
315							
400							
500							
630							
800							
1000							
1250							
1600							
2000							
2500							
3150							

Experiment 5.3   Experiment to study the effect of masking sound on speech intelligibility

Apparatus

Two tape recorders, tape of white noise or other suitable constant sound, tape of monosyllabic words, sound level meters.

Method

Lists of suitable words are recorded on tape at the rate of about one every 4 s using a normal voice. Sets of suitable words are listed in Table 5.7 pages 221–224. It is important to use a series of unrelated monosyllabic words because if sentences are used, words not heard can be worked out.

The tape recorder with the words recorded is set up with its loudspeaker at a height of about 1.5 m, and the one with the white noise nearby. With the sound level meter(s) the sound level in dB(A) is measured with no white noise at each listening position. The first list of recorded words is then played at the level of a normal teaching voice and written down by the subjects, after which they are checked against the originals. The percentage correct is noted.

The white noise is run so that the sound level is approximately 45 dB(A). It will vary throughout the room and the level at each subject position noted. The second list is now played at the same level as before. Again the percentage correct is noted.

The experiment is repeated for levels of masking noise of approximately 50 dB(A), 55 dB(A), 60 dB(A), 65 dB(A) and the percentage intelligibility determined.

Theory

Background sound can interfere with the hearing of speech. Table 5.6 gives a guide to the maximum background levels before speech interference starts.

Results

List 1.   Background level = _____ dB(A)

          % intelligibility = _____

          Distance from words = _____ m

List	2	3	4	5	6
White-noise sound level (dB(A))					
% intelligibility					

Distance from speaker = m
   (masking sound)

Table 5.6 Speech interference levels

Distance from speaker to hearer (m)	Normal voice (dB)
0.1	73
0.2	69
0.3	65
0.4	63
0.5	61
0.6	59
0.8	56
1.0	54
1.5	51
2.0	48
3.0	45
4.0	42

Raised voice: add 6 dB to each
Very loud voice: add 12 dB to each
Shouting: add 18 dB to each
In the case of a female voice, all levels should
be reduced by 5 dB

Table 5.7 Monosyllabic word lists

List 1

1. in	21. gang	41. eye
2. roost	22. hump	42. cart
3. theme	23. fair	43. beard
4. sigh	24. soak	44. brass
5. web	25. get	45. cork
6. ace	26. skid	46. joke
7. duke	27. rouge	47. crate
8. salve	28. slush	48. puss
9. slice	29. ramp	49. clog
10. rout	30. through	50. click
11. quip	31. lid	
12. did	32. flash	
13. retch	33. seed	
14. tilt	34. robe	
15. pew	35. judge	
16. base	36. fast	
17. pad	37. walk	
18. pack	38. mow	
19. wash	39. souse	
20. gob	40. wise	

Table 5.7 Monosyllabic word lists (*Cont.*)

List 2

1. shave	21. lip	41. lunge
2. chill	22. loud	42. his
3. note	23. grab	43. lynch
4. gush	24. rob	44. weed
5. chain	25. art	45. axe
6. flare	26. dot	46. rose
7. trod	27. thine	47. bale
8. fat	28. camp	48. cat
9. grew	29. freeze	49. bless
10. claws	30. thorn	50. claw
11. grew	31. dice	
12. debt	32. fool	
13. hush	33. cub	
14. hide	34. lime	
15. sieve	35. thaw	
16. sash	36. chip	
17. aims	37. sack	
18. fade	38. got	
19. ouch	39. waste	
20. chaff	40. crab	

List 3

1. fright	21. part	41. three
2. turn	22. had	42. dub
3. aid	23. sang	43. rye
4. wield	24. knee	44. cheese
5. gab	25. hash	45. kind
6. rogue	26. house	46. next
7. droop	27. pump	47. closed
8. map	28. pitch	48. gas
9. hose	29. crews	49. drape
10. stress	30. tuck	50. nap
11. rug	31. ton	
12. book	32. rock	
13. leash	33. suit	
14. cliff	34. dame	
15. fifth	35. tire	
16. thresh	36. thou	
17. barge	37. sheep	
18. lay	38. stab	
19. din	39. ink	
20. sheik	40. soar	

Table 5.7   Monosyllabic word lists (*Cont.*)

List 4

1. dead	21. shook	41. stuff
2. wrist	22. heat	42. rid
3. waif	23. darn	43. quack
4. grate	24. clip	44. foam
5. thy	25. life	45. at
6. shoot	26. muss	46. coax
7. hunk	27. news	47. nick
8. cute	28. prude	48. me
9. tell	29. kick	49. lathe
10. vote	30. nod	50. howl
11. wag	31. fife	
12. curve	32. douse	
13. oft	33. soil	
14. shrug	34. sing	
15. group	35. tent	
16. barn	36. vague	
17. dung	37. purge	
18. dash	38. slab	
19. isle	39. tray	
20. car	40. bust	

List 5

1. ball	21. laugh	41. reek
2. gnash	22. jaw	42. out
3. chink	23. set	43. depth
4. chafe	24. fought	44. bluff
5. dime	25. lash	45. shut
6. vine	26. ripe	46. hear
7. cling	27. clutch	47. sod
8. wove	28. hunch	48. cad
9. tile	29. chair	49. park
10. sky	30. loose	50. throb
11. jazz	31. cave	
12. and	32. fed	
13. rove	33. hug	
14. greet	34. flog	
15. frill	35. romp	
16. wage	36. sledge	
17. priest	37. knife	
18. ass	38. jolt	
19. foot	39. chap	
20. flood	40. done	

Table 5.7 Monosyllabic word lists (*Cont.*)

List 6

1. back	21. page	41. clothe
2. tree	22. nudge	42. tag
3. thug	23. rape	43. plus
4. bug	24. slug	44. force
5. jay	25. goose	45. line
6. bash	26. hurt	46. cue
7. earth	27. wade	47. put
8. gull	28. chance	48. etch
9. ears	29. bob	49. youth
10. snipe	30. real	50. ail
11. rush	31. daub	
12. staff	32. pink	
13. those	33. flaunt	
14. maze	34. lap	
15. flight	35. champ	
16. cow	36. mope	
17. fir	37. void	
18. value	38. scrub	
19. hat	39. cord	
20. rip	40. wake	

226

Experiment 5.4 Measurement of sound absorption coefficients in a reverberation room

Apparatus

Beat-frequency oscillator, 75–100 W amplifier, two 50 W loudspeakers, omnidirectional microphone, audio-frequency spectrometer, high-speed level recorder, 10 m² absorbent material, reverberation room (ideally this should be specially constructed with a volume of

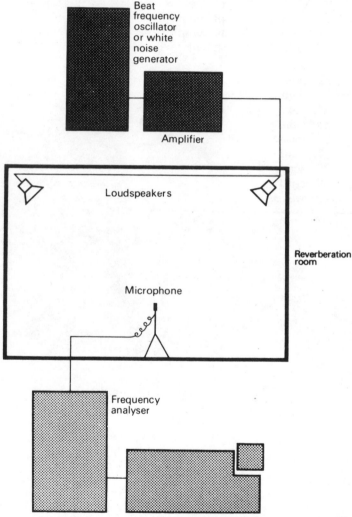

Fig. 5.13 Measurement of reverberation time using a beat-frequency oscillator or white-noise generator as sound source

about $200\,\text{m}^3$, but for the purpose of this experiment a fairly reverberant room of about this volume should be satisfactory).

Method

The reverberation time (t_1) is first measured for the room without the absorbent material as described in Experiment 5.1. Six measuring positions are chosen for each octave interval with mean frequencies at 125, 250, 500, 1000, 2000 and 4000 Hz. The absorbent material is then placed in the room, preferably covering a single area as near as possible to $10\,\text{m}^2$. The reverberation time measurements are repeated at each frequency for six different room positions. The amount of absorption added is then calculated for each bandwidth and hence from the superficial area the absorption coefficients are calculated.

Theory

Sabine's formula states that for a room,

$$t = \frac{0.16\,V}{A}$$

where

t = reverberation time (s)

V = volume of the room (m^3)

A = area of absorption (m^2)

Thus for the empty room,

$$A = \frac{0.16\,V}{t_1}$$

and for the room containing the extra absorption, δA

$$A + \delta A = \frac{0.16\,V}{t_2}$$

where

δA = the extra absorption

t_1 = original reverberation time

t_2 = reverberation time with the absorbent material added

$$\therefore \quad \delta A = 0.16\,V\left\{\frac{1}{t_2} - \frac{1}{t_1}\right\}$$

Hence the absorption coefficient,

$$\alpha = \frac{\delta A}{s} = \frac{0.16\,V}{s}\left\{\frac{1}{t_2} - \frac{1}{t_1}\right\}$$

where s = superficial area of the material.

(*Reference:* BS 3638:1963 Method for the Measurement of Sound Absorption Coefficients (ISO) in a Reverberation Room.)

228

Results

Area of absorbent = _____ m²

Centre frequency of octave band (Hz)	Reverberation time (s)							Extra absorption (m²)	Absorption coefficient (α)
	1	2	3	4	5	6	Mean		
125									
250									
500									
1000									
2000									
4000									

Experiment 5.5   Determination of absorption coefficients using a standing-wave tube

Apparatus

Loudspeaker, 100 mm and 30 mm diameter measuring tubes, microphone and probe, pure-tone oscillator, measuring amplifier and third-octave filters.

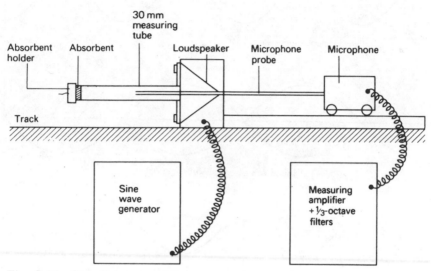

Fig. 5.14   Standing-wave tube

Method

The apparatus is set up as shown in Fig. 5.14. but with the larger standing-wave tube. A 100 mm diameter sample of the absorbent material is inserted in the end of this tube by means of the holder. The pure-tone oscillator is adjusted to 100 Hz at a suitable amplitude and the microphone and probe moved until a maximum is measured by the measuring amplifier. This value $(A + B)$ is noted (see Fig. 5.15). The microphone and probe are then moved to obtain the minimum $(A - B)$, which is noted. Hence $n = (A + B)/(A - B)$ is calculated, and by use of the graphs shown in Fig. 5.16 (a) or (b) the absorption coefficient is determined. This is repeated for frequencies up to 1600 Hz.

The large tube is then changed for the smaller tube with a 30 mm diameter sample of the same absorbent material. The process is repeated for frequencies from 800 to 6300 Hz. It will be noticed that

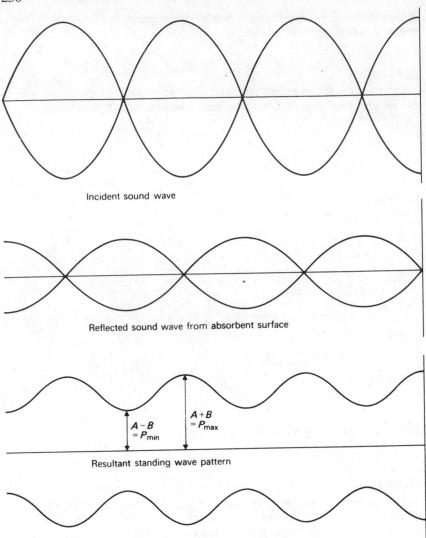

Incident sound wave

Reflected sound wave from absorbent surface

$$A - B = P_{min}$$

$$A + B = P_{max}$$

Resultant standing wave pattern

Fig. 5.15 Formation of the standing wave pattern in the impedance tube

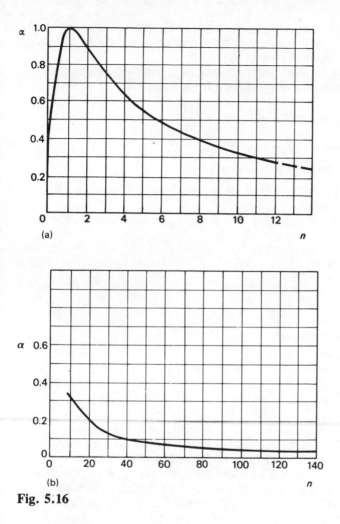

Fig. 5.16

the overlap enables the results from the two different tubes to be compared.

Theory

This method is convenient for the measurement of absorption coefficients at normal incidence for fibrous materials. It is not suitable for absorbers which rely on resonance. It works on the basis of comparing the incident and reflected pure tones on a small sample of the absorbent. As there is a quarter wavelength phase shift on reflection from an absorbent material, this is relatively easy. It means that the maximum amplitude for the reflected wave coincides with the position of minimum for the incident wave. Similarly, the maximum for the incident wave coincides with the minimum for the reflected wave. The result is illustrated in Fig. 5.15.

The definition of absorption coefficient is the proportion of the incident energy which is not reflected from the surface.

$$\text{Thus absorption coefficient } \alpha = 1 - \left(\frac{B}{A}\right)^2$$

where A = amplitude of reflected wave

         B = amplitude of incident wave

Note the fact that the energy is proportional to the square of amplitude or pressure.

The maximum measured is $A + B$ while the minimum is $A - B$

$$\therefore \quad \frac{\text{maximum pressure}}{\text{minimum pressure}} = \frac{P_{max}}{P_{min}}$$

$$= \frac{A + B}{A - B}$$

Now let $\qquad n = \dfrac{A + B}{A - B}$

i.e. the ratio of maximum to minimum pressure.

It can be shown that $\alpha = \dfrac{4n}{n^2 + 2n + 1}$

With some equipment, the voltage scales on the measuring amplifier are calibrated to enable the coefficient to be read directly. If not, the graphs shown in Fig. 5.16 (a) or (b) may be used.

The maximum diameter of standing-wave tube that can be used is approximately half the wavelength under investigation (0.586λ). The shortest length possible is a quarter wavelength. Thus to make measurements down to 63 Hz, a tube of about 1.25 m minimum is needed. At the same time, a 100 mm diameter tube can only be used up to 1800 Hz. A 30 mm diameter tube of length at least 100 mm can be used from about 800 Hz to 6.5 kHz.

Results

Frequency (Hz)	Max $(A + B)$	Min $(A - B)$	$\therefore \ n = \dfrac{A + B}{A - B}$	Absorption coefficient (α)
100				
125				
160				
200				
250				
315				
400				
500				
630				
800				
1000				
1250				
1600				
2000				
2500				
3150				
4000				
5000				
6300				

Experiment 5.6 Determination of the NR and NC ratings of a noise from its octave band spectrum

Apparatus

Sound level meter with octave band filters, calibrator, noise source, tripod stand.

Method

A suitable noise source is chosen. This may be an electric drill, extractor fan fixed in a room, or any other convenient source. A suitable position in the room is chosen for the measurements. The sound level meter batteries are checked and the meter calibrated. Readings of sound pressure level are taken in octave bands with centre frequencies from 63 Hz to 8 kHz.

The values obtained are then superimposed upon the noise rating (NR) and noise criteria (NC) curves (Figs. 5.17 and 5.18). The lowest NR or NC curve just above the values obtained determines the NR number or NC number. This is best found by plotting the results graphically together with the appropriate NC and NR curves.

Theory

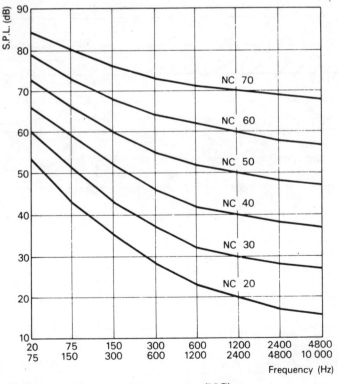

Fig. 5.17 Noise criteria curves (NC)

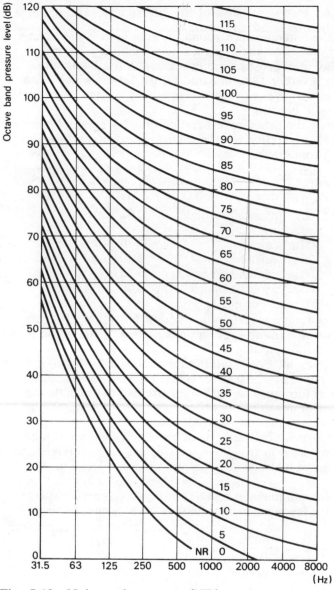

Fig. 5.18 Noise rating curves (NR)

Results

Noise source:

NR =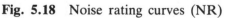

NC = ‗‗‗‗‗‗‗

Questions

1. (a) State the units in which sound pressure level is most commonly measured and explain why a doubling of sound pressure level causes a meter reading to rise by 6 units.
   (b) Define reverberation time and list the factors which affect it.
2. (a) Explain why the rear wall of large halls is often treated with perforated board.
   (b) Calculate the actual reverberation time for a hall of volume $4000 \, \text{m}^3$ at $500 \, \text{Hz}$ given the following surface finishes and seating conditions:

Item	Absorption coefficient at 500 Hz
$520 \, \text{m}^2$ brick wall	0.02
$700 \, \text{m}^2$ plaster on solid backing	0.02
$300 \, \text{m}^2$ carpet	0.30
$70 \, \text{m}^2$ curtain	0.40
$200 \, \text{m}^2$ canvas scenery	0.30
$100 \, \text{m}^2$ acoustic board	0.70
300 seats occupied	$0.4 \, \text{m}^2$ units/seat

   (c) If the optimum reverberation time is $1.14 \, \text{s}$, calculate the number of extra absorption units needed to adjust the actual value to the optimum.
3. (a) The optimum reverberation time at $500 \, \text{Hz}$ for a hall is given by $t = r(0.012 \sqrt[3]{V} + 0.1070)$ where $r = 6$ for choral music. Use this formula and the Sabine formula to calculate the number of absorption units required for a hall of $4000 \, \text{m}^3$ capacity when used for choral music.
   (b) Describe practical methods of introducing the absorption units required in (a) above, making particular reference to the type of material and the recommended locations in the hall.
4. Calculate the optimum and actual reverberation times for speech in a hall with the following surface finishes and seating conditions:

Item	Absorption coefficient at 500 Hz
$750 \, \text{m}^2$ brick walls	0.02
$535 \, \text{m}^2$ plaster on solid backing	0.02
$63 \, \text{m}^2$ glass	0.10
$70 \, \text{m}^2$ curtain	0.40
$130 \, \text{m}^2$ acoustic board	0.70
$300 \, \text{m}^2$ wood block floor (allow 40 per cent for shading)	0.05
500 occupied seats	$0.4 \, \text{m}^2$ units/seat

5. St Paul's Cathedral has a volume of approximately $160\,000\,\text{m}^3$. If its reverberation time when empty is 11 s and with a full congregation 6 s, find how many people are present when the Cathedral is full.
(Assume that each person has an absorption of $0.4\,\text{m}^2$ units.)

6. State the factors which determine the reverberation time (t) of a hall and explain how they influence the value of t. Using the data provided, calculate the optimum and actual reverberation times at 500 Hz of a hall intended for orchestral music and say what acoustical treatment you would recommend.
   Data: Volume of hall $= 3000\,\text{m}^3$
            Number of absorption units at 500 Hz $= 312$
            The constant r in the Stephens' and Bate formula $= 5$ for orchestral music.

7. (a) A hall is to be built to hold about 600 people. Assuming the main use will be for speech, suggest the main points to be considered in order to obtain a suitable acoustic environment. (You should assume that amplification systems are not to be used.)
   (b) A hall has a volume of about $10\,\text{m}^3$ per person when full. It was designed to hold up to 1100 audience, 50 choir and 35 orchestral players. If the ratio of the volume/absorption is $10/1$, how does the actual reverberation time compare with the optimum suggested by the Stephens' and Bate formula? Assume $r = 6$ for choral music.

8. Describe the acoustic properties required for a lecture room (of volume $216\,\text{m}^3$) near a railway line. (Stephens' and Bate formula $t = r(0.012\ ^3\!\!\sqrt{V} + 0.1070)$ where V is in m^3.)

9. The reverberation time of a concert hall holding 300 people was measured in 1964 and found to be 1.5 s when full. The average absorption per person in that year was $0.4\,\text{m}^2$ units. It has been shown that an average girl in a miniskirt only contributes $0.24\,\text{m}^2$ units of absorption. If the volume of the hall is $19\,500\,\text{m}^3$, what would be the reverberation time of the hall if it were full of girls in this attire?

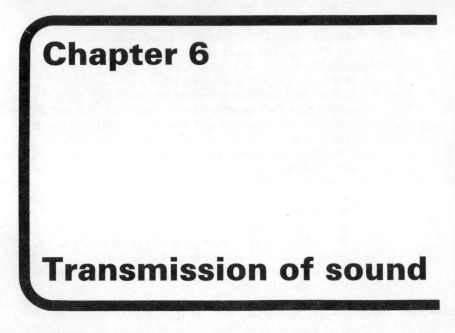

Chapter 6

Transmission of sound

It was explained in Chapter 4 that sound may be transmitted through solids, liquids and gases. The transmission through air and the solids of buildings is clearly very important. The aim of this chapter is to consider the propagation of sound in order to understand how best to reduce noise levels.

Propagation of sound in open air

Point source

Air cannot sustain a shear force and only a longitudinal type of sound wave is possible. It can be seen that if a point source of sound is radiating in the unobstructed atmosphere then the intensity (I) at a distance (r) is such that:

$$I = \frac{W}{4\pi r^2}$$

where W = sound power of the source

$4\pi r^2$ = surface area of the sphere

Thus it can be seen that:

$$I \propto \frac{1}{r^2}$$

or the intensity is inversely proportional to the square of the distance (inverse square law).

The effect of increasing the distance from a point source can be determined (Fig. 6.1).

$$\frac{I_r}{I_R} = \frac{R^2}{r^2}$$

where I_r = intensity at distance r from the source

I_R = intensity at distance R from the source

Now if L_r = sound pressure level at distance r from the source

then $L_r = 10\log_{10}\dfrac{I_r}{I_0}$

where I_0 = reference intensity

and if L_R = sound pressure level at distance R from the source

then $L_R = 10\log_{10}\dfrac{I_R}{I_0}$

∴ the reduction

$$L_r - L_R = 10\log_{10}\frac{I_r}{I_0} - 10\log_{10}\frac{I_R}{I_0}$$

$$= 10\log_{10}\frac{I_r}{I_R}$$

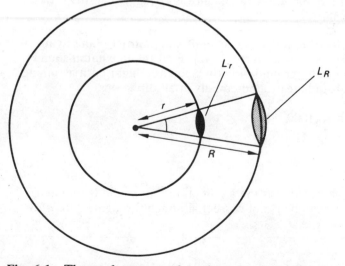

Fig. 6.1 The surface area of a sphere is proportional to the square of its radius. The sound intensity will therefore decrease in proportion to the square of the distance from the point source.

But $\dfrac{I_r}{I_R} = \dfrac{R^2}{r^2}$

$\therefore\quad L_r - L_R = 10\log_{10}\left(\dfrac{R}{r}\right)^2$

$\qquad\qquad = 20\log_{10}\dfrac{R}{r}$

Thus the reduction in sound pressure level for a point source will be 6 dB for each time the distance is doubled.

$L_r - L_{2r} = 20\log_{10}\dfrac{2r}{r}$

$\qquad\qquad = 20 \times 0.3010$

$\qquad\qquad = \underline{6\,dB}$

It will be remembered that in Chapter 4 sound pressure level (SPL) was calculated from the sound power level (PWL).

SPL = PWL $- 20\log_{10} r - 11$

or in the case of hemispherical radiation

SPL = PWL $- 20\log_{10} r - 8$

In each case the reduction with distance is $20\log_{10} r$ or 6 dB for each doubling of distance.

Example 6.1 The sound power level of a certain jet plane flying at a height of 1 km is 160 dB (*re* 10^{-12} W). Find the maximum sound pressure level on the ground directly below the flight path assuming that the aircraft radiates sound equally in all directions.

$L_r = 160 - 20\log_{10} 1000 - 11$

$\quad = 160 - 20 \times 3 - 11$

$\quad = \underline{89\,dB}$

It should be noted that the assumption that sound is radiated equally in all directions is not correct although the figures are of the correct order.

Line source

A line source could be considered as being made up from a line of point sources for each of which the inverse square law applies. It is easier to consider the line radiating in cylindrical form (Fig. 6.2). As

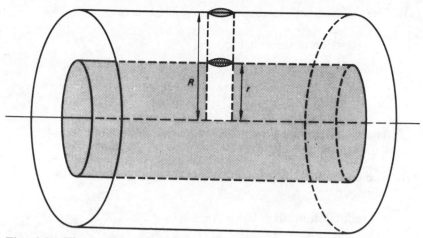

Fig. 6.2 The surface area of a cylinder is proportional to its radius. Sound intensity will therefore decrease directly with distance from a line source

the surface area of a cylinder is $2\pi r l$ it can be seen that the effective reduction is such that:

$$\frac{I_r}{I_R} = \frac{R}{r}$$

It should be noted that in this case R and r represent the perpendicular distance from the line.

$$L_r - L_R = 10 \log_{10} \frac{R}{r}$$

$$\text{If} \quad R = 2r$$

$$\text{Then} \quad L_r - L_{2r} = 10 \log 2$$
$$= \underline{3 \, \text{dB}}$$

It can be seen that there is only 3 dB reduction for a doubling of the distance.

Example 6.2   A policeman measures the sound pressure level at a distance of 7.5 m from the line of traffic on a road and finds it to be 80 dB. What would the level be at a distance of 75 m from the line of traffic if the policeman's reading was from:

(a)   an isolated vehicle?
(b)   a continuous line of closely packed identical cars? (Assume that the ground is flat and unobstructed.)

(a) The isolated vehicle constitutes a point source

$$\therefore \quad \text{reduction in dB} = 10\log_{10}\frac{75}{7.5}$$
$$= 20\log_{10} 10$$
$$= 20 \times 1$$
$$= \underline{20\,\text{dB}}$$

The sound pressure level at 75 m from the vehicle would therefore be $80 - 20 = \underline{60\,\text{dB}}$

(b) The continuous line of closely packed cars could be taken as a line source.

$$\therefore \quad \text{reduction in dB} = 10\log_{10}\frac{75}{7.5}$$
$$= 10\log_{10} 10$$
$$= \underline{10\,\text{dB}}$$

The level at 75 m from the line of cars would be $80 - 10 = 70\,\text{dB}$

Area source

In the two cases considered of point and line sources assumptions have been made. In practice most sources of sound are not point sources but have a finite area. In the far field where the distance from the source is several times the larger linear dimension of the source it may be considered as a point. At closer distances this assumption is not true and for large sources of sound there may be little or no attenuation with distance in the near field. This must be considered when measurements of noise sources are made as a basis for predictions at other distances. It is clearly better to carry out predictions from a knowledge of sound power levels of sources.

Reduction by barriers

The reduction of sound by means of a wall or fence is only effective where the barrier is large compared with the wavelength of the noise. It has been shown that approximately:

$$x = \frac{H^2}{D_s}$$

where H = 'effective' height of barrier
   D_s = distance of source from the barrier

as shown in Fig. 6.3.

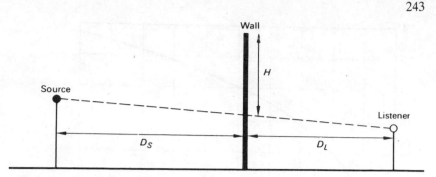

Fig. 6.3   Relation of source and listener to wall

x is related to the sound reduction as shown in Fig. 6.4. The formula applies provided:

$$D_L \gg D_s$$
$$D_s > H$$

Example 6.3   A school is situated close to a furniture factory with a flat roof on top of which is a dust extractor plant as shown in Fig. 6.5. The highest window in the school facing the factory is at the same height as the noise source, which can be taken as 1 m above the factory roof. The extractor plant is 6 m from the edge and produces a note of 660 Hz. Find the minimum height of wall to be built on the edge of the factory roof to give a reduction of 15 dB.

From Fig. 6.4 the value of x required to give 15 dB reduction is 3.

$$3 = \frac{H^2}{\dfrac{V}{f} \cdot D_s}$$

$$= \frac{H^2}{(330 \times 6)/660}$$

$$= \frac{H^2}{3}$$

$$H^2 = 9$$

$$\therefore \quad H = \underline{3 \text{ metres}}$$

$\therefore$ total height of wall = 4 metres.

It is unlikely that this is the best solution to this particular problem. These calculations are based on the amount of sound diffracted over the barrier. It is necessary for the barrier to be sufficiently long. It is also essential that the barrier itself does not

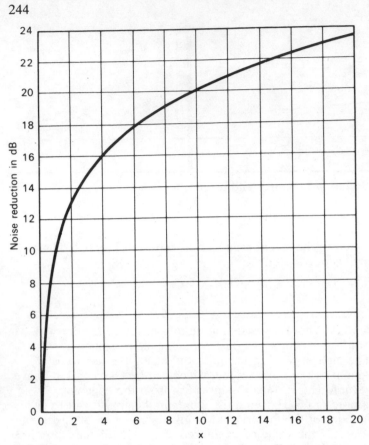

Fig. 6.4 Sound reduction by means of a wall or screen

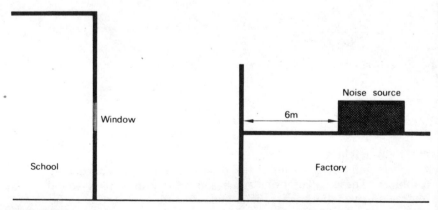

Fig. 6.5

transmit sound. In effect this means that the insulation of the barrier must be greater than the screening effect needed. Gaps in a fence would make the screen useless. Barriers can be light in weight, a mass of $20\,kg/m^2$ usually being sufficient.

Bushes and trees

The noise attenuation achieved by shrubs and trees can only be marginal. Measurements made under jungle conditions allowing for the normal loss by distance alone are shown approximately in Fig. 6.6.

Considering the marginal improvement, trees are not an economic means of achieving sound insulation. However, there is some absorption effect which would not be present if the ground were paved. If trees are used they must be of a leafy variety and be thick right to the ground. A point sometimes forgotten is that the rustle of leaves may occasionally produce a noise level as high as

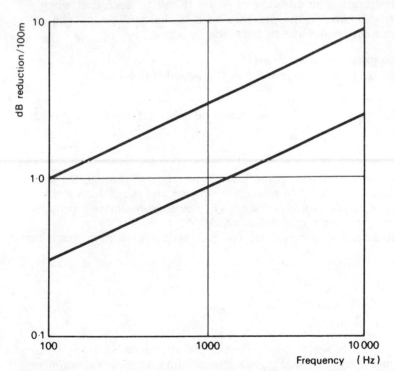

Fig. 6.6 Attenuation by means of trees. The upper line represents maximum attenuation under conditions of very dense growth of leafy trees and bushes where penetration is only possible by cutting. The lower line shows the attenuation under conditions where penetration is easy and visibility up to 100 m, but still very leafy undergrowth

50 dB(A). Trees undoubtedly have a very beneficial psychological effect probably enhanced by the masking effect provided by the rustle of the leaves.

Factors which influence noise measurements

Noise is defined as unwanted sound. When measurements are being made of noise various external factors may affect the readings such as air temperature, wind speed and other background sounds.

Background noise

There are many situations where it is necessary to measure the sound level of one noise in the presence of one or more others. It may be realised from Chapter 4 that the result will be the sum of the noises. If however the background sound is 10 dB or more lower than the sound to be measured then it has no significant effect on the readings. If the background noise were equal to the one whose level is needed then the result is an addition of 3 dB. It can be seen that when measuring a noise in the presence of a background sound two measurements may have to be made:

(a)   the background level; and
(b)   the result of background and the sound concerned.

Thus it is necessary to arrange to measure the background without the other noise. This may sometimes be achieved by switching off the sound.

Air absorption

Energy is absorbed as a sound wave is propagated through the air. These losses are due to a relaxation process and depend upon the amount of water vapour present. They are approximately proportional to the square of the frequency.

The attenuation per metre (α) has been shown to be such that:

$$\alpha = kf^2 + \alpha_2$$

where $k = 14.24 \times 10^{-1}$

f = frequency in Hz

α_2 is humidity dependent

Typical values are about 3 dB per 100 m at 4000 Hz dropping to 0.3 dB per 100 m at 1000 Hz. Air attenuation becomes very important for ultrasonic frequencies and is greater than 1 dB/m at 100 Hz, but is of comparatively little significance in architectural acoustics (Fig. 6.7). The exception to this will be with model studies where frequencies up to 80 kHz are frequently used. It is important to control the humidity for such studies.

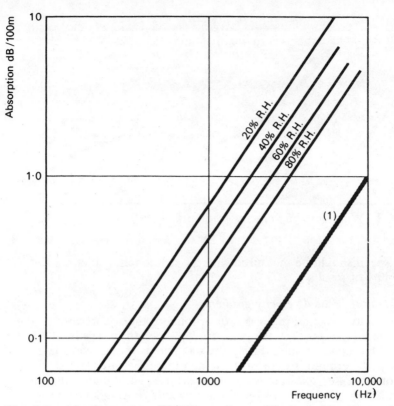

Fig. 6.7 Air absorption dB/100 m. Curve (1) shows the absorption due to the kf^2 part of the equation. The other lines show the total contributions

Wind created velocity gradients

For propagation close to the ground, sound velocity gradients can have a big influence on the levels received from a distance. These velocity gradients may be caused by wind or temperature.

Friction between the moving air and the ground results in a decreased velocity near ground level. This causes a distortion of the wave front. Downwind from the source, sound rays are refracted back towards the ground and the received level is not affected. Upwind the sound is refracted up and away from the ground causing acoustic shadows in which the level is considerably reduced (Fig. 6.8). It has been suggested that a motorway or industrial development built east (downwind) of a town would have an advantage of about 10 dB extra attenuation over one to the west.

It is important when making noise measurements to make allowance for wind speed. In many cases, such as for traffic noise, the

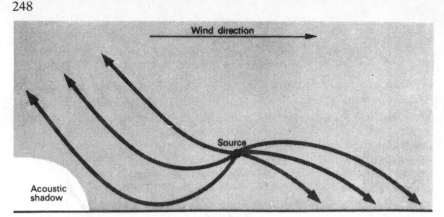

Fig. 6.8 Wind-created velocity gradients

wind speed should be measured with an anemometer to check the reliability of results.

Temperature created velocity gradients
The velocity of sound in air is proportional to the square root of the absolute temperature. A temperature lapse condition arises where the ground has been warmed during the day and as the air cools later the air near the ground is kept warmer than the upper air. The result is that the speed of sound in the ground layer of air is higher than that above. The sound rays are thus bent upwards causing an acoustic shadow around the source (Fig. 6.9).

While the acoustic shadow effect may be beneficial to some the temperature lapse condition can cause sound to be transmitted over obstructions causing worse disturbance to others. At least one major

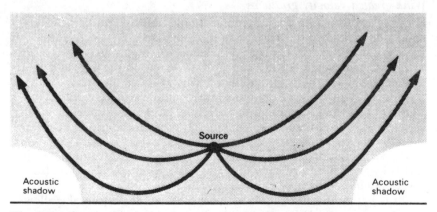

Fig. 6.9 Temperature-lapse-created velocity gradients

international firm attempts to forecast temperature lapse conditions near its plant in order to avoid the extra evening annoyance to local residents.

Sound insulation

Sound insulation is defined as the prevention of transmission of sound. It must not be confused with sound absorption which is the prevention of reflection of sound. There is no direct relationship between the sound insulation value of a partition and the sound absorption of its surfaces. The amount of absorbent in a room does however affect the amount of reverberant sound which in turn affects the amount of transmitted sound. The use of absorbents within cavity construction can assist in improving sound insulation. The lining of ducts can significantly reduce transmission along the ducts.

Airborne and impact sounds

Before any sound can reach the ear it must become airborne. For the purpose of sound insulation airborne and impact sounds are differentiated according to origin. Hence those initiated in the air such as speech, a radio, etc., before reaching a structure would be termed airborne. Those originated directly on a structure by blows or vibration would be termed impact sound. The definition is complicated because it is dependent upon the position of the listener. As an example footsteps on a floor would be heard as airborne sound in that same room but as impact sound in the receiving room below. In practice there is no great difficulty in understanding the difference. It is necessary to appreciate that the difference is important in that materials may differ in their ability to reduce the two types of sound. For instance increasing the mass of a floor gives a significant improvement in its ability to reduce airborne sound but only a very small reduction in impact sound transmission.

The term structure-borne sound is often used to distinguish sound which is predominantly transmitted by the structure. It has no precise definition in acoustic terminology.

Airborne sound insulation by partitions

Airborne sound can be transmitted in a receiving room by some or all of the methods shown in Fig. 6.10.

1. Airborne sound in the source room excites the separating partition into vibration which directly radiates the sound into the receiving room. The amount of attenuation will depend upon the frequency of sound, the mass, fixing conditions and thus the resonant frequencies of the partition.
2. Airborne sound in the source room may excite walls other than the separating one into vibration. The energy is then transmitted

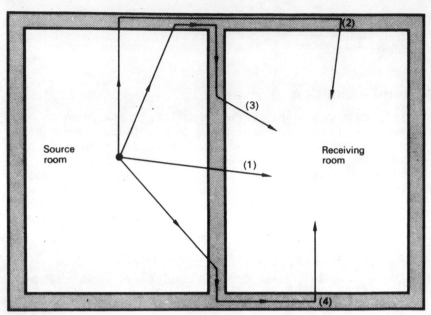

Fig. 6.10 Four different transmission paths from source room to receiving room

through the structure and reradiated by some other partition into the receiving room.

3. Any wall other than the separating one may be excited. The sound is transmitted to the separating wall and then reradiated by it.

4. Sound energy from the separating partition is radiated into the receiving room by some other wall.

The flanking transmission means that a laboratory measurement of the sound insulation provided by a partition, which is mounted in massive side walls, may give results different from those achieved in actual buildings. There is a limit to the insulation obtained by improving only the adjoining partition. Where a partition has a low insulation value of 35 dB or less, flanking transmission is of little consequence, but when values of 50 dB reduction for the partition are reached, further improvement is limited by the indirect sound paths.

The sound transmission properties of a partition can be divided into a number of distinct regions (Fig. 6.11). For convenience here they are divided into three regions.

Region 1: where resonance and stiffness conditions control the insulation. At low frequencies it is the stiffness rather than mass which governs the behaviour of a partition. For low frequencies

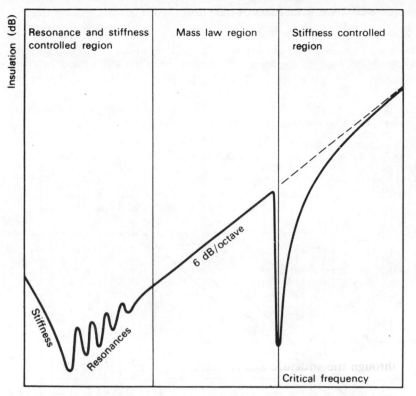

Fig. 6.11 Three distinct regions showing the way a panel will react to different frequency sounds

inertia forces are low because the velocity of motion is small. The inherent 'spring' in the wall resists movement. Unfortunately within this region, which is stiffness controlled, there are certain frequencies which cause resonance due to the stiffness. The lowest resonant frequency is known as the fundamental where the sound reduction is at its lowest. In addition there are higher resonant frequencies. Alteration to mass, stiffness and fixing conditions changes the resonant frequencies.

Region 2: which is mass controlled and where the partition can be considered as a large number of masses free to slide over each other.

The higher the mass the greater is the resisting inertia force. At the higher frequencies it is the mass rather than stiffness which controls resistance to vibration. In this section which is known as the mass law region the sound insulation is:

$$R = 20 \log mf - 43 \, \text{dB}$$

252

where m = superficial mass of partition (kg/m^2)
         f = frequency (Hz)

The mass law equation is only strictly applicable for normal angles of incidence (0°) which shows a 6 dB increase for each doubling of frequency. For moderately reverberant rooms where the angles of incidence mainly vary from 0° to 80° incidence or field incidence the insulation is lower by about 6 dB. Where a room is highly reverberant the insulation is lower and the gradient nearly 5 dB (Fig. 6.12).

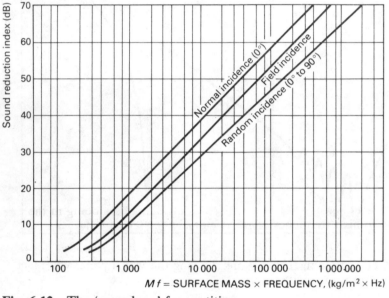

Fig. 6.12   The 'mass laws' for partitions

Region 3:   where again the partition becomes stiffness controlled due to an effect known as coincidence. If a partition is vibrated in a direction normal to its surface then a bending wave will move along the panel. It can be seen from Fig. 6.13 that the panel vibrates in a corrugated form. If a sound wave strikes the panel at a certain angle the situation can arise where the projection of its wavelength is exactly equal to the bending wave in the panel. The effect is similar to low frequency resonance. The mass law suddenly no longer applies.

The coincidence problem is unfortunately not confined to a single frequency. The lowest frequency at which the effect takes place is where the velocity of the bending wave equals the velocity of sound in air. This is called the critical frequency and the sound wave is at grazing incidence. As the bending wave velocity increases with

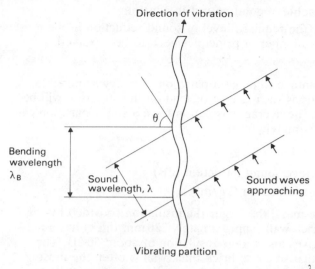

Direction of vibration

θ

Bending
wavelength
λ_B

Sound
wavelength, λ

Sound waves
approaching

Vibrating partition

Fig. 6.13 Coincidence effect occurs when $\lambda_B = \dfrac{\lambda}{\sin \theta}$

frequency the result is that above the critical frequency the sound reduction index will stay below the mass law prediction. The coincidence problem is more important than low frequency resonance which is normally below 100 Hz. The coincidence loss being in the middle to high frequency range is more important.

The ideal material for good sound insulation has a very high mass and low stiffness. Unfortunately some of the most convenient building materials have low mass and relatively high stiffness. Details of some of these are shown in Table 6.1.

Table 6.1 Critical frequencies and densities for some common materials

Material	Density (kg/m^3)	Mass/unit area × coincidence frequency $(kg/m^2 \times Hz)$
Plywood	580	13 000
Hardboard	810	31 000
Aluminium	2700	35 000
Plasterboard	750	35 000
Glass	2500	39 000
Concrete	2300	44 000
Brick	1900	47 000
Steel	8100	98 000
Lead	11 200	600 000

Requirements to achieve good sound insulation

In order to achieve the required level of sound reduction in the most economic manner a number of principles need to be observed.

Mass

It is clearly an advantage to make a partition as heavy as possible. Practical considerations such as cost or load on the structure will be the limiting factor. The average insulation from a single partition can be calculated approximately from:

$$R_{AV} = 10 + 14.5 \log_{10} m$$

where R_{AV} = average sound reduction (dB)

  m = mass/unit area (kg/m^2)

The greater the mass, the larger the insulation provided by the partition. A one brick wall (approximately 230 mm thick) has a mass of 415 kg/m^2 and gives an average insulation of about 50 dB. The use of mass up to that of a one brick thick wall is often the most economic method of providing sound insulation. Above 50 dB average sound reduction other methods must be considered as the cost of extra mass alone becomes prohibitive.

Completeness

For any type of construction requiring good sound reduction it is important that there are no gaps. A very small air hole in a brick wall can easily reduce insulation from 50 dB to as low as 20 dB. This is frequently an unappreciated problem in party walls between dwellings where incomplete mortar joints reduce the insulation. These may

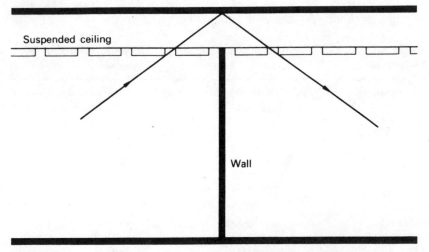

Fig. 6.14 An example of incomplete construction where a good sound path is present above a perforated false ceiling

go unseen as the plaster covers the defects. Ducting for hot air heating and suspended ceilings (Fig. 6.14) are other examples where failure to appreciate that a gap exists results in poor sound insulation.

Multiple or discontinuous construction

This type of construction can achieve much higher levels of sound reduction than a single construction of the same total mass per unit area. It is however much more expensive and a number of details of construction are important for success. The most common use of multiple partitions is where lightweight construction is necessary, for example double glazing. This form of construction is used for lightweight stud partitions, and occasionally with heavy masonry construction where the requirements of very high standards of insulation are more important than cost, such as in broadcasting and recording studios. Speed of erection and ability to use dry construction may influence the choice of double lightweight construction.

The main requirements are as follows:

1. The gap is large, below 50 mm having no advantage.
2. The two panels are of different superficial weight.
3. The gap contains sound absorbent material.
4. There are no air paths through the panels.
5. The panels are not coupled together by the method of construction. It will be fairly clear that if it is necessary to tie the panels together, then to get the greatest impedance mismatch and hence the highest insulation, very light ties should be used with heavy panels, and massive ties with light panels.

If average figures for sound reduction are compared for a massive construction such as 230 mm brick with double stud plasterboard construction with 225 mm gap the lightweight one may seem at least as good. An average figure can however be deceptive. On the whole lightweight construction gives better insulation at higher frequencies. Figures 6.15 and 6.16 which apply to glazing show the great improvement at high frequencies with little improvement at the lower end.

Timber frame house construction for a party wall would require the sort of mass and gap shown in Fig. 6.17. Many details of construction have been omitted for clarity. The second sheet of plasterboard is fixed by adhesive to the first layer. This minimises gaps and adds mass. It can be seen that to some extent mass and the gap are interchangeable. If the gap is reduced the mass must be increased. A third layer of plasterboard may well be an advantage. A problem exists on site involving accuracy. It is easy to specify a 225 mm gap. Construction problems may result in the actual spacing being reduced. Such a problem needs to be appreciated and suitable tolerances allowed for in design.

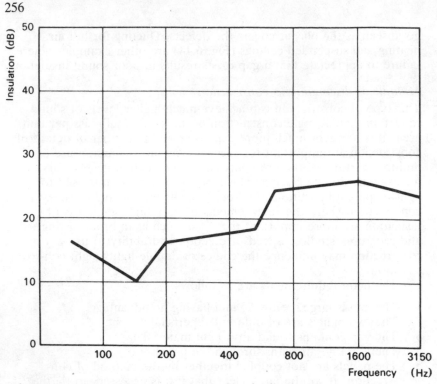

Fig. 6.15 A sound insulation curve for single glazing

A normal cavity wall with about 50 mm gap does not give any improvement over the equivalent mass solid construction. This is largely due to the size of the spacing and the need for ties. Reductions of over 80 dB average can be achieved by massive constructions with a suitably large gap (e.g. 300 mm) incorporating absorbent material.

Examples of sound reduction index by single and double glazing are shown in Table 6.2.

Apparent insulation by the use of absorbents
While absorbent materials should not be confused with insulating ones some benefit can often be achieved by reducing the reverberant sound in an enclosure.

The actual reduction expected using absorbents can be approximately calculated from:

$$\text{Reduction in sound pressure level} = 10\log_{10}\frac{A + a}{A}$$

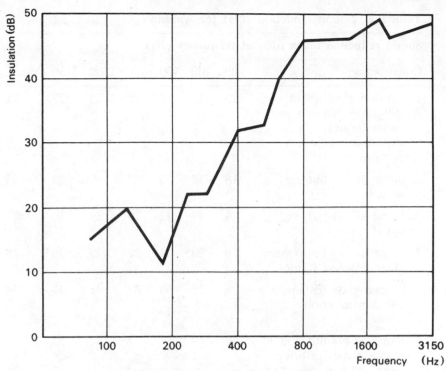

Fig. 6.16 A sound insulation curve for double glazing

where A = original area of absorption (m^2)

$\quad a$ = additional area of absorption (m^2)

Thus it can be seen that doubling the amount of absorption can be expected to achieve 3 dB reduction. A 10-fold increase in area of absorption would give 10 dB reduction. This calculation is oversimplified because in practice the placing of absorbent near to noisy machines can make it more effective (Fig. 6.18). However it should be realised that a 10 dB reduction by the use of absorbers would normally be the limit.

Sound transmission coefficient (T)

This is the ratio which the sound energy of a given frequency transmitted through and beyond a surface or partition bears to that incident upon it.

Sound reduction index

Sound reduction index = $10 \log_{10} \dfrac{1}{T}$

where T is the sound transmission coefficient.

258

Table 6.2 Sound reduction (dB) for windows

Sound reduction index (dB) at frequency (Hz)

Construction	100	200	400	800	1600	3150	Mean
1. 2.5 mm glass about 500 mm × 300 mm in metal frames, closed openable sections	23	12	18	22	29	27	21
2. Same as (1) but sealed	18	12	20	27	30	28	23
3. Same as (1) but wood frames	14	17	22	26	29	30	23
4. Same as (1) but 6 mm glass in wood frames.	16	20	24	29	29	36	26
5. 25 mm glass 650 mm × 540 mm in wood frames	28	25	30	33	41	47	34
6. Same as (1) but double wood frames in brickwork, 50 mm cavity between glazing	19	19	27	37	47	52	33
7. Same as (6) but 100 mm cavity	22	20	35	43	51	53	37
8. Same as (6) but 180 mm cavity	28	25	36	43	50	53	39
9. Same as (8) but with acoustic tiles on reveals	25	34	41	45	53	57	42

It can be seen that the term sound reduction index is simply related to the proportion of sound energy transmitted. It also only applies to the separating partition and does not include flanking transmission. Sound reduction index and transmission coefficient can be determined for a partition from laboratory measurements. In the case of actual constructions the measurement includes flanking sound and a level difference is found. This may be adjusted to allow for the absorption in or reverberation of the receiving room to find the normalised level difference.

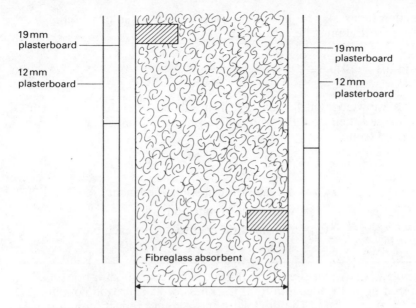

Fig. 6.17 Double stud construction (not to scale)

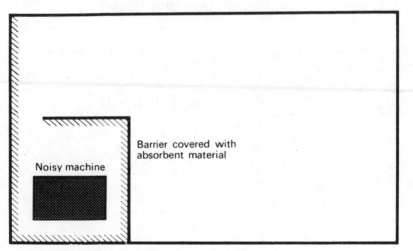

Fig. 6.18 Use of absorbent covered screen can help localise noise. Absorbent material is necessary to prevent sound being reflected back to the operator of the machine, raising the level there even more

Sound reduction of composite partitions

Sound reduction indices are available from measurements made on standard partitions. In practice a partition may include several different items such as brickwork, doors and windows. The sound reduction index may be known for each but the value of the composite partition must be calculated.

$$\text{SRI in dB} = 10 \log_{10} \frac{1}{T}$$

From a knowledge of the values of transmission coefficients for each material and their area the average transmission coefficient may be calculated from:

$$T_{AV} \times A = T_1 \times A_1 + T_2 \times A_2 + T_3 \times \cdots$$

where
$$T_{AV} = \text{average transmission coefficient}$$
$$A = \text{total area of partition}$$
$$T_1, T_2, \text{etc.} = \text{transmission coefficients of each part}$$
$$A_1, A_2, \text{etc.} = \text{areas of each part}$$

Example 6.4 A partition of total area $10 \, \text{m}^2$ consists of a brick wall plastered on both sides to a total thickness of 250 mm and contains a window of area $2 \, \text{m}^2$. The brickwork has a sound reduction index of 51 dB and the window 18 dB at a certain frequency. Calculate the sound reduction of the complete partition at this frequency.

Brickwork:

$$51 = 10 \log_{10} \frac{1}{T_B}$$

where
$$T_B = \text{transmission coefficient of the brick}$$

$$\therefore \quad 5.1 = -\log_{10} \frac{1}{T_B}$$

$$\text{Log}_{10} T_B = \bar{6}.9$$

$$\therefore \quad T_B = \text{antilog} \, \bar{6}.9$$
$$= \underline{0.000\,008} \, (8 \times 10^{-6})$$

Window:

$$18 = 10 \log_{10} \frac{1}{T_W}$$

where
$$T_W = \text{transmission coefficient of the window}$$
$$\therefore \quad 1.8 = -\log_{10} T_W$$
$$\text{Log}_{10} T_W = \bar{2}.2$$
$$\therefore \quad T_W = 0.015\,85 \, (1.585 \times 10^{-2})$$

Now $T_{AV} \times 10 = 8 \times 10^{-6} \times 8 + 1.585 \times 10^{-2} \times 2$

$$\therefore \quad \underline{T_{AV} = 3.1764 \times 10^{-3}}$$

Actual sound reduction index in dB

$$= 10 \log_{10} \frac{1}{T_{AV}}$$
$$= -10 \log_{10} T_{AV}$$
$$= -10 \log_{10} (3.1764 \times 10^{-3})$$
$$= -10 \times \overline{3}.5019$$
$$= 30 - 5.019$$
$$= \underline{25 \text{ dB}}$$

It can be seen that the poor insulation of the window of small area reduces the overall insulation very considerably. If the window had fitted badly the insulation would be even lower.

Impact sound insulation

As explained earlier in the chapter, impact sound is sound originating by impact or vibration at the structure. The main principles which apply to minimising impact sound also apply to reducing other structure-borne sounds. These may originate from mechanical plant such as lifts, compressors, pumps, etc, which are fixed to the building. There are three ways in which structure-borne sound may be reduced:

1. Use of mass to increase inertia. This is only of limited effectiveness by itself in reducing the amplitude of the vibration. Seldom will it cure an impact or structure-borne noise problem. It may be useful in conjunction with a resilient mounting.
2. Use of resilient mountings to avoid coupling the source to the structure. The same basic principles apply as applied to discontinuous construction for airborne sound. The main difference is that a resilient mount rather than air is needed to obtain a high impedance mismatch between source and structure.

   Unfortunately a resilient mounting can actually make the situation worse if its resonant frequency coincides with the driving frequency of the source (Fig. 6.19). The resonant frequency of the mount should be much lower than the source frequency. The resonant frequency of the mount can be found from its static deflection by:

$$f = \sqrt{\frac{250}{h}}$$

where f = resonant frequency of mount in Hz

h = static deflection in mm under the load

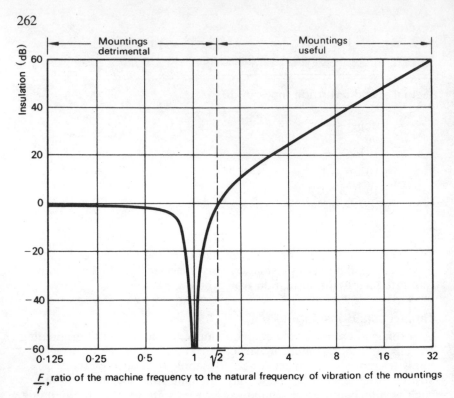

Fig. 6.19 Energy reduction in dB by the use of resilient machine mountings

In simple terms, soft mounts with a big deflection are better than stiff ones.

3. Damping layers. These are of limited use in building.

Floating floors

The best way to provide good impact sound insulation for floors is to isolate the walking surface from the structure of the building. While specially designed mountings can be used for floors it is more usual to use a resilient blanket laid between the structure and the walking surface. A layer of fibreglass is ideal as it is inexpensive and highly compressible giving a large static deflection for small loads. This large deflection means that it can be effective in isolating low frequency sound from impact. Typical floor constructions are shown in Figs. 6.20 to 6.24. It will be seen that in each case a minimum mass is needed to provide adequate inertia. Details of the construction are very important to ensure that the walking surface is completely isolated from the rest of the building.

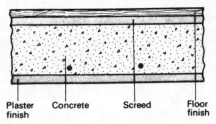

Plaster
finish   Concrete   Screed   Floor
                                       finish

Fig. 6.20 This gives **Grade** 1 insulation for impact and airborne
sound if concrete + plaster + screed weighs more than 365 kg/m^2
and a soft floor finish is used. If a hard floor finish is used airborne
insulation is still Grade 1, but the floor fails Grade 2 for impact
sound

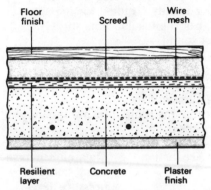

Floor
finish   Screed   Wire
                          mesh

Resilient   Concrete   Plaster
layer   finish

Fig. 6.21 This gives **Grade** 1 insulation for airborne and impact
sound if concrete + plaster weigh more than 220 kg/m^2

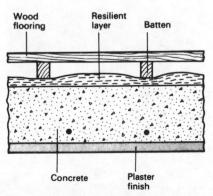

Wood   Resilient
flooring   layer   Batten

Concrete   Plaster
              finish

Fig. 6.22 This gives **Grade** 1 insulation for airborne and impact
sound if concrete + plaster weigh more than 220 kg/m^2

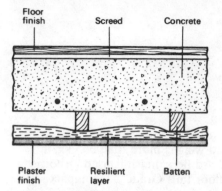

Fig. 6.23 This gives **Grade** 1 insulation for airborne and impact sound if concrete + screed weigh more than 220 kg/m² and a soft floor finish is used. If a medium floor finish is used airborne insulation is still Grade 1 but the floor is only Grade 2 for impact sound. If a hard floor finish is used airborne insulation is still Grade 1 but the floor fails Grade 2 for impact sound

Table 6.3 Airborne and impact sound insulation standards required by the Building Regulations:
Airborne sound insulation is the minimum reduction.
Impact sound is the maximum in the receiving room from a standard tapping machine on the floor of the source room.

Third octave band centre frequency (Hz)	Airborne sound, reduction (dB)		Maximum octave band sound pressure level (dB) for impact sound
	Party walls	Floors	
100	40	36	63
125	41	38	64
160	43	39	65
200	44	41	66
250	45	43	66
315	47	44	66
400	48	46	66
500	49	48	66
630	51	49	65
800	52	51	64
1000	53	53	63
1250	55	54	61
1600	56	56	59
2000	56	56	57
2500	56	56	55
3150	56	56	53

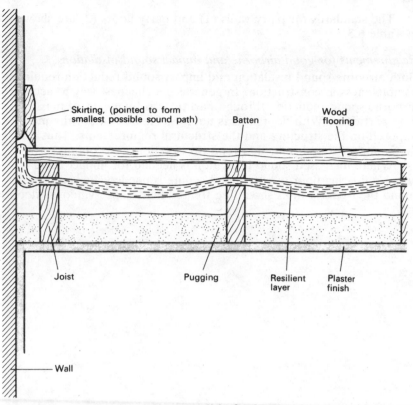

Fig. 6.24 Grade 1 insulation possible for both airborne and impact sound if heavy lath and plaster with 80 kg/m² of pugging or heavy lath and plaster with 15 kg/m² of pugging on heavy walls or heavy lath and plaster with no pugging but with very heavy walls or plasterboard and one coat plaster with 15 kg/m² pugging and very heavy walls

Standards for impact and airborne sound insulation

The Building Regulations set standards of airborne and impact sound insulation for partitions and floors between dwellings. In effect three standards are implied.

1. The airborne insulation which can be achieved by the horizontal attenuation provided by a 230 mm brick party wall, which is considered the desirable minimum.
2. That which can be achieved economically by floors which is set by the vertical attenuation provided by a concrete construction with a floating floor.
3. Other situations where the adjacent part of the building is non-inhabited and contains no machinery.

The standards for party walls (1) and party floors (2) are shown in Table 6.3.

Requirements for good airborne and impact sound insulation

Both airborne sound insulation and impact sound reduction require completeness in construction. In general the cheapest way to achieve airborne sound reduction through partitions is by means of massive limp partitions. With floors this is not possible because of the mass imposed on the structure and the structural requirements. Thus for floors impact sound reduction requires isolation for economy.

Where increased sound insulation is needed for impact noise either a floating floor or suspended ceiling is needed. For airborne sound the weakest links should be treated first, gaps, windows, doors, etc.

Experiments

Experiment 6.1   Measurement of the airborne sound insulation of a partition

Apparatus

White-noise generator or tape recorder with white noise recorded, amplifier (about 100 W), two 50 W loudspeakers, 2 microphones, extension leads, switchbox, third-octave filters, microphone stands, level recorder.

Methods

The apparatus is set up as shown in Fig. 6.25 with the two microphones at position 1 in each room. Each of the microphone positions is chosen so as to be at least 1 m from any large object. White noise is produced at a level of approximately 100 dB in the source room. The filters are adjusted to one-third octave centred around 100 Hz and the levels in the source and receiving room recorded on paper. This is then repeated for each third-octave bandwidth up to 3150 Hz.

The whole is then repeated for a total of six microphone positions in each room up to and including 400 Hz, and three positions each above this frequency.

The mean sound reduction at each frequency is calculated.

Reverberation time is measured as described in Experiment 5.1 and the correction term applied to obtain the sound reduction index for the partition for each bandwidth.

The results may be compared with the airborne sound insulation requirements of the Building Regulations (Part G) for dwellings.

Theory

Normalised level difference in dB

  = sound pressure level in the source room

    − sound pressure level in the receiving room

    + correction for the absorption provided by the receiving room

or

Normalised level difference = $L_S - L_R$ + correction term

For dwellings, the correction term is: $+10\log_{10}\dfrac{T}{0.5}$

where T = reverberation time of the receiving room in seconds

∴   Normalised level difference = $L_S - L_R + 10\log_{10}\dfrac{T}{0.5}$

(*Reference:* Building Regulations, BS 2750:1956.)

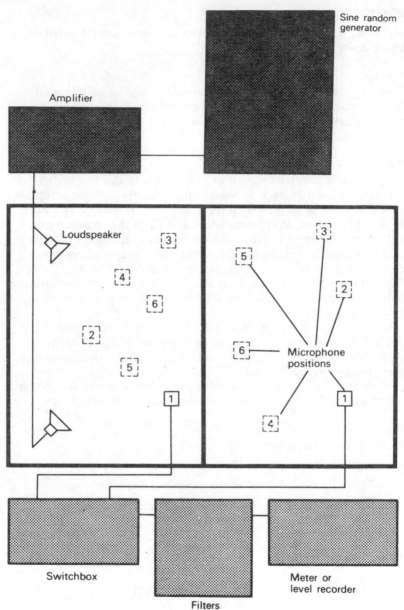

Fig. 6.25 Measurement of airborne sound insulation

Results

⅓-octave centre frequency (Hz)	dB difference			Mean diff. (dB) (1)	Reverberation time (s)	$10\log_{10}\dfrac{T}{0.5}$ (2)	Normalised sound level difference (dB) (1) + (2)
100							
125							
160							
200							
250							
315							
400							
500							
630							
800							
1000							
1250							
1600							
2000							
2500							
3150							

Experiment 6.2 Measurement of the impact sound insulation of a floor

Apparatus

Tapping machine, sound level meter, third-octave filters, calibrator, tripod stand, level recorder.

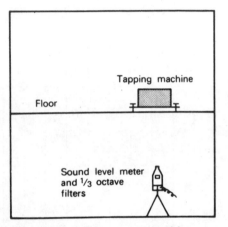

Fig. 6.26 Measurement of impact sound insulation of a floor

Method

The apparatus is set up as shown in Fig. 6.26 with the tapping machine near the centre of the floor whose insulation is to be measured. If the floor is liable to be damaged by the metal feet they are removed and replaced by the rubber ones and then adjusted for height. The sound level meter with attached filters is calibrated using the acoustic calibrator, and set up on the tripod stand. The microphone should be more than 1 m from any large objects or people.

The level produced in the receiving room by the tapping machine is then measured in third-octave bandwidths from 100 to 3150 Hz centre frequencies. This is repeated for six different microphone positions up to 400 Hz, and three positions above this. If the range is greater than 6 dB at any frequency then further readings are taken to reduce the standard deviation.

The reverberation time of the receiving room is then measured as described in Experiment 5.1 to enable the correction term to be applied.

Theory

Normalised sound pressure level in the receiving room in dB = measured sound pressure level in dB − correction terms.

There are two correction terms. The first to allow for the absorption of the receiving room and the second to enable the results to be compared to octave measurements.

1. *Allowance for absorption*

For dwellings the correction is $10\log_{10}\dfrac{T}{0.5}$

2. *Allowance for third-octave measurements*

This is $+10\log 3 = 4.77$
$$\simeq 5\,\text{dB}$$

$\therefore$ normalised SPL in dB $= L - 10\log_{10}\dfrac{T}{0.5} + 5$

where
$L =$ measured SPL in receiving room
$T =$ receiving room reverberation time in seconds

(*Reference:* Building Regulations Part G, BS 2750:1956.)

272

Results

Frequency (Hz)	Readings (dB)			Mean (dB)	Reverberation Time (T)	$10\log_{10}2T$	Rec. room level Corr. $\frac{1}{3}$-oct.	Rec. room level
100							+5	
125							+5	
160							+5	
200							+5	
250							+5	
315							+5	
400							+5	
500							+5	
630							+5	
800							+5	
1000							+5	
1250							+5	
1600							+5	
2000							+5	
2500							+5	
3150							+5	

Questions

1. Explain briefly how the insulation of a partition to airborne noise is measured and how the mass law relates to this.
2. Explain the limitation of obtaining 'apparent' sound insulation by means of the use of sound absorbent materials.
3. A room in a building has an external window; there is an office above it and a light-machine workshop below it. Describe the methods available for reducing sound entering from outside the room.
4. (a) Explain what is meant by the term 'sound reduction index'.
   (b) An external wall, of area 4 m by 2.5 m in a house facing a motorway is required to have a sound reduction of 50 dB. The construction consists of a 275 mm cavity wall containing a double glazed (and sealed) window. The sound reduction indices are 55 dB for the 275 mm cavity wall and 44 dB for the sealed double glazed window with a 130 mm cavity. Calculate to the nearest $0.1 \, \text{m}^2$ the maximum size of window to achieve the required insulation.
5. An external wall of a house is of area $10 \, \text{m}^2$ and is to have a sound reduction of 45 dB. The available construction is 275 mm cavity wall (sound reduction index 50 dB) and double glazed window (sound reduction index 40 dB). Calculate, to the nearest tenth of a square metre, the area of window that can be incorporated.
6. A room in a building used for light industry and in a district of heavy traffic is to be converted and used for meetings and lectures. Explain:
   (a) How the noise level, due to external sources, can be reduced.
   (b) How the reverberation time can be reduced.
   (Within each suggested treatment explain the reasons for this choice.)
7. A policeman measuring motor vehicle noise is exactly 7.5 m from the line of traffic on a straight road. The reading shown on his meter is 86 dB(A). An isolated house faces the road and is exactly 30 m from the line of traffic. A front room of the house has an external 225 mm brick wall of area $10 \, \text{m}^2$, a window $3 \, \text{m}^2$ and a door $2 \, \text{m}^2$.
   Insulation values are: brick 50 dB, window 18 dB, door 24 dB. You may assume that air gaps for the window and door have already been allowed.
   What level would you expect the sound of traffic to be in the room if:
   (a) The policeman's reading was from an isolated vehicle?
   (b) The reading were taken from a continuously closely packed line of identical cars?

8. Calculate the sound reduction of a partition of total area $20\,m^2$ consisting of $15\,m^2$ of brickwork, $3\,m^2$ of windows and $12\,m^2$ of door. The sound reduction indices are $50\,dB$, $20\,dB$ and $26\,dB$ for the brickwork, windows and door respectively.

9. A workshop of volume $1000\,m^3$ has a reverberation time of 3 s. The sound pressure level with all the machines in use is 105 dB at a certain frequency. If the reverberation time is reduced to $0.75\,s$, what would the sound pressure level be?

Chapter 7

The effects of noise on man

In the previous chapter the physical aspects of sound and its measurement was discussed. In this chapter the effects of noise on man are examined.

Hearing sound

What is heard as sound does not always bear a simple relationship to the sound which is present. This is partly due to the mechanism by which the ear works and hence its response. The ear is a transducer converting sound pressure waves into signals which are sent to the brain (Figs. 7.1–7.3).

Sound first reaches the outer and visible part of the ear known as the concha. A concave shape of a certain size will act as a focusing device only for wavelengths up to the same order of size, and so the concha will tend to scatter the longer wavelengths while reflecting shorter ones into the meatus. The meatus is the tube connecting the outer ear to the eardrum, and because of its size it resonates to a frequency of about 3 kHz. The eardrum separates the outer from the inner ear. Major atmospheric pressure changes can be equalised on either side of the eardrum through the Eustachian tube by the act of swallowing. The problem at this stage is the high impedance mismatch due to the outer and middle ear being filled with air and the inner ear filled with liquid. The small bones which connect the eardrum with the oval window, effectively make the middle ear

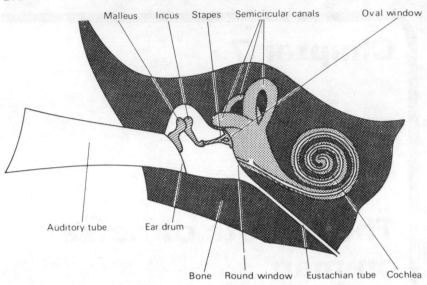

Fig. 7.1 Main component of the ear, showing outer, middle and inner parts

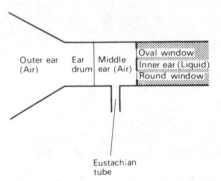

Fig. 7.2 Three main sections of the ear shown schematically

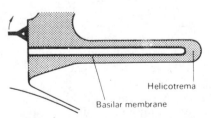

Fig. 7.3 Diagram of the cochlea unwound to show its main components

equivalent to a 'step-up transformer' of about 20 times. This just compensates for the theoretical loss of approximately 30 dB between air and fluids of the inner ear.

The lever system consists of three bones when in theory only one is needed (as in the case with birds). It does give three important advantages:

1. Minimum bone conduction.
2. More linear response for different frequencies.
3. A protective overload device possible as the ossicles change their mode of operation above 140 dB sound pressure level.

The middle ear also possesses another protective device consisting of two small muscles which adjust the eardrum and stapes for levels of sound above 90 dB which last more than 90 ms.

The inner ear is a system of liquid-filled canals protected both mechanically and acoustically by being located inside the temporal bone of the skull. The cochlea is a hollow coil of bone filled with liquid, with a total length of 40 mm (Fig. 7.3). This is divided along its length by the basilar membrane with a small gap at its far end known as the helicotrema. Acoustic energy is converted into impulses transmitted to the brain at the basilar membrane on which about 24 000 nerve endings terminate.

Intense sounds can damage or even destroy any of the moving parts of the ear. In the more common case of hearing damage because of prolonged exposure to high levels of noise it is the hair cells that are damaged.

Pitch

Frequency is an objective measure of the number of vibrations per second whereas the term pitch is subjective, and although dependent mainly on frequency is also affected by intensity.

Audiometry

Audiometry is the term used to describe the measurement of hearing sensitivity. The instrument used for this purpose is an audiometer which produces pure tones of various frequencies at different known sound pressure levels. The subject is required to indicate for each ear which level at each frequency he can just detect. With no hearing defect the audiometer results follow the bottom curve of Fig. 7.4. If the subject has some hearing defect the audiometer levels would need to be raised at some or all frequencies above the normal threshold. The amount the level needs to be increased gives the hearing loss. Examples of audiometric tests for left and right ears are given in Fig. 7.5.

It is important that audiometric testing is carried out only in a suitably quiet background to avoid masking effects. This is particularly important for industrial audiometry where the subject

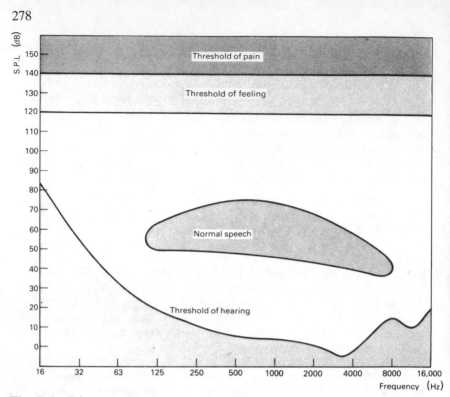

Fig. 7.4 Diagram showing the appropriate threshold of hearing for young people in the age range 18–25. The thresholds of feeling and pain occur at about 120 and 140 dB respectively. The range of levels and frequencies of normal speech are shown. Frequency limits for the piano and singers are shown

may have pre-employment tests at a young age, when there should not be any hearing loss. For these tests to be valid there must be no interference due to masking sound. Suggested maximum levels are given in Table 7.1.

Hearing defects

These may conveniently be divided into two general types. First, those which are not related to noise exposure; and second, those which are directly attributable to damage caused by noise.

Defects not normally caused by noise

1. *Presbyacousis:* This is a loss which is normally associated with age. It is manifest in a deterioration in high frequency acuity and tends to progress with age. It can be distinguished from noise-induced hearing loss because losses are greater with higher frequencies.

Noise-induced hearing loss shows a larger dip at around 4 kHz. Typical presbyacousis losses are shown in Fig. 7.6. These should apply to both men and women in the absence of noise-induced hearing loss.

There have been suggestions that presbyacousis is not due to age alone and that everyday noises in a Western-type civilisation are the cause. Audiometric tests on one primitive tribe show that their hearing at 70 years of age is comparable with that of Americans at 30 years of age. In practice this is probably an academic point for most of us.

2.   Tinnitus: This defect which is experienced by most of us from time to time is usually in the form of a high pitched ringing in the ears. This ringing effect may accompany a serious complaint known as Ménières disease or may be the aftermath of an exposure to a very high level of sound. In the former case it usually accompanies attacks of dizziness with perhaps nausea and vomiting. The second situation

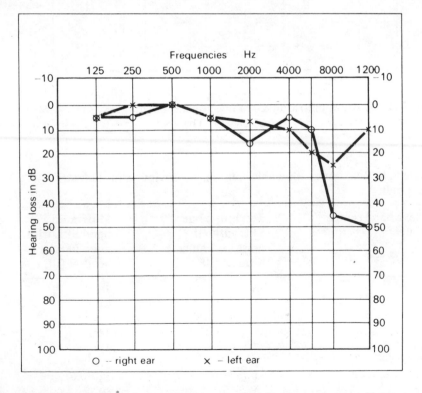

Fig. 7.5(a)   Audiometric test result showing some hearing loss at high frequencies

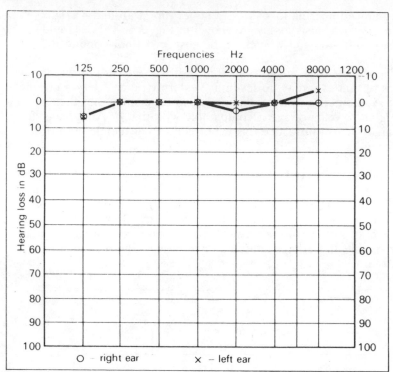

Fig. 7.5(b) Audiometric test result showing normal hearing

Table 7.1 Maximum allowable sound pressure levels for industrial audiometry

Octave band centre frequency (Hz)	Maximum band SPL without noise excluding headset (dB)	Maximum band SPL with noise barrier headset (dB)
31.5	76	76
63	61	61
125	46	55
250	31	44
500	7	31
1000	1	31
2000	4	43
4000	6	50
8000	9	44

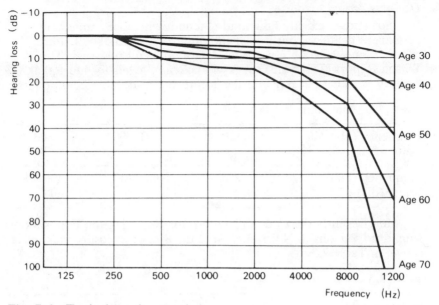

Fig. 7.6   Typical presbyacousis loss

is experienced by most people at some time in their lives after exposure to high levels of noise. It is due to the discharge of nerve impulses in the fibres of the auditory nerve.

3.   Deafness: There are three main types of deafness–conductive deafness, nerve deafness and cortical deafness.

Conductive deafness is due to defects in those parts of the ear (external canal, eardrum and ossicles) which conduct the sound waves in the air to the inner ear. Examples are a thickening of the eardrum, stiffening of the joints of the ossicles or a blocking of the external canal by wax. These affect all frequencies evenly but the loss is limited to between 50 and 55 dB, due to conduction through the head.

Conductive hearing loss causes a loss despite amplitude. A 20 dB loss would mean that a sound of 20 dB is needed for threshold, 40 dB to hear a level of 20 dB, 60 dB to hear 40 dB, etc. (at a particular frequency). People afflicted may not hear normal speech but may hear loud speech in a noisy factory.

Nerve deafness is either due to loss of sensitivity in the sensory cells in the inner ear or to a defect in the auditory canal. There is no medical remedy and the hearing loss is usually different for different frequencies.

Cortical deafness chiefly affects old people, and is due to a defect in the brain centres.

Deafness caused by noise

It is not known exactly how loud a sound will cause immediate permanent deafness, but it is of the order of 150 dB. The harmful effect of long duration noise is even more difficult to assess. The Wilson Report (Table 7.2) suggests maximum levels for more than 5 hours' working per day in order to prevent noticeable hearing loss.

Table 7.2 Maximum levels of noise

Frequency (Hz)	37.5–150	150–300	300–600
Value (dB)	100	90	85

Frequency (Hz)	600–1200	1200–2400	2400–4800
Value (dB)	85	80	80

More recently the Code of Practice for Reducing the Exposure of Employed Persons to Noise suggested the 8 hour exposure maxima shown in Table 7.3.

Table 7.3

Octave band centre frequency (Hz)	63	125	250	500	1000	2000	4000	8000
Maximum levels (dB)	97	91	87	84	82	80	79	78

It can be seen that noise-induced hearing loss depends on frequency. Narrow band noise is far more serious than broad band noise.

Temporary threshold shift

When a person of normal hearing is exposed to intense noise for a few hours he suffers a temporary loss of hearing sensitivity known as temporary threshold shift. The threshold of his hearing has been raised. After a sufficiently long rest from noise he usually recovers. If the noise is of a pure tone of low level the threshold shift is greatest at the same frequency as the noise. However for pure tones of high intensity the shift of threshold is greatest at about a half octave above (at about 1.4 kHz for a noise of 1 kHz). The most susceptible frequency range however is from 3 to 6 kHz. With broad band noise deterioration first occurs in this region, with 4 kHz being often worst affected. The amount of temporary threshold shift is related to the exposure duration.

Permanent threshold shift

In the case of someone exposed to intense occupational noise during the working day for a matter of years the stage is often reached

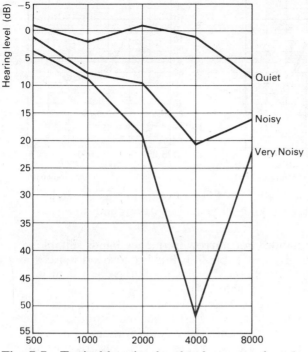

Fig. 7.7 Typical hearing levels of groups of workers exposed to different noise conditions for 10 years

where the temporary threshold shift has not completely recovered overnight before the next exposure. It appears that 'persistent threshold shift' is followed eventually by 'permanent threshold shift'. That permanent damage is being done to hearing may often be indicated by signs of dullness of hearing often combined with tinnitus after exposure to noise at work.

Noise-induced hearing loss is characterised by a 4 kHz drop in sensitivity as can be seen in Fig. 7.7. Continued exposure leads to a widening of the dip as can be seen in Fig. 7.8(a) and (b).

The actual hearing level of a person will be worse than might have been expected because it is the sum of the permanent threshold shift and presbyacousis loss. Hearing level becomes a function of three variables; noise level, exposure time and age.

Noise susceptibility

Susceptibility to hearing damage varies from person to person. It is not enough to protect the average person. It is not possible in all situations to ensure standards which will protect the most sensitive. Clearly maximum levels of noise must ensure protection for the

284

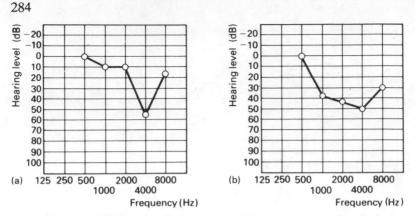

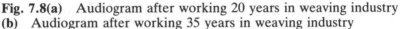

Fig. 7.8(a) Audiogram after working 20 years in weaving industry
(b) Audiogram after working 35 years in weaving industry

majority of the population. For narrow frequency bands suitable maximum levels are given in Table 7.3. For broad band noise a maximum of 90 dB(A) for 8 hours' exposure or the equivalent is usually considered suitable (Table 7.4).

Protection for the more sensitive involves a screening process. Thus a pre-employment audiometric test is needed. Routine testing can then be carried out to determine any deterioration, of hearing which is not due to presbyacousis. Where noise levels are close to or exceed those shown control of noise exposure is important. This may include: reduction of noise at source; reduction of noise transmission; use of ear protection (ear plugs or ear muffs); or in some cases a change of job.

Equivalent continuous noise levels (L_{eq})

The maximum sound levels given in Table 7.4 were continuous levels. In practice for most circumstances levels are not continuous. For the

Table 7.4 Maximum noise exposure to continuous broad band noise at work

Continuous sound level in dB (A)	Maximum exposure time per day
90	8 h
93	4 h
96	2 h
99	1 h
102	30 min
105	15 min

purpose of estimating exposure to noise it is convenient to consider continuous levels. Thus the term equivalent continuous level is used. The equivalent continuous level of sound in dB(A) is that continuous sound level which has the same total energy as the actual fluctuating sound over the time considered.

Noise criteria

Noise is simply unwanted sound. As can be seen above it may be unwanted because it could cause damage to hearing. There are however various other reasons why a sound is unwanted; it may interfere with our hearing of speech, disturb what we are trying to do (such as going to sleep), or it may be annoying. It can be appreciated that any criteria must be dependent upon what we are doing. Various criteria have been devised to ensure safety, avoid speech interference or minimise disturbance. In some cases the criteria will need to be frequency dependent.

Deafness risk criteria

It has already been explained that sound levels above 150 dB(A) are almost certain to cause instantaneous hearing damage and must always be avoided. In practice no one should ever be exposed to more than 135 dB(A) without ear protectors. One is at risk for sound levels above 90 dB(A) and Table 7.4 shows maximum exposure above which ear protectors should be used. It is often not appreciated that it is not only noise exposure at work which can cause damage. Many hobbies, particularly shooting, are damaging. Attendance at discos and even use of some domestic power tools could damage hearing. The same criteria still apply.

Speech interference criteria

Communication by speech is important in many different situations. Clarity is dependent upon the level of the speaker's voice, how close he is to the listener, the masking noise level and the frequency distribution of this masking noise. The most important frequencies are in the three octave bands from 600 to 4800 Hz(600–1200 Hz, 1200–2400 Hz, 2400–4800 Hz). Sound in this frequency band is most effective in masking speech due to the coincidence with speech frequencies.

The actual criteria or maximum levels of intruding noise to avoid speech interference depend upon how loud one can speak and how close one can be to the listener. This can be seen approximately in Table 7.5. These speech interference criteria do not ensure avoidance of disturbance and some sounds, particularly of lower frequency, may be very disturbing.

Table 7.5 Speech interference levels

Distance from speaker to hearer	Maximum level of background noise to ensure speech intelligibility			
(m)	Normal voice (dB)	Raised voice (dB)	Very loud voice (dB)	Shouting (dB)
0.1	73	79	85	91
0.2	69	75	81	87
0.3	65	71	77	83
0.4	63	69	75	81
0.5	61	67	73	79
0.6	59	65	71	77
0.8	56	62	68	74
1.0	54	60	66	72
1.5	51	57	63	69
2.0	48	54	60	66
3.0	45	51	57	63
4.0	42	48	54	60
8.0	35	41	47	53
16.0	29	35	41	47

Noise rating and noise criteria curves

The fact that speech interference is partially dependent upon the frequency spectrum of the intruding noise suggests that any criteria should be frequency dependent. As a result noise rating (NR) curves and noise criteria (NC) curves have been devised. These are shown in Figs. 7.9 and 7.10. The aim behind both sets of curves is the same and the differences are fairly small. Provided speech interference is avoided disturbance should also be minimised at the most economic cost. Thus if a speech interference level of 50 dB is required then NC 50 or NR 50 should be specified.

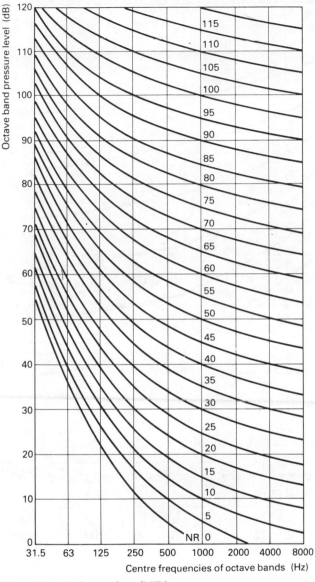

Fig. 7.9 Noise rating (NR) curves

288

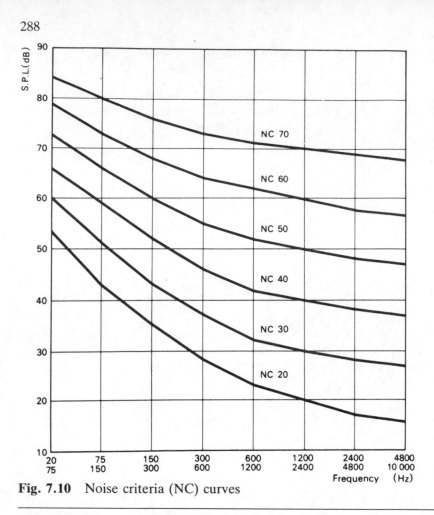

Fig. 7.10 Noise criteria (NC) curves

Example 7.1 A noise on the site of a proposed school classroom has the following characteristics:

Octave band centre frequency (Hz)	63	125	250	500	1000	2000	4000	8000
Sound pressure Level (dB)	75	71	70	69	65	62	61	60

Calculate:

(a)   the NC level;
(b)   the NR level;
(c)   the sound insulation needed to achieve NC 25.

(a)   The noise spectrum is drawn on the graph of NC curves (Fig. 7:11) and it can be seen that it just touches the NC 65 curve at the worst frequency.

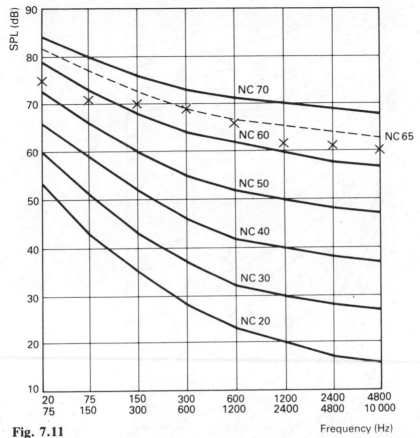

Fig. 7.11

Frequency (Hz)

Frequency (Hz)	Level (dB)	NC 25	Insulation (dB)
63	75	57	18
125	71	47	24
250	70	39	31
500	69	32	37
1000	65	28	37
2000	62	25	37
4000	61	22	39
8000	60	21	39

This could be achieved by means of a 115 mm brick wall.

290

(b) Similarly to the above the noise spectrum is drawn on the graph of noise rating (NR) curves. It just touches the NR 66 curve at the worst frequency (Fig. 7.12).

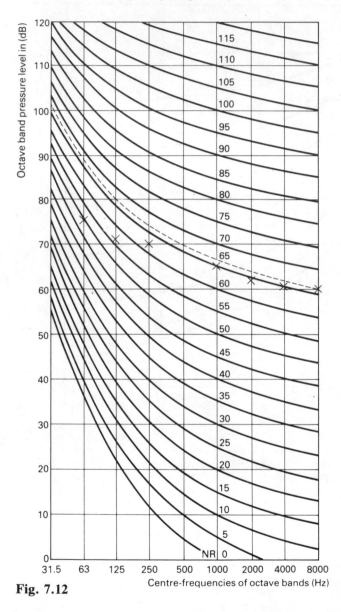

Fig. 7.12

It is not always easy to determine the optimum NC figure for each situation, but Table 7.6 gives some suggested figures

Table 7.6 Suggested noise criteria (NC) for different rooms

	NC
Concert halls	< 20
Large churches	20–25
Boardroom	20–25
Hospital theatre	25
Classroom/lecture room	25–35
Bedrooms	25–35
Private offices	25–40
Hospital dayrooms/treatment rooms	35
Restaurant	35–45
Hospital kitchens	40
Typing pool	50
Workshops	55–75

which should be suitable. These can only be average figures and there is some variation between the ideal and the cheapest solution.

It should be appreciated that it is uneconomic to insist on too low a noise level. In many situations background sound will not be heard above the activity noise in the room. In other situations a background noise can be of positive benefit as masking sound thus providing audible privacy.

Example 7.2 The recommended maximum sound level for the sleeping area of a house is the NC 35 curve. The external bedroom wall of a house is 4 m by 2.5 m in area, facing a motorway at a distance of 30 m from the line of traffic. An octave band analysis at 7.5 m from the line of traffic gave the following results:

Frequency (Hz)	20–75	75–150	150–300	300–600
SPL (dB)	99	95	93	92

Frequency (Hz)	600–1200	1200–2400	2400–4800	4800–10 000
SPL (dB)	93	90	88	85

The construction consists of a 280 mm cavity wall containing a double glazed (and sealed) window. The sound reduction indices are 55 dB for the 280 mm cavity wall and 44 dB for the sealed double window both at 1000 Hz. Assume for the purpose of this question that the insulation of both the window and brickwork improves by 5 dB for each doubling of the frequency. Calculate, for 1000 Hz, the maximum area of window to the nearest $0.1 \, m^2$. State what the insulation is expected to be at the other seven octave values and whether they are likely to be adequate.

The level of the sound will drop 3 dB for each doubling of the distance for noise from a line source. In this example the distance has been doubled twice so that $2 \times 3 = 6$ dB must be subtracted to find the level immediately outside the window.

Frequency (Hz)	20–75	75–150	150–300	300–600
Motorway (dB)	99	95	93	92
Outside window (dB)	93	89	87	86

Frequency (Hz)	600–1200	1200–2400	2400–4800	4800–10 000
Motorway (dB)	93	90	88	85
Outside window (dB)	87	84	82	79

To find the necessary insulation the maximum recommended level given by the NC 35 curve must be subtracted from the values outside the window. Hence:

Frequency (Hz)	20–75	75–150	150–300	300–600
Outside window (dB)	93	89	87	86
NC 35	63	55	47	42
Insulation needed (dB)	30	34	40	44

Frequency (Hz)	600–1200	1200–2400	2400–4800	4800–10 000
Outside window (dB)	87	84	82	79
NC 35	37	35	33	32
Insulation needed (dB)	50	49	49	47

Thus it can be seen that the necessary insulation at 1000 Hz must be 50 dB.

Now the reduction in dB $= 10 \log_{10}(1/T)$ where T is the transmission coefficient

$$\therefore \quad 50 \, dB = 10 \log_{10} \frac{1}{T_{AV}}$$

(T_{AV} = average transmission coefficient for the window and brickwork combined)

$$\therefore \quad 5 = \log_{10} \frac{1}{T_{AV}}$$

$$= -\log_{10} T_{AV}$$

$$\therefore \quad \bar{5}.0000 = \log_{10} T_{AV}$$

$$T_{AV} = 0.00001$$

For the 280 mm cavity wall:

$$55 \, dB = 10 \log_{10} \frac{1}{T_B}$$

(T_B = transmission coefficient for the brickwork)

$$\therefore \quad 5.5 = -\log_{10} T_B$$

$$\bar{6}.5 = \log_{10} T_B$$

$$T_B = 0.000003162$$

Similarly for the window:

$$44 = 10 \log_{10} \frac{1}{T_W}$$

(T_w = transmission coefficient for the window)

$$\therefore \quad 4.4 = -\log_{10} T_w$$

$$\bar{5}.6 = \log_{10} T_w$$

$$T_w = 0.00003981$$

Now:

(Total area) $\times T_{AV}$ = (area of brick) $\times T_B$ + (area of window) $\times T_w$

Let the window area = $a \, m^2$

$$\text{Total area} = 10 \, m^2$$

$$\text{Area of brick} = (10 - a) \, m^2$$

$$\therefore \quad 10 \times 1.0 \times 10^{-5} = (10 - a) \times 3.162 \times 10^{-6} + a \times 3.981 \times 10^{-5}$$

$$1.0 \times 10^{-4} = 3.162 \times 10^{-5} - 3.162 \times 10^{-6} a + 3.981 \times 10^{-5}$$

$$\therefore \quad 6.838 \times 10^{-5} = 3.665 \times 10^{-5} a$$

$$a = \frac{6838}{3665}$$

$$= 1.9 \, m^2 \text{ to the nearest 0.1 square metre}$$

A 5 dB improvement in insulation for each doubling of frequency also means a reduction of 5 dB for each halving of the frequency below 1000 Hz. Thus the expected insulation would be:

Hz	Octave Band	Insulation (dB)
2000	1200–2400	55
4000	2400–4800	60
8000	4800–10 000	65
500	300–600	45
250	150–300	40
125	75–150	35
63	20–75	30

Comparing these values with those from the NC curve the insulation should be adequate for all frequencies.

Traffic noise criteria

Noise rating (NR) and noise criteria (NC) curves were primarily intended to assist in avoiding speech interference. There are many situations particularly in dwellings where it is important to have criteria which avoid disturbance. This does not mean that NC or NR figures cannot be specified to avoid nuisance but they are frequently not the most convenient.

Traffic noise is the largest single source of noise nuisance, not only in Britain but all the Western world. In practice traffic noise levels have been creeping up by an average of 1 dB(A) per year. While the annual increase is not noticed as such it is found that levels are being reached which are unacceptable to householders. A problem of measurement arises due to the fluctuating levels. Various options for units of measurement of traffic noise have been explored including:

L_{eq} – the equivalent continuous sound level in dB(A).

L_{10} – the sound level in dB(A) exceeded for 10 per cent of the measurement time.

L_{50} – the sound level in dB(A) exceeded for 50 per cent of the measurement time.

L_{90} – the sound level in dB(A) exceeded for 90 per cent of the measurement time.

L_{eq} is favoured in certain countries but in Britain an 18 hour L_{10} is used. This is the sound level exceeded for 10 per cent of the time over the 18 hours from 6 a.m. until midnight. There appears to be reasonable correlation between the 18 hour L_{10} value and the amount of nuisance caused. It can be seen that the L_{10} is a measure of the peak noise levels. An L_{10} of 68 dB(A) is considered the limit of acceptability outside dwellings. In practice 70 dB(A) was chosen, but due to an expected $\pm 2\,\mathrm{dB(A)}$ accuracy in measurement $68\,\mathrm{dB(A)}$ has been used.

Industrial noise affecting residential areas

Apart from the problem already considered of industrial noise affecting hearing it may cause noise nuisance to the people in nearby dwellings. A satisfactory criterion can only be achieved after making allowance for the character of the noise and the neighbourhood. BS 4142 suggests a possible method:

1. Basic criteria $50\,\mathrm{dB(A)}$
2. Add:
    (a) $0\,\mathrm{dB(A)}$ for new factories;
    (b) $0\,\mathrm{dB(A)}$ for existing factories to which changes of structure or process are being made;
    (c) $5\,\mathrm{dB(A)}$ for existing factories out of character with the neighbourhood;
    (d) $10\,\mathrm{dB(A)}$ for long-established factories in keeping with the neighbourhood.
3. To allow for type of district:
    (a) Subtract $5\,\mathrm{dB(A)}$ for a rural area.
    Add:
    (b) $0\,\mathrm{dB(A)}$ for suburban area;
    (c). $5\,\mathrm{dB(A)}$ for urban areas;
    (d) $10\,\mathrm{dB(A)}$ for mainly residential urban area with some light industry or main roads;
    (e) $15\,\mathrm{dB(A)}$ for evenly mixed residential and industrial area;
    (f) $20\,\mathrm{dB(A)}$ for a predominantly industrial region with few dwellings.
4. To allow for times of day that the factory is in use:
    (a) Weekdays 8.00–18.00 hours only, add $5\,\mathrm{dB(A)}$;
    (b) To include night time 22.00–07.00 hours, subtract $5\,\mathrm{dB(A)}$;
    (c) If use to include the periods 18.00–22.00 hours or 07.00– 08.00 hours, but not 22.00–07.00 hours then no change is needed.

Provided the corrected noise level (CNL) is below the criterion the sound should not cause much nuisance. If the CNL is more than $10\,\mathrm{dB(A)}$ above the criterion complaints are almost certain. If it is more than $10\,\mathrm{dB(A)}$ below then complaints are very unlikely.

The corrected noise level (CNL) of the sound is found as follows:

1. Find the noise level in $\mathrm{dB(A)}$.
2. Apply a tonal character adjustment of $+5\,\mathrm{dB(A)}$ where the noise has a definite continuous note such as a whine or hiss.
3. Add $5\,\mathrm{dB(A)}$ if the sound is impulsive, e.g. bangs.
4. Apply a correction to allow for the intermittent nature of the sound. This is done by finding the percentage of the time the noise is produced and applying a correction from Fig. 7.13 or Fig. 7.14 depending whether it is day or night.

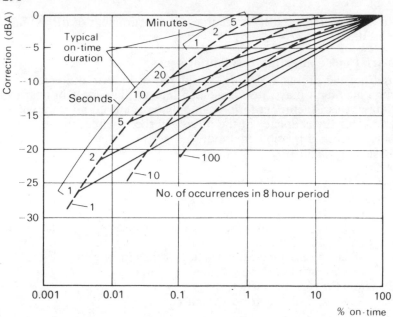

Fig. 7.13 Intermittency duration correction for night time

Example 7.3 A new factory is to be built to operate 24 hours per day and it is estimated that the sound level at nearby houses will be 57 dB(A). Predict the possibility of complaints.

	Day time		Night time	
	Old noise	New noise	Old noise	New noise
Measured noise level				57 dB(A)
Tonal character correction				5
Impulsive character correction				—
CNL				62 dB(A)
Basic criteria	50	50	50	50
Correction for type of installation				0
Correction for type of district				+10
Correction for time of day				− 5
Corrected criterion				55

As the corrected noise level is higher than the criterion, complaints can be expected.

Construction site noise

Noise from building sites can have a number of effects including deafness to workers, speech interference and nuisance to neighbours. The first two have been dealt with earlier but the question of nuisance has special implications for contractors because of the provisions within the Control of Pollution Act 1974. This gives local authorities power to deal with noise from construction and demolition sites. These powers can extend from before work commences in requiring noise predictions. It may be to the contractor's advantage to discover the local authority's noise requirements at the tender stage. The aim of any local authority is to avoid noise nuisance or if that is impossible to minimise it. This may affect the contract price and the plant to be used.

Criteria used by local authorities will normally include L_{eq} limits as recommended in BS 5228, Noise Control on Construction and Demolition Sites. In some cases L_{10} levels may also be used to avoid excessive peak levels. In addition limits on hours of noise producing work may be given. It is thus advisable for contractors to

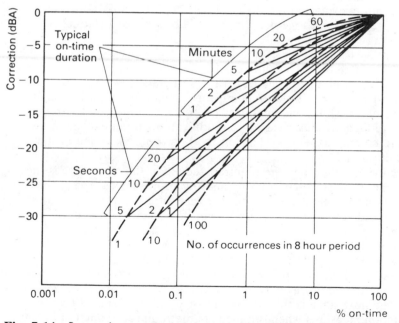

Fig. 7.14 Intermittency and duration correction for other than night time

predict likely noise levels. The prediction methods are explained in BS 5228. They are based on calculating the sound level (dB(A)) at a given position such as the site boundary from a knowledge of the A-weighted sound power level of the equipment. The relationship was explained in Chapter 4. It is sound level (dB(A)) = sound power level $- 20\log_{10} r - 8$, where r = distance from the noise source.

It is unlikely in practice that nuisance will be common on smaller construction jobs but it can be serious for contracts of £1000 000 and above.

Motor cars

In Britain the approach to noise control is in general one of avoiding nuisance and suitable criteria are usually derived for each situation as explained for motor traffic, construction, etc. Many countries have adopted a 'fixed standard' approach to avoiding nuisance. This is used in certain situations in Britain and the limit for motor cars is one of them as shown in Table 7.7.

The steps in a noise control programme

It is important to adopt a logical approach to noise control problems. The first step, for an existing noise problem is the measurement of the noise levels and the diagnosis of the most important sources, where several exist together. In the case of proposed sources, the levels must be predicted.

The next step is the establishment of a target noise level for the particular situation. The amount of noise reduction which is required can then be estimated. The target level or standard will be set to meet a particular criterion relating for example to employee health and safety, customer acceptability or public annoyance. The required noise reductions may necessitate modifications to an existing noise source or may have to be built in to the proposed source at the planning stage.

Having established the required reduction the next stage is the application of noise control engineering principles to design the reduction treatment. At this stage several other factors have to be considered such as the cost, safety and fire hazard implications of the proposed noise reduction measures and their possible effects on working procedures and production processes. Any adverse environmental conditions likely to be encountered by the materials to be used, such as the presence of dust and grit, corrosive chemicals and high temperatures must be taken into account at this stage.

The next stage is the assessment of the effectiveness of noise reduction measures when installed. If the required reduction is not achieved the process has to be repeated, starting at the design stage. It must be emphasised that many noise sources may contribute to

Table 7.7 Maximum sound level for vehicles

Type of vehicle	Maximum sound levels in dB (A)		
	New vehicles (lst registered after 1 April 1970)	Vehicles in use (lst registered before 1 Nov 1970)	Vehicles in use (1st registered after 1 Nov 1970)
1. Motor cycle up to 50 cc	77	80	80
2. Motor cycle 50 to 125 cc	82	90	85
3. Motor cycle above 125 cc	86	90	89
4. Motor car	85	88	88
5. Passenger vehicle for more than 12 passengers	89	92	92
6. Any other passenger vehicle	84	87	87
7. Heavy goods vehicle	89	92	92
8. Tractor, locomotive works truck or engineering plant	89	92	92
9. Any other vehicle	85	92	89

a particular problem, or there may be many transmission paths associated with one particular source. This means that the effective solution of a noise problem may involve more than one noise control measure (e.g. absorption as well as screening , or isolation as well as insulation). Therefore if a particular treatment does not give the required reduction it does not necessarily mean that the diagnosis and design have been completely wrong–they may just have been incomplete and require other additional measures as well.

The above sequence is shown schematically in Fig. 7.15.

The noise chain: source–path–receiver

When considering various noise control options it is often convenient to think of the noise problem as consisting of a three link chain: the

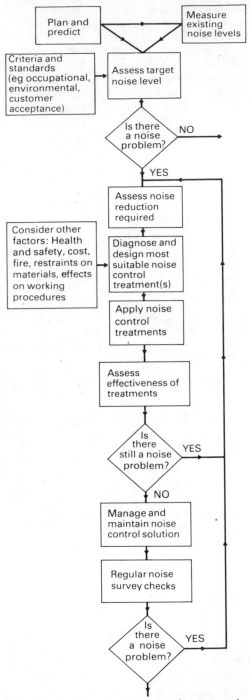

Fig. 7.15 The steps in a noise control programme

noise source; the transmission path; and the receiver. The noise reduction measures may then be applied to either one or more of these links. Noise control at source is obviously desirable if it is practical. In extreme cases this might mean the complete removal of the noise producing process and its substitution by a quieter one. An alternative might be the re-siting of the source (e.g. a plant room) to a less noise-sensitive area. In cases where such complete solution is inappropriate it might be possible to reduce the noise at source. Noise sources mainly fall into two broad categories – first of all noise emitted from vibrating surfaces, e.g. machine panels, and second, 'aerodynamic noise sources' in which there is a direct disturbance of the air (or other medium, such as water) by, for example, the presence of a fan, a jet or a pump.

The noise is transmitted from the source to the receiver by one or more transmission paths. This might be by airborne sound transmission, in which case the use of sound absorbing and sound insulating materials will be useful as noise reduction measures. Alternatively the path may be via a solid structure, in which case techniques appropriate to vibration reduction, such as isolation and damping may be the appropriate noise control measures. The transmission path for fan noise will be via the ductwork as well as direct radiation from the fan, and, possibly vibration transmission from both fan and ductwork.

As a last resort noise control measures may be applied at the

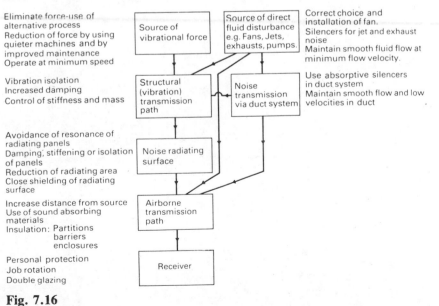

Fig. 7.16

reception point. For example this might be the provision of hearing protection for a worker in industry or the provision of double glazing for a household near to a main road or airport.

Figure 7.16 shows schematically the various stages in the transmission of noise from source to receiver and lists some of the standard noise control measures appropriate to each stage.

Questions

1. Explain why a decibel noise reading may have no relation to loudness.
2. Write an account of the determination of loudness level.
3. Sketch an audiogram for a typical case of noise-induced hearing loss and give the probable cause in terms of the damage to the structure of the ear.
   Evaluate the permitted daily exposure time for a person subjected to a steady sound level of 109 dB(A).
4. Define L_{10} and explain the criterion which is applied to dwellings near to new roads.
5. Explain what is meant by L_{eq} and its use.

Section III

Light

Chapter 8

Artificial lighting

Illumination is the process of lighting an object. This may be a visual task or the total visual scene inside or outside a building. Light is provided in a building either by the sun (natural lighting) or by the use of artificial sources (artificial lighting). The sun provides light within a restricted portion of the day. In order to extend the working day, supplement daylight or enable recreational activities to continue after sunset, a number of artificial sources have been invented. Although candles, oil lamps and gas lamps fall into this category, the most important sources today are electric ones.

Sir Humphrey Davy produced light from a carbon arc early in the nineteenth century but this source proved to be too large and powerful to be used conveniently to light building interiors. The quest to 'subdivide the carbon arc' led Swan in the United Kingdom and Edison in America to invent incandescent lamps by the 1880s. Today we have two main groups of electric lamps. There are those based on the heating effect of an electric current to produce light by incandescence and those producing light by electrical discharge in a gas.

Some of the important types of lamps and their efficacies (the amount of light flux (in lumens) produced per watt of electricity consumed) are listed in Table 8.1.

Terminology and units of illumination

The following basic terms provide a starting point – other units will be introduced as the need arises. A full list of terms used in

Table 8.1 Lamp types and efficacies

Lamps	Designation	Efficacy (lumens/watt)
Incandescent or filament		
1. Tungsten	GLS – coiled coil	11–14
	– mushroom bulb	10–12
	single coil	9.6–13.6 to 17.3 for high wattages
2. Tungsten Halogen (TH)		17.5–22
Discharge		
3. Low pressure mercury vapour (fluorescent tubes)	MCFE/U and MCFA/U	35–80
4. High pressure mercury vapour	MB	40–50
	MBF (fluorescent)	36–54
	MBFR (reflector)	42–48
	MBTF (tungsten ballast)	16–23
Metal halide	MBI	60–78
	MBIF (fluorescent)	64–85
	MBIL	72–78
	CSI	67
5. Low pressure sodium vapour	SOX	122–160
	SLI	125–143
6. High pressure sodium vapour	SON	84–100
	SONT	90–105
	SONTD	105

Note:
These efficacies are based on lighting design lumens per watt at the lamp.
Efficacies based on circuit watts for discharge lamps are lower by 5–10%.

lighting can be found in the IES Code for Interior Lighting or CIE Publication No. 17 – International Lighting Vocabulary.

Steradian (sr)

This is the unit for measuring three-dimensional (or solid) angles. It is the three-dimensional equivalent of the two-dimensional radian. One steradian is defined as the solid angle subtended at the centre of a sphere of radius r by a surface element of area r^2 (Fig. 8.1) The total surface area of a sphere of radius r is $4\pi r^2$ and therefore there are 4π steradians at the centre of the sphere ($4\pi r^2/r^2 = 4\pi$ steradians). For large values of r, 1 steradian is subtended at a point by unit plane area, unit distance from the point.

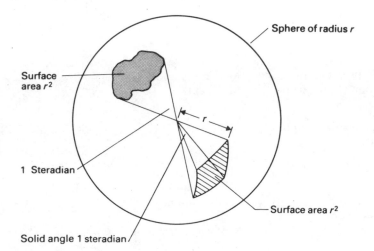

Fig. 8.1 The steradian

Candela (cd)

This is defined as the SI unit of luminous intensity, in a given direction, of a source which emits monochromatic radiation of frequency 540×10^{12} Hz and of which the radiant intensity in that direction is 1/683 watts per steradian.

Lumen (lm)

This is the unit of luminous flux which is the light emitted by a source or received by a surface. A small source which has a uniform intensity of 1 candela emits a total of 4π lumens in all directions and emits 1 lumen within unit solid angle (1 steradian) (Fig. 8.2).

Where a source is not of uniform intensity it is possible to calculate an average value or mean spherical intensity. The total number of lumens given out by the source would then equal $4\pi \times$ mean spherical intensity. The light emitted by a lamp is stated in

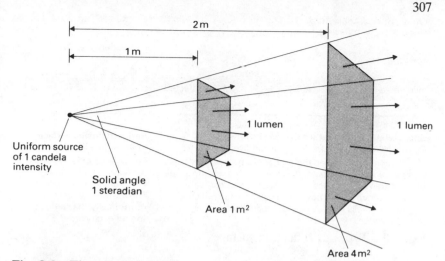

Fig. 8.2 The relationship between the lumen and the candela

lumens; e.g. a 40 W Warm White fluorescent tube gives out 2950 lumens when new. Methods for calculating the flux from a non-uniform source are given in *Lighting Fittings Performance and Design*, by A. R. Bean and R. H. Simons.

Illuminance (*E*)

Illuminance is the luminous flux density received at a surface; i.e. the number of lumens received per square metre of surface. The unit is the *lux* (lx) where:

1 lumen per square metre = 1 lux

The illuminance is the lighting level in lux on a surface.

Luminance

Luminance is the physical measure of a stimulus such as a self-luminous, transmitting or reflecting surface which produces the subjective sensation of luminosity (brightness). The unit is the candela per square metre; i.e. the intensity in a given direction divided by unit area of the emitting surface.

The luminance of a reflecting surface in a particular direction is given by:

$$\frac{\text{Illuminance } (lx) \times \text{luminance factor } (\beta)}{\pi} = \text{luminance in cd/m}^2$$

(An alternative form is:

illuminance × luminance factor = luminance in 'apostilbs')

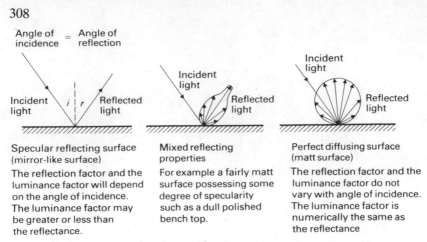

Fig. 8.3 Types of reflecting surface

The luminance factor (β) is the ratio of the luminance of a reflecting surface viewed in a given direction to that of a perfect white uniform diffusing surface identically illuminated. The need to use a luminance factor is due to the range of ways light may be reflected from surfaces (Fig. 8.3).

For most interior lighting design calculations it is convenient to assume that the luminance factor of a surface is the same as its reflection factor where the reflection factor is the ratio of flux reflected from a surface to that incident on it (Fig. 8.4). The product illuminance × reflection factor is then the luminous excitance in lumens per square metre, or apostilbs.

In SI units:
$$\frac{\text{illuminance} \times \text{reflection factor}}{\pi} = \text{luminance in cd/m}^2$$

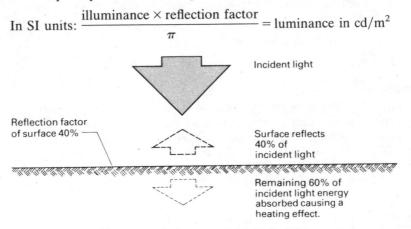

The reflection factor or reflectance (ρ) of the surface is the ratio of the quantity of light reflected by the surface to the quantity of light incident on the surface expressed as a percentage.

Fig. 8.4 Reflection factor

Artificial lighting calculations

An important quantity for task lighting in a building interior is the illuminance (in lux) measured on the task or on the working plane on which the task is carried out.

Calculation techniques are used in order to predict the number of lamps of a certain type necessary to produce a required lighting level (illuminance) in a particular room. Although it is not often the practice to produce calculations for domestic lighting, when planning the lighting of larger buildings it is necessary to quantify the number, type and positions of the lights and to be able to assess their effect.

The artificial sources mentioned earlier are used as bare lamps or more usually are placed within light fittings (luminaires). The light that is produced reaches the task either directly or by reflection and inter-reflection from the interior surfaces of the room. The total illuminance thus comprises both a direct component and an indirect component.

Direct component of illuminance

The direct illuminance produced by a source on a surface can be calculated using the following two illuminance laws.

The inverse square law of illuminance

The illuminance (E) at a point on a surface varies inversely as the square of the distance (d) between the surface and the point source

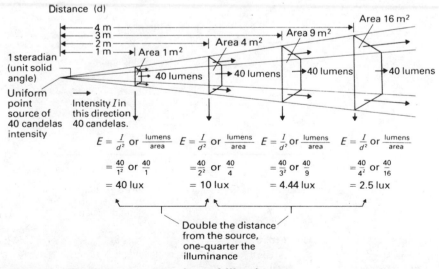

Fig. 8.5 The inverse square law of illuminance

of light which illuminates it; i.e. the illuminance produced by a source falls off inversely with the square of the distance from it.

For a surface normal to the direction of the incidence light:

$$E = \frac{I}{d^2} \text{ lux}$$

where I = the intensity (in candelas) in the direction of the surface

d = the distance (in metres) between the source and the surface (Fig. 8.5).

The cosine law of illuminance

The illuminance on any surface varies as the cosine of the angle of incidence of the surface to the direction of the light. The illuminance on a plane normal to the incident light calculated by the inverse square law must now be corrected for the angle of incidence on to any other surface because the flux is distributed over a larger area. Referring to Fig. 8.6:

$$E_h = \frac{E}{1/\cos\theta} \qquad E_h = E\cos\theta \qquad E_h = \frac{I}{d^2}\cos\theta$$

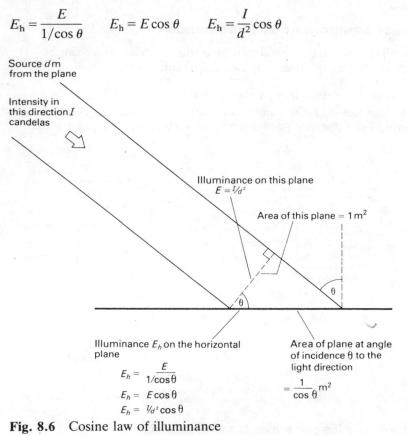

Source d m from the plane

Intensity in this direction I candelas

Illuminance on this plane $E = 1/d^2$

Area of this plane $= 1\,\text{m}^2$

Illuminance E_h on the horizontal plane

$$E_h = \frac{E}{1/\cos\theta}$$
$$E_h = E\cos\theta$$
$$E_h = 1/d^2\cos\theta$$

Area of plane at angle of incidence θ to the light direction

$$= \frac{1}{\cos\theta}\,\text{m}^2$$

Fig. 8.6 Cosine law of illuminance

Example 8.1 A uniform source giving out 25 133 lumens is placed 3 m directly above point A on a working plane. Point B is on the same plane but 4 m away from A. Calculate the direct illuminance at each point (Fig. 8.7).

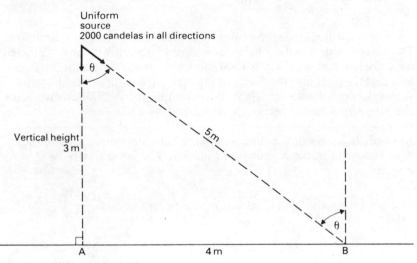

Uniform source
2000 candelas in all directions

θ

Vertical height 3 m

5 m

A   4 m   B

Fig. 8.7 (Example 8.1)

The source is uniform and will have an intensity in all directions of 2000 candelas.

$$\text{Intensity} = \frac{\text{total lumens}}{4\pi} \text{ for a uniform source}$$

$$= \frac{25\,133}{4\pi}$$

$$= 2000 \text{ candelas}$$

$$\text{Illuminance at point A} = \frac{I}{d^2}$$

$$= \frac{\text{intensity towards A}}{d^2}$$

$$= \frac{2000}{3^2}$$

$$= 222.22 \text{ lux}$$

Illuminance at point B $= \dfrac{I}{d^2} \cos \theta$

$$= \frac{2000}{5^2} \cdot \frac{3}{5}$$

$$= 48 \, \text{lux}$$

For most luminaires the intensity is not the same in all directions. The intensity distribution for a luminaire in a vertical plane is plotted on a polar curve scaled to a 1000 lumen source. To find the intensity in a certain direction the intensity at that angle is multiplied by the source lumens divided by 1000. For example for a 2500 lumen source the intensity in any direction is multiplied by 2.5.

Example 8.2　An opal cylinder replaces the luminaire in Example 8.1. The polar curve is shown in Fig. 8.8. The lamp flux is 1000 lumens.

Intensity towards A $= 112 \, \text{cd} \left(\text{from the polar curve} \times \dfrac{1000}{1000} \right)$

Illuminance at A $\quad = \dfrac{I}{d^2}$

$$= \frac{112}{3^2}$$

$$= 12.44 \, \text{lux}$$

Intensity towards B $= 75 \, \text{cd}$ ($\theta = 53°8'$ from the vertical)

Illuminance at B $\quad = \dfrac{I}{d^2} \cos \theta$

$$= \frac{75}{5^2} \cos 53°8'$$

$$= 1.8 \, \text{lux}$$

An alternative expression for the horizontal illuminance on a plane beneath the light source is:

$$E_\text{h} = \frac{I_\theta \cos^3 \theta}{h^2} \, \text{lux}$$

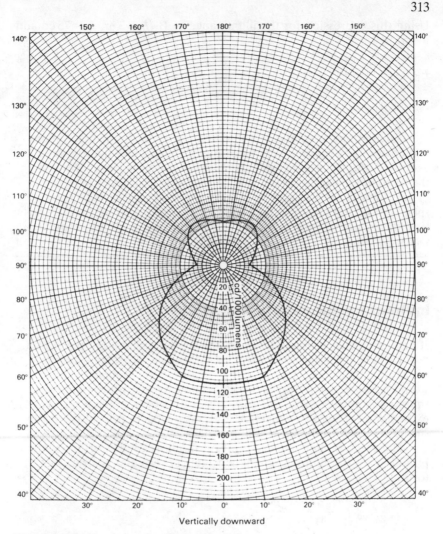

Fig. 8.8 Polar curve for Example 8.2

where I_θ = the intensity in the direction $\theta°$ from the downward
vertical

     $\theta°$ = the angle of incidence

     h = the vertical height of the light above the horizontal
plane

Using this expression for the illuminance at B ($h = 3\,\text{m}$, $I_\theta = 75\,\text{cd}$ and $\theta = 53° 8'$):

$$E_B = \frac{75 \cos^3 53° 8'}{3^2}$$

$$= \underline{1.8\,\text{lux}}$$

Calculations of total direct illuminance using the inverse square and cosine laws become very laborious if the general lighting area is extensive and the number of luminaires required is large. However, the technique is appropriate to local lighting situations where one or two luminaires are used to light a very restricted area or some specific task and also in areas where the reflecting properties of the boundaries are unknown such as some high bay installations in large factory spaces.

Indirect component of illuminance

If measurements of illuminance on the working plane are made in a room where the reflection factors of the various surfaces – ceiling, walls and floor – are not zero, the measured illuminance values would be higher than the values calculated by just using the inverse square and cosine laws. This is because the working plane receives an extra bonus of light by reflection and inter-reflection from the room surfaces to add to the direct illuminance.

The average illuminance (E_ind) on the horizontal working plane provided by the indirect component can be calculated from the formula:

$$E_\text{ind} = \frac{\Phi_u \rho_u + \Phi_d \rho_d}{A(1 - \rho)}$$

where Φ_u = the amount of luminous flux emitted upwards from the luminaire

   Φ_d = the amount of luminous flux emitted downwards from the luminaire

   ρ_u = the average reflectance of the room surfaces above the plane of the luminaire

   ρ_d = the average reflectance of the room surfaces below the plane of the luminaire

   ρ = the average reflectance of all the room surfaces

   A = the total area of all the room surfaces

The lumen design method

The lumen design method takes into account both the direct and indirect components of illuminance in calculating the average illuminance on the working plane. The method can be applied only to square or rectangular rooms with a regular array of luminaires.

In a regular array there are equal spaces between luminaires along a row and half spaces to the walls. If the luminaires were essentially point sources the spacing would be the same on the other room axis (Fig. 8.9). For linear fittings the spacing may be different on the other room axis but the rule about half spaces to the walls still applies (Fig. 8.10).

Full spaces, S between luminaires, half spaces, $\frac{1}{2}S$ to walls

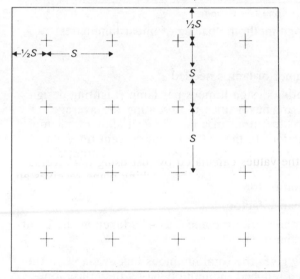

Fig. 8.9 Lumen design method: spacing of point sources

The method is intended primarily for task lighting situations. Other forms of treatment are more appropriate in situations such as hotel foyers, restaurants and lounges.

The lumen design method calculates the number of lumens required from the sources to produce a given average illuminance.

Lumen design method formula

$$E_{\mathrm{av}} = \frac{n \times \Phi \times \mathrm{UF} \times \mathrm{MF}}{A}$$

where E_{av} = the average illuminance required, usually on a horizontal working plane

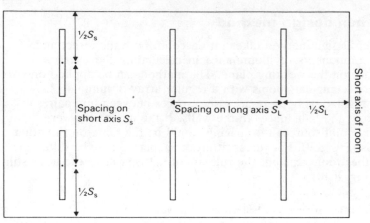

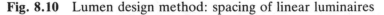

Long axis of room

Fig. 8.10 Lumen design method: spacing of linear luminaires

n = the number of lamps needed

Φ = the lighting design lumens per lamp (Lighting design lumens are flux values representing the average through life output of a lamp; i.e. it allows for lamp depreciation. In the case of a fluorescent tube the lighting design lumens are taken as the output after 2000 hours burning.)

UF = utilisation factor

MF = maintenance factor

A = area of the working plane – usually taken as the plant area of the room

The product $n \times \Phi$ gives the total luminous flux available in the luminaires. This is then reduced by multiplying by the utilisation factor (UF) and the maintenance factor (MF) to give the lumens reaching the working plane. A division by the area of the working plane yields the illuminance (in lux).

A rearrangement of the formula enables the number of lamps to be calculated:

$$n = \frac{E_{av} \times A}{\Phi \times UF \times MF}$$

Data for terms in the formula

E_{av} – the value of the illuminance on the working plane can be selected by reference to the CIBS/IES Code. Standard service illuminances are scheduled appropriate to the room use; i.e. the types of visual task to be lit (Table 8.2). The standard service

Table 8.2 Examples of standard service illuminance, limiting glare index and lamp colour appearance recommendations from the IES Code

	Standard service illuminance (lux)	Limiting glare index	Colour appearance of lamps
Offices			
General offices	500	19	Intermediate or warm
Deep-plan general offices	750	19	Intermediate or warm
Drawing offices	750 (on boards)	16	Cool, intermediate or warm
Shops			
Conventional shops	500	19	Cool, intermediate or warm
Supermarkets	500	22	Intermediate or warm
Industrial			
Assembly shops			
casual work	200	25	Intermediate or warm
rough work	300	25	Intermediate or warm
medium work	500	22	Intermediate or warm
fine work	1000	19	Cool, intermediate or warm
very fine work	1500	16	Cool, intermediate or warm
Structural steel fabrication plants			
general	300	28	Intermediate or warm
marking off	500	28	Intermediate or warm
Woodwork shops rough sawing fine bench work	300 750	22	Intermediate or warm
Warehouse–racks	200	25	Cool, intermediate or warm

Table 8.2 (*Cont.*)

	Standard service illuminance (lux)	Limiting glare index	Colour appearance of lamps
Hospitals			
Wards	100 (floor)		Warm
	30–50 (bedhead)		Warm
	150 (reading)		Intermediate or warm
Operating theatres	400–500 (general)		Intermediate
	10 000–50 000 (on table)		Warm
Further education establishments			
Teaching spaces	500	19 (16 in direction of chalkboard)	Intermediate or warm
Laboratories	500	19	Cool, intermediate or warm
Homes			
Kitchen	300		Intermediate or warm
Reading areas	300		Intermediate or warm

illuminance may be varied in steps to allow for particular peculiarities such as very low reflectances or contrasts, the seriousness of errors, duration of task and lack of windows. The standard service illuminance steps are: 150 lux; 200 lux; 300 lux; 500 lux; 750 lux; 1000 lux; 1500 lux; and 3000 lux. The steps being large enough to produce noticeable differences in the subjective impression of the illuminance. The Code contains a flow chart (Fig. 8.11) to use to amend the standard service illuminance to obtain a final design value.

A – the area of the working plane is the room area in square metres.

Φ – manufacturers quote values for the lighting design lumens of electric lamps in their handbooks.

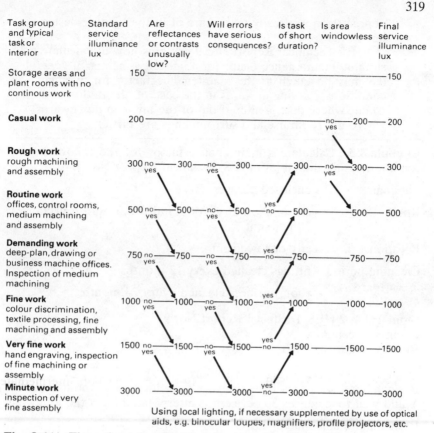

Task group and typical task or interior	Standard service illuminance lux	Are reflectances or contrasts unusually low?	Will errors have serious consequences?	Is task of short duration?	Is area windowless	Final service illuminance lux
Storage areas and plant rooms with no continous work	150					150
Casual work	200				no / yes — 200	200
Rough work rough machining and assembly	300 no / yes	300 no / yes	300	300 no / yes	300	300
Routine work offices, control rooms, medium machining and assembly	500 no / yes	500 no / yes	500 yes / no	500	500	500
Demanding work deep-plan, drawing or business machine offices. Inspection of medium machining	750 no / yes	750 no / yes	750 yes / no	750	750	750
Fine work colour discrimination, textile processing, fine machining and assembly	1000 no / yes	1000 no / yes	1000 yes / no	1000	1000	1000
Very fine work hand engraving, inspection of fine machining or assembly	1500 no / yes	1500 no / yes	1500 yes / no	1500	1500	1500
Minute work inspection of very fine assembly	3000	3000	3000 yes / no	3000	3000	3000

Using local lighting, if necessary supplemented by use of optical aids, e.g. binocular loupes, magnifiers, profile projectors, etc.

Fig. 8.11   Flow chart for deciding final service illuminance

Some typical values of lumen output are:

Lamp	Lighting design (lumens)	
Tungsten GLS 100 W	1200	
Fluorescent tubes:		
1500 mm 65 W White	4750	lighting design lumens at 2000 hours
1500 mm 65 W Natural	3400	
1500 mm 65 W Warm White	4600	

MF – the maintenance factor is a coefficient less than 1.0 to allow for the average light reduction due to dirt accumulation on the luminaire and deterioration of the reflectances of interior surfaces during a maintenance period. (Note that lamp depreciation is already taken care of in the lighting design lumen figure.)

IES Technical Report No. 9 explains how to calculate a suitable maintenance factor for a lighting scheme. In general 0.8 may be taken as the maintenance factor for average conditions, varying upwards in the case of air conditioned rooms where dust is filtered out of the air, and downwards where the room use and surroundings are dirtier.

Example 8.3 Calculate a maintenance factor for the following conditions using the tables in IES Technical Report No. 9.

Luminaire — enclosed diffuser, BZ 6

Room — clean laboratory, room index 2.0, wall reflectances 30 per cent

Location — city or town centre

Cleaning period — fittings cleaned every 12 months

— room surfaces cleaned every 24 months

From Table 2 (IES Technical Report No. 9)
Luminaire category = C/D
Room category   = Y

From Table 4.6 (IES Technical Report No. 9)
Basic maintenance factor for category C/D luminaire, room category Y and wall reflectances 30 per cent is interpolated from the following:

Basic maintenance factor for C   = 0.75
Basic maintenance factor for D   = 0.69
∴   Basic maintenance factor for C/D = 0.72

From Table 3 (IES Technical Report No. 9)

$k = 1.02$ for a BZ6 luminaire, room index 2, and wall reflectance $\rho_w = 30$ per cent

Maintenance factor = basic maintenance factor $\times k$
$$= 0.72 \times 1.02$$
$$= 0.73$$

UF – the utilisation factor is a coefficient less than 1.0 which takes into account both the fall off of illuminance with distance due to the inverse square law and the lighting bonus derived from inter-reflection for the whole scheme.

$$\text{Utilisation factor} = \frac{\text{total flux reaching the working plane}}{\text{total lamp flux}}$$

The main points which affect the value of the utilisation factor are as follows:

(a) light output ratio of the luminaire;
(b) the flux distribution of the luminaire;
(c) room proportions;
(d) room reflectances;
(e) suspension;
(f) spacing/mounting height ratio.

(a) *The light output ratio of the luminaire*

The amount of light emitted by the source is reduced by the luminaire because light energy is lost both inside and by transmission through the fitting. The light output ratio (LOR) of a luminaire is the percentage of light emitted by a luminaire compared with the flux produced by the source(s) inside.

The light output ratio is further subdivided into light available in the upper hemisphere (above a horizontal plane through the fitting) compared with the total source flux – the upward light output ratio (ULOR) and its downward counterpart, the downward light output ratio (DLOR).

$$ULOR + DLOR = LOR$$

A luminaire with the following specification:

Upward light output ratio (ULOR)   = 20 per cent
Downward light output ratio (DLOR) = $\underline{50}$ per cent
∴   the light output ratio (LOR)   = $\underline{70}$ per cent

Of the total lamp lumens available inside the fitting 20 per cent are radiated upwards, 50 per cent downwards and 30 per cent are absorbed as heat in the luminaire.

The upper and lower flux fractions (UFF and LFF) for the luminaire are derived from the light output ratios. The flux fractions being the percentage of light coming out upwards or downwards compared with the total emitted, e.g.

Light output ratios   Flux fractions

DLOR = 50 per cent
ULOR = $\underline{20}$ per cent
  LOR = $\underline{70}$ per cent

$$LFF = \frac{50}{70} = 71 \text{ per cent}$$

$$UFF = \frac{20}{70} = 29 \text{ per cent}$$

$$\text{Flux fraction ratio} = \left(\frac{29}{71}\right)$$

$$= 0.41$$

Lower flux fraction (LFF)	Upper flux fraction (UFF)	Distribution	Classification	Examples
>90%	<10%		direct	Luminous ceilings Recessed luminaires Downlighters
90%–60%	10%–40%		semi-direct	Bare tubes Opal diffuser fluorescent tube luminaire Prismatic controller
60%–40%	40%–60%		general diffusing	Opal sphere
40%–10%	60%–90%		semi-indirect	Uplighters with translucent base Portable drum shade luminaires
<10%	>90%		indirect	Uplighters, cornice lighting

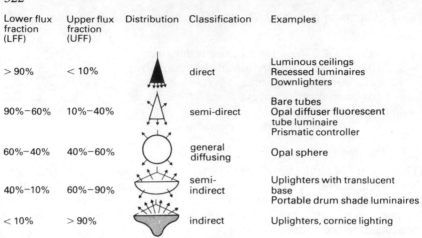

Fig. 8.12 Luminaire classification

The flux fractions give a basis for luminaire classification as shown in Fig. 8.12. Figure 8.13(a) and (b) gives examples of light output ratios for direct and semi-direct luminaires.

(b) *Flux distribution of the luminaire*
Data tables can be simplified if:

- (i) the upward light is all assumed to fall on the ceiling so that it is only the quantity of upward light and not its distribution which is important;
- (ii) the downward light distribution can be based on a set of standard light distributions.

The intensity distribution pattern of the light radiated from a luminaire in the lower hemisphere will affect how much of the downward flux falls directly on the working plane and how much will be available for reflection from the walls in a given room. It affects the direct ratio which is the ratio of flux received directly on the working plane to the total downward flux (see Fig. 8.16). The downward light distribution characteristics of luminaires can be defined by the British Zonal classification.

In the BZ classification 10 theoretical lower hemisphere intensity distributions (Fig. 8.14) are used to evaluate the direct ratio zone limits that would be produced in rooms of different size (see room index, page 326) with appropriate spacing to mounting height ratios (see page 328) which would give a uniformity ratio (E_{min}/E_{av}) equal or greater than 0.8. Zone limits for variation of direct ratio with BZ number and room index are shown in Fig. 8.15.

A luminaire is classified as a certain BZ number if it produces a variation of direct ratio with room index for a given spacing to

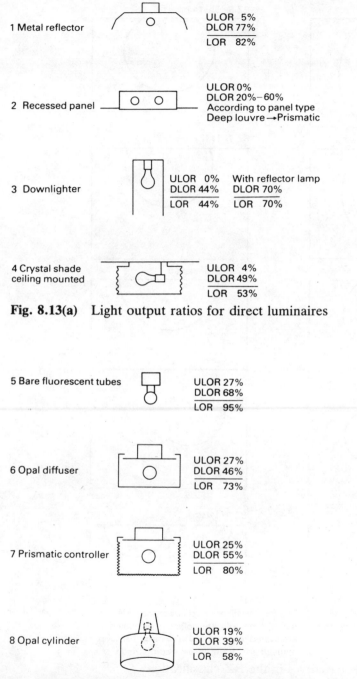

1 Metal reflector

ULOR 5%
DLOR 77%
LOR 82%

2 Recessed panel

ULOR 0%
DLOR 20%–60%
According to panel type
Deep louvre →Prismatic

3 Downlighter

ULOR 0%   With reflector lamp
DLOR 44%   DLOR 70%
LOR   44%   LOR 70%

4 Crystal shade
ceiling mounted

ULOR 4%
DLOR 49%
LOR 53%

Fig. 8.13(a) Light output ratios for direct luminaires

5 Bare fluorescent tubes

ULOR 27%
DLOR 68%
LOR 95%

6 Opal diffuser

ULOR 27%
DLOR 46%
LOR 73%

7 Prismatic controller

ULOR 25%
DLOR 55%
LOR 80%

8 Opal cylinder

ULOR 19%
DLOR 39%
LOR 58%

Fig. 8.13(b) Light output ratios for semi-direct luminaires

324

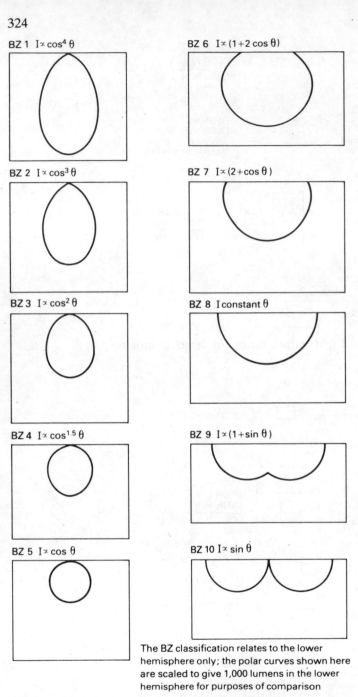

BZ 1 $I \propto \cos^4 \theta$

BZ 2 $I \propto \cos^3 \theta$

BZ 3 $I \propto \cos^2 \theta$

BZ 4 $I \propto \cos^{1.5} \theta$

BZ 5 $I \propto \cos \theta$

BZ 6 $I \propto (1 + 2\cos \theta)$

BZ 7 $I \propto (2 + \cos \theta)$

BZ 8 I constant θ

BZ 9 $I \propto (1 + \sin \theta)$

BZ 10 $I \propto \sin \theta$

The BZ classification relates to the lower hemisphere only; the polar curves shown here are scaled to give 1,000 lumens in the lower hemisphere for purposes of comparison

Fig. 8.14 Polar curves in the BZ classification

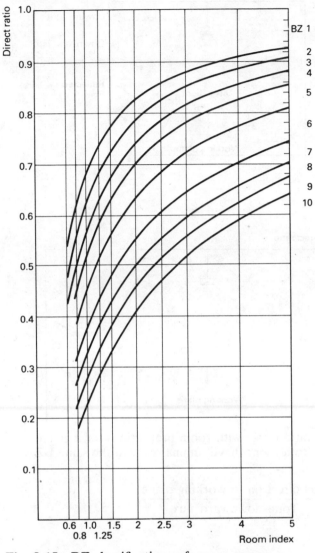

Fig. 8.15 BZ classification reference curves

mounting height ratio within a zone defined by one of the 10 theoretical intensity distributions. (See IES Technical Report No. 2.)

The lower the BZ number of a fitting the narrower the distribution of light from it. The higher the BZ number the greater the proportion of sideways flux.

The BZ classification not only serves to aid the calculation of utilisation factors but also the computation of discomfort glare index.

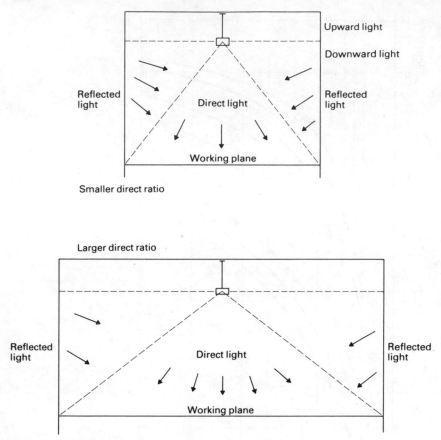

Fig. 8.16 Direct ratio varies with room proportions for a given BZ classification. Both rooms have luminaires with the same BZ classification.

$$\text{Direct ratio} = \frac{\text{Flux direct on to working plane}}{\text{Total downward flux}}$$

(c) *Room proportion*

How much light is received directly by the working plane and how much arrives by reflection from the walls for a particular BZ number luminaire will depend on the room proportions (Fig. 8.16).

The room proportions are described by a room index (RI). The room index is the ratio of the areas of two bounding horizontal planes, one through the luminaire and the other the working plane, to the area of the walls between them.

In Fig. 8.17 the area of the two horizontal planes is $2 \times l \times w$ and

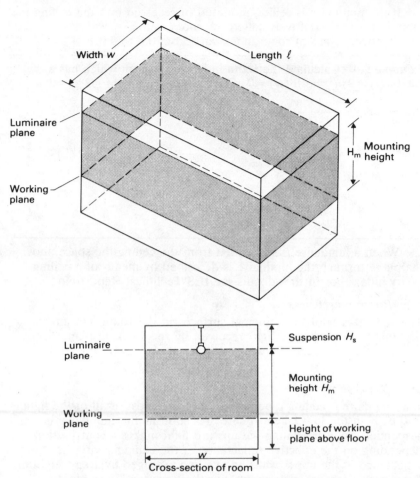

Width w

Length ℓ

Luminaire plane

H_m Mounting height

Working plane

Luminaire plane

Suspension H_s

Mounting height H_m

Working plane

Height of working plane above floor

w

Cross-section of room

Fig. 8.17 Room proportions for room index calculation

the vertical area is $H_m \times 2(l + w)$ where:

$$l = \text{room length}$$

$$w = \text{room width}$$

$$H_m = \text{the mounting height of the luminaire above the working plane}$$

The room index (RI) $= \dfrac{2 \times l \times w}{H_m \times 2(l + w)}$

$$\therefore \quad \text{RI} = \dfrac{l \times w}{H_m(l + w)}$$

(k_r was also used as a symbol for room index).

If the luminaire is ceiling mounted or recessed into the ceiling the room index is alternatively called the room ratio.

Practical values of room index vary from 0.6 to 5.0.

Example 8.4 Calculate the room index for a room which has a length $(l) = 9\,\text{m}$, width $(w) = 7\,\text{m}$ and mounting height $(H_m) = 2\,\text{m}$.

$$\text{Room index} = \frac{l \times w}{H_m(l + w)}$$

$$\therefore \quad \text{RI} = \frac{9 \times 7}{2(9 + 7)}$$

$$= \frac{63}{32}$$

$$\simeq 2$$

When a luminaire is suspended from the ceiling the space above the plane through the luminaire is described by means of a ceiling cavity index. For further details see IES Technical Report No. 2.

(d) *Room reflectances*

The reflectances of the principal surfaces in the room will affect the quantity of reflected light received by the working plane. The important surfaces are the walls, ceiling, effective working plane or floor.

(e) *Suspension* (H_s)

If a luminaire is ceiling mounted or recessed most or all of the light output is in the lower hemisphere. As soon as the luminaire is suspended from the ceiling the upward light makes a contribution depending on the effective reflectance of the ceiling cavity; i.e. the reflectance of the upper walls and ceiling averaged by area weighting. Some manufacturers' data sheets give three suspension conditions expressed as a suspension (H_s) to mounting height (H_m) ratio (Fig. 8.17).

$\dfrac{H_s}{H_m} = 0$ for ceiling mounted luminaire $(H_s = 0)$

$\dfrac{H_s}{H_m} = \begin{array}{l} 0.3 \text{ if the luminaire is suspended a quarter of the way down} \\ \text{from the ceiling to the working plane} \\ \quad H_s = 1x \ \ H_m = 3x \end{array}$

$\dfrac{H_s}{H_m} = \begin{array}{l} 1.0 \text{ when the luminaire is half way between the ceiling and} \\ \text{working plane} \\ \quad H_s = 1x \ \ H_m = 1x \end{array}$

(f) *Spacing/mounting height ratio* (S/H_m)

The illuminance produced on the working plane under a regular array of luminaires varies with position in a room. Higher values of illuminance occur directly beneath luminaires and lower levels between rows of luminaires.

The ratio of the minimum illuminance (E_{min}) to the average illuminance (E_{av}) is called the uniformity ratio for the installation, (the measurement of average illuminance is given in Appendix 2 of the IES Code). It has been shown that people prefer a uniformity ratio of 0.8 or greater for the type of work interiors for which the lumen design method is suitable. (The uniformity ratio was formerly defined as the ratio of minimum (E_{min}) to maximum (E_{max}) illuminance and the criterion 0.7 or greater was used.)

For a given BZ distribution and room index the uniformity ratio will depend on the spacing of the luminaires and their height above

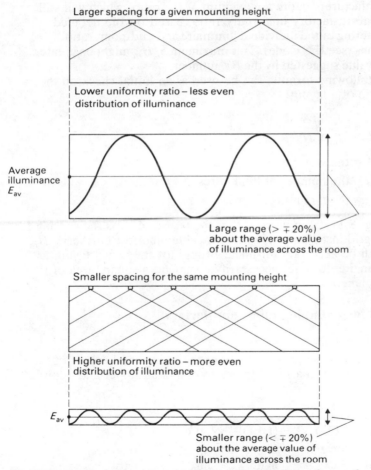

Fig. 8.18 Influence of spacing/mounting height ratio on the distribution of illuminance across the working plane

the working plane. The closer spaced the luminaires for a given mounting height the higher the uniformity ratio (Fig. 8.18), or the greater the mounting height for a given spacing the greater the uniformity ratio.

In order that the minimum uniformity ratio criterion is obtained limits must be set on the spacing (S)/mounting height (H_m) ratio.

For BZ classified luminaires these are:

Maximum S/H_m

BZ 1–3	1:1
BZ 4	1.25:1
BZ 5–10	1.5:1

Manufacturers quote a maximum S/H_m ratio, which will still produce a satisfactory uniformity ratio, based on the detailed photometering of the individual luminaire and mid-point ratio calculations (see IES Code). This maximum S/H_m might be greater than the value suggested by the BZ number.

The following formula can be used as an initial check on the S/H_m ratio of a layout:

$$S/H_m = \frac{1}{H_m} \sqrt{\frac{A}{N}}$$

where A = the working plane area

N = the number of luminaires

H_m = the mounting height

In Fig. 8.19 it can be seen that the term $\sqrt{A/N}$ is the spacing for a square grid installation of symmetrical luminaires. Further S/H_m checks will be necessary for linear fittings for the actual layout as each luminaire does not necessarily command a square area of the working plane.

Methods for obtaining utilisation factors

IES Technical Report no. 2
Basic photometric information about the fittings and details of the room are applied to a set of tables to obtain an upper and a lower flux utilance.

Then:

(DLOR × LFU)	+ (ULOR × UFU)	= UF
downward × lower light flux output utilance ratio	+ upper × upper light flux output utilance ratio	= utilisation factor

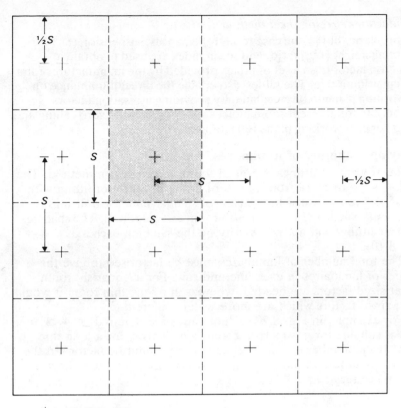

+ Luminaire positions

Shaded area $= \dfrac{A}{N} = \dfrac{\text{room area}}{\text{number of luminaires}}$

Spacing, $S = $ side of square $= \sqrt{\dfrac{A}{N}}$

Fig. 8.19

Lower flux – the proportion of the downward flux from the lumin-
utilance aires which reaches the working plane, some directly
and some after inter-reflection.

Upper flux – the proportion of the total upward flux from the lumin-
utilance aires which reaches the working plane after inter-
reflection.

IES Code 1977

The Code gives a method for obtaining utilisation factors based
on flux transfer factors. A full treatment is now available in CIBS
Technical Memorandum No. 5 'Calculation and Use of Utilisation
Factors', which supersedes IES Technical Report No. 2.

Manufacturer's recipe sheet method

Practice varies but in one case room reflectances, suspension (H_s) to mounting height (H_m) ratio, and room index are used to obtain a utilisation factor from a set of tables provided by the manufacturer for a particular luminaire. The tables also enable the direct illuminance on the working plane to be calculated by providing utilisation factors appropriate for different room indices against zero wall, zero ceiling and 0–100 per cent working plane reflectances.

Arraying a number of luminaires

A regular array of fittings is central to the lumen design method. The spacing is the distance from centre of fitting to centre of fitting with half spaces to the walls. In the case of small symmetrical fittings the spacing will be the same on each room axis (Fig. 8.9). The spacing of linear fittings will not necessarily be the same on each axis (Fig. 8.10).

The total number of luminaires must be factorised to give the number of luminaires in each line and row. For a 'squarish' room similar sized factors are needed, whereas in a long thin room it would be a pair of factors which are more widely different.

For example, in Fig. 8.20 12 luminaires are required. A 4×3 array is suitable for a 'squarish' room, a 6×2 array for a long thin room. If the number of luminaires required cannot be factorised the number should be varied until an array is possible and the effect on illuminance recalculated.

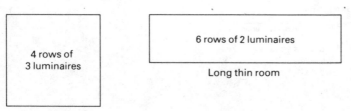

Fig. 8.20 Twelve luminaire array – influence of room shape

The physical length of linear fittings in a row must also be checked to see if they will fit into the room size.

Alternatively, a layout may be determined first and a lamp and luminaire sought out which will give the required illuminance and uniformity ratio.

Example 8.5 A room measures 20 m × 9 m and the light fittings are mounted in the ceiling 2.5 m above the working plane. The required illuminance is 300 lux with a maintenance factor of 0.8.

Calculate:
(a)   the room index;
(b)   the utilisation factor;
(c)   the number of light fittings;
(d)   whether the spacing/mounting height ratio is acceptable.

Data:
BZ 4 luminaire max $S/H_m = 1.25:1$
Lighting design lumens of lamps 6134 lumens – 1 lamp per luminaire.

Lower flux utilance (LFU)	0.88
Upper flux utilance (UFU)	0.42
Downward light output ratio (DLOR)	50 per cent
Upper light output ratio (ULOR)	20 per cent

(a)   Room index (RI) $= \dfrac{l \times w}{H_m(l + w)}$

$$= \frac{20 \times 9}{2.5(20 + 9)}$$

$$= \underline{2.48}$$

(b)   Utilisation factor (UF)

$$= (DLOR \times LFU) + (ULOR \times UFU)$$
$$= (0.5 \times 0.88) + (0.2 \times 0.42)$$
$$= \underline{0.524}$$

(c)   Number of fittings $(n) = \dfrac{E_{av} \times A}{\Phi \times UF \times MF}$

$$= \frac{300 \times 20 \times 9}{6134 \times 0.524 \times 0.8}$$

$$= \underline{21} \text{ lamps and luminaires}$$

(d)   Spacing/mounting height ratio

$$S/H_m = \frac{1}{H_m}\sqrt{\frac{A}{N}}$$

$$= \frac{1}{2.5}\sqrt{\left(\frac{20 \times 9}{21}\right)}$$

$$= \underline{1.17:1} < 1.25:1 \text{ checks as shown i.e. } S/H_m \text{ ratio is}$$
acceptable.

(For BZ 4 luminaires the maximum $S/H_m = 1.25:1$)

Example 8.6 (Using a manufacturer's data sheet) Design a lighting installation for a college seminar room so that the average illuminance is 500 lux on the horizontal working plane, using the equipment listed below. Sketch the layout and give appropriate spacing to mounting height checks if the luminaires are mounted crosswise to the long axis of the room.

Data:

Room dimensions: 12 m long × 8 m wide × 3.20 m high
Working plane at 0.850 m above floor.

Reflection factors: Ceiling   70 per cent
Walls   50 per cent
Working plane 20 per cent

Maintenance factor: 0.74

Luminaires: 1800 mm twin tube with opal diffuser (photometric data sheet Table 8.3)
Ceiling mounted
Downward light output ratio 44 per cent
Maximum S/H_m 1.73:1
Dimensions: 1812 mm long × 200 mm wide

Lamps: 1800 mm 75 W Plus White; 5500 lighting design lumens per lamp.
Two lamps per luminaire

(a) Calculate the room index

$$\text{Room index, RI} = \frac{l \times w}{H_m(l + w)}$$

$$= \frac{12 \times 8}{2.35(12 + 8)}$$

$$\therefore \quad \text{RI} = 2.04$$

(b) Calculate $H_s/H_m = \dfrac{0}{2.35} = 0$

($H_s = 0$, ceiling mounted fitting)

(c) Select a utilisation factor from the manufacturer's photometric data sheet

Utilisation factor = 0.59
(or 0.5932 by interpolation for room index = 2.04)

(d) Using the lumen design formula:

$$\text{No. of fittings, } n = \frac{E_{av} \times A}{\Phi \times UF \times MF}$$

$$\therefore \quad n = \frac{500 \times 12 \times 8}{5500 \times 0.59 \times 0.74}$$

$$= 19.989$$

$$= 20 \text{ lamps}$$

Two lamps per luminaire gives 10 luminaires.

(e) Initial check on S/H_m ratio

$$S/H_m = \frac{1}{H_m}\sqrt{\frac{A}{N}}$$

$$\therefore \quad S/H_m = \frac{1}{2.35}\sqrt{\left(\frac{12 \times 8}{10}\right)}$$

$$= 1.32:1$$

Maximum $S/H_m = 1.73:1$ (recommended by manufacturer)

Therefore it should be possible to use 10 luminaires.

(f) Trial layout:

Room shape	$12\,m \times 8\,m$
Factors of 10	10×1, 5×2
Try 5×2 layout	(Fig. 8.21)

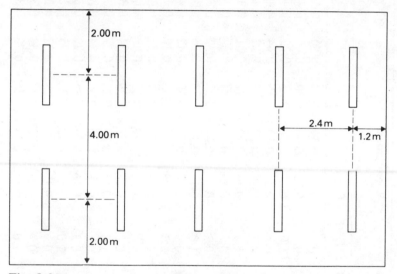

Fig. 8.21

(g) It is important now to check that the luminaires fit in the space:

$2 \times 1.812\,m = 3.624 < 8\,m$ (width of room)

Therefore they will fit.

Usually the only need is to check the linear dimension of the fitting for space. The other dimension is small (200 mm).

(h) S/H_m checks – NB for spaces and half-spaces (ends) the total number of spaces = the number of luminaires in a row.

Table 8.3 Typical manufacturer's data for an 1800 mm twin tube fluorescent luminaire with opal diffuser pre Technical Memorandum N:5

Utilisation factors

Room reflectances

Work plane	Ceiling	Wall	H_s/H_m	Room index									
				0.6	0.8	1.0	1.25	1.5	2.0	2.5	3.0	4.0	5.0
20%	70%	50%	0	0.31	0.39	0.44	0.50	0.54	0.59	0.63	0.66	0.70	0.72
			0.3	0.25	0.33	0.39	0.44	0.48	0.54	0.59	0.62	0.67	0.70
			1.0	0.21	0.28	0.33	0.37	0.41	0.48	0.52	0.56	0.61	0.65
		30%	0	0.25	0.33	0.38	0.44	0.48	0.54	0.58	0.62	0.66	0.69
			0.3	0.19	0.27	0.32	0.37	0.41	0.48	0.53	0.57	0.62	0.65
			1.0	0.16	0.22	0.27	0.31	0.35	0.41	0.45	0.49	0.55	0.59
		10%	0	0.20	0.28	0.34	0.39	0.43	0.50	0.54	0.58	0.63	0.66
			0.3	0.15	0.22	0.27	0.32	0.36	0.43	0.48	0.52	0.58	0.62
			1.0	0.13	0.19	0.23	0.26	0.30	0.35	0.40	0.44	0.50	0.54
	50%	50%	0	0.26	0.34	0.38	0.42	0.46	0.50	0.53	0.56	0.59	0.61
			0.3	0.25	0.30	0.35	0.39	0.43	0.48	0.51	0.54	0.57	0.59
			1.0	0.20	0.27	0.31	0.35	0.38	0.43	0.47	0.50	0.54	0.57
		30%	0	0.21	0.29	0.33	0.38	0.41	0.46	0.50	0.53	0.56	0.58
			0.3	0.18	0.25	0.29	0.34	0.37	0.42	0.46	0.49	0.53	0.56
			1.0	0.16	0.22	0.26	0.29	0.33	0.38	0.41	0.43	0.49	0.52
		10%	0	0.18	0.25	0.30	0.34	0.38	0.43	0.47	0.50	0.54	0.56
			0.3	0.16	0.21	0.25	0.29	0.33	0.38	0.42	0.46	0.50	0.53
			1.0	0.13	0.19	0.22	0.25	0.28	0.33	0.37	0.40	0.45	0.48
	30%	30%	0	0.19	0.25	0.29	0.32	0.35	0.39	0.42	0.44	0.47	0.49
			0.3	0.17	0.23	0.27	0.30	0.33	0.38	0.41	0.43	0.46	0.48
			1.0	0.16	0.21	0.25	0.28	0.31	0.35	0.38	0.41	0.44	0.46

		0.16	0.22	0.26	0.30	0.32	0.37	0.40	0.42	0.45	0.48
10%		0.14	0.20	0.23	0.27	0.30	0.34	0.37	0.40	0.43	0.46
	0.3	0.13	0.18	0.21	0.25	0.27	0.31	0.34	0.37	0.40	0.43
	1.0	0.16	0.17	0.19	0.22	0.24	0.27	0.29	0.31	0.33	0.35
0%		6	6	6	6	6	6	6	6	6	6

0–100%   0%
BZ classification

Recommended maximum S/H_m ratio = 1.73 : 1   ULOR 40%   DLOR 44%   LOR 84%

(i) Long axis check

$$S = 12/5$$
$$= 2.4 \, \text{m}$$

$$S/H_m = \frac{2.4}{2.35}$$
$$= 1.02 : 1 < 1.73 : 1 \text{ checks}$$

(ii) Short axis check (in the direction of the length of the luminaire)

$$S = 8/2$$
$$= 4 \, \text{m}$$

$$S/H_m = \frac{4.0}{2.35}$$
$$= 1.70 : 1 < 1.73 : 1 \text{ checks}$$

If the checks had not worked out it would be necessary to reconsider the number of luminaires used and to recalculate the illuminance produced.

For example, a 3×3 array would lower the lux level or
a 4×3 array would increase the lux level.

Permanent supplementary artificial lighting of interiors (PSALI)

Historically the day-time lighting of buildings using natural lighting penetrating through windows and/or roof lights has principally concerned the architect. The use of artificial lighting for illuminating interiors at night became the province of electrical engineers and later lighting engineers. The two areas of design remained separate for many years, the artificial lighting often being designed after the major architectural decisions had been made.

More recently it became obvious that the electric lights designed primarily for night-time conditions were also being used regularly during the day even in some south-facing rooms and in June when daylight and sunlight would be expected to provide more than enough illumination.

Referring to the chapter on daylighting it can be appreciated that even with large windows the fall-off in daylight factor across a room is rapid. This effect is increased as the percentage fenestration is reduced. The eye adapts to the brightest part of the visual field and even though at the back of the room remote from the windows a physical measurement might indicate a reasonable illuminance, the eye at a high adaptation level conditioned by the areas near the windows registers the area as relatively gloomy.

The object of PSALI schemes is to integrate artificial lighting with daylight in order to compensate for the relative lack of daylight in those parts of a room which are remote from windows. At night the scheme should either be capable of alteration by switching to provide night-time illuminance levels or be switched off in favour of a separate scheme altogether. The intention of a PSALI scheme is indicated in Fig. 8.22. Daylight provides the major contribution to illuminance near the windows with artificial lighting at the rear of the room.

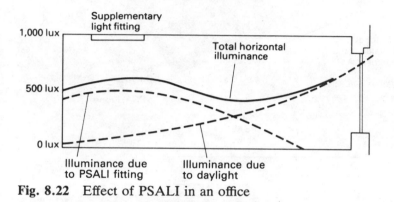

Fig. 8.22   Effect of PSALI in an office

PSALI schemes range from a simple system with just a single supplementary luminaire through to highly sophisticated schemes in which the artificial lighting is graded in light output across the ceiling towards the windows and the lamps are on dimmers so that the illuminance produced can be controlled by photocells on the roof Solar Activated control (SAC's). Artificial lighting near the windows is stepped up as daylight fades.

Deep rooms with large areas remote from windows need a permanent artificial lighting (PAL) scheme along the lines of a night-time installation.

In rooms less than 15 m deep the two questions that have to be answered are: What area of the room should be regarded as a PSALI zone? What illuminance level should be provided in it? It should also be borne in mind that quite high levels of artificial light would be needed to match the daylight near the window and the colour rendering properties of the lamps need to be good to blend with daylight.

The guidance given in IES Technical Report No. 4, 'Daytime Lighting in Buildings', is as follows:

1. PSALI Zone – Referring to Fig. 8.23 the PSALI zone starts where the daylight factor has fallen to 10 per cent of that at a

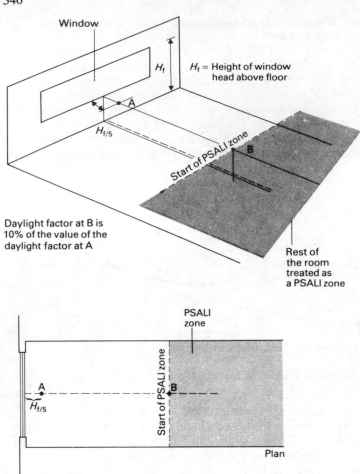

Window

H_f

H_f = Height of window
head above floor

A

$H_{f/5}$

Start of PSALI zone

B

Daylight factor at B is
10% of the value of the
daylight factor at A

Rest of
the room
treated as
a PSALI zone

PSALI
zone

Start of PSALI zone

A

$H_{f/5}$

B

Plan

Fig. 8.23 PSALI zone

distance $H_f/5$ from the window where H_f is the height of the
window head above the floor.

2. Illuminance level (E) to be provided by artificial lighting in the
   PSALI zone is based on an empirical formula:

$$E = 500\,D \text{ lux}$$

where D = the average daylight factor in the PSALI zone.

Daylight factors at working plane height are obtained for a room by
one of the various methods and plotted on a vertical section of the
room. The position of the start of the PSALI zone and the illuminance
required from the supplementary luminaires can then be calculated and
a lumen design method approach used to calculate the scheme.

Experiments

Experiment 8.1   Determination of the reflectance of a surface (Method 1)

Apparatus
EEL galvanometer and reflectometer head, standard magnesium carbonate block (98.3 per cent reflectance).

Method
The galvanometer is switched on and zeroed by means of the zero adjustment knob. The reflectometer is then plugged into the galvanometer and with its filter set to neutral (grey spot) placed on top of the magnesium carbonate block. (It is important that the block is clean, if not it should be scraped clean before use.) Using the sensitivity control the galvanometer is set to 98.3. The reflectometer is then placed on the surface whose reflectance is desired and the value obtained directly from the galvanometer.

Results

Surface	Reflectance (%)

Experiment 8.2 Determination of surface reflectance (Method 2)

Apparatus

Hagner meter or similar, magnesium carbonate block.

Method

The meter is set to measure luminance in candelas per square metre using 1° acceptance angle to the internal photocell. The luminance of the surface whose reflectance is required is then measured. The magnesium carbonate block is then placed over the surface and its luminance read.

Reflectance is then calculated from:

reflectance

$$= \frac{\text{luminance of surface}}{\text{luminance of magnesium carbonate block at the same point}}$$

$\times 98.3$ per cent

Results

Surface	Luminance of surface (cd m^{-2})	Luminance of magnesium carbonate block (cd m^{-2})	Reflectance $(\%)$

Experiment 8.3   Luminance survey of room surfaces

Apparatus

Hagner meter or similar.

Method

A list is made of all the major surfaces within the room such as each
wall, floor, ceiling, cupboards, doors, radiators, blackboard, etc. The
meter is switched on to read the luminance or brightness in candelas
per square metre using the internal photocell. The particular surface
is sighted through the viewfinder and the average luminance read
from the scale seen in the viewfinder. This is repeated for each
surface and the values noted.

Results

Surface	Luminance (cd m^{-2})

Experiment 8.4   Gray's light box apparatus to study some interior illumination parameters in model rooms

Apparatus

Gray's light box with polar curve apparatus and integrating box, Hagner meter or other means of measuring illuminance, variable resistance, voltmeter, low-voltage supply, model light fittings.

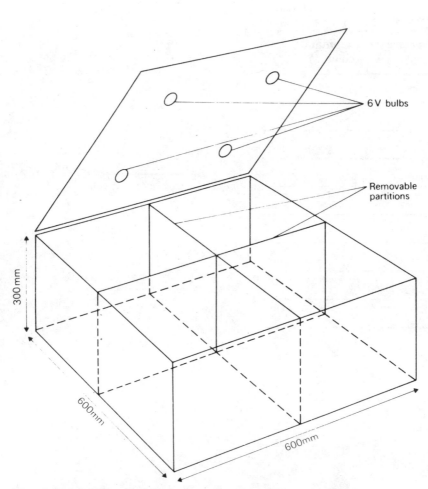

Fig. 8.24   (a) Gray's light box

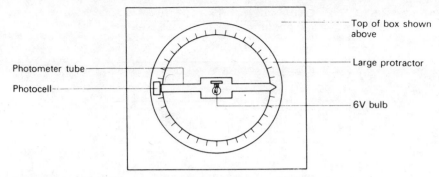

Fig. 8.24 **(b)** Polar curve apparatus on top of light box

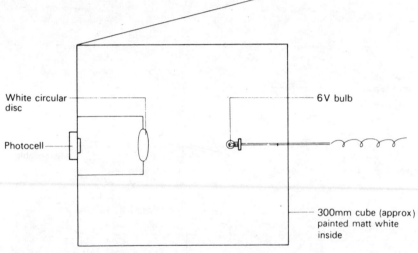

Fig. 8.24 **(c)** Integrating cube

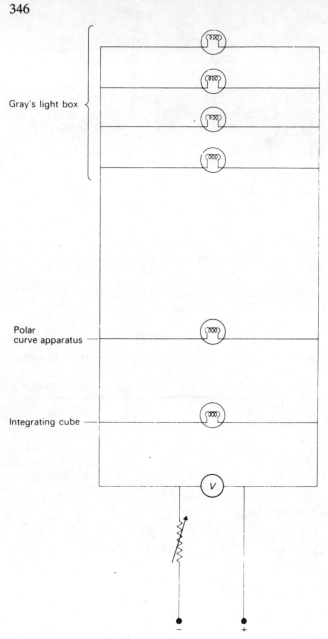

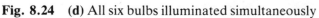

Fig. 8.24 **(d)** All six bulbs illuminated simultaneously

Experiment 8.4(a) Plotting a polar curve using a model lamp fitting

Method

The apparatus is set up as shown in Fig. 8.24 (b), (c) and (d). All six 6 V bulbs are connected in parallel to the low-voltage supply as shown in Fig. 8.24 (d). It is important that the voltage is kept constant throughout to avoid variation in light output. It is for this reason that the four bulbs in the light box, the one on the photo-meter and the one for the integrating cube are on simultaneously.

An external cell from the meter is connected to the end of the photometer tube shown in Fig. 8.24 (b). Readings of illuminance are taken at 10° intervals around the circle. The intensity (in candelas) at each interval is calculated by multiplying each illuminance value by d^2, where d is the distance in metres from the centre of the lamp to the photocell. The intensities may then be plotted on polar graph paper.

It is then possible to use the procedures outlined in IES Technical Report No. 2 to obtain the BZ number for the fitting.

Results: $d = m$

Angle	Illum. (lux)	Intensity cd (lux × d²)	Angle	Illum. (lux)	Intensity cd (lux × d²)	Angle	Illum. (lux)	Intensity cd (lux × d²)
0°			130°			260°		
10°			140°			270°		
20°			150°			280°		
30°			160°			290°		
40°			170°			300°		
50°			180°			310°		
60°			190°			320°		
70°			200°			330°		
80°			210°			340°		
90°			220°			350°		
100°			230°			360°		
110°			240°					
120°			250°					

348

Experiment 8.4(b) Determination of the light output ratio (LOR) of a fitting

Method

Suitable model fittings are made to fit on to the lamp in the integrating cube as shown in Fig. 8.24(c). A table-tennis ball is ideal as an opal diffusing sphere (approx. BZ 8 distribution). Other types are fairly easily modelled. The integrating cube may be made out of a large 'biscuit' tin or in wood, about 300 mm cube. The lamp is fixed on a small rod through one side with the photocell on the other. As it is important not to measure direct light, a white circular disc is mounted some 75 mm in front of the cell.

The illuminance may then be read directly on the meter with the fitting on the lamp (E_1). This is then repeated without the fitting (E_2). The experiment is repeated for different fittings.

Theory

The light output ratio is calculated by dividing the illuminance value with the fitting by the illuminance value without the fitting, or

$$LOR = \frac{\text{illuminance with fitting}}{\text{illuminance without fitting}} = \frac{E_1}{E_2}$$

Results

Type of fitting	Illuminance with fitting(E_1) (lux)	Illuminance without fitting(E_2) (lux)	LOR $= \dfrac{E_1}{E_2}$

Experiment 8.4(c) Measurement of illuminance for rooms with different room indices

Method

With all the removable partitions in position the room index is calculated for one of the room spaces and the illuminance measured at the appropriate number of positions.

One section of partition is removed producing a model room 600 mm × 300 mm × 300 mm high and the experiment repeated.

All partitions are removed and the experiment repeated. In each case the average illuminance is calculated.

Theory

Illuminance (E) is defined as the luminous flux density at a surface and is measured in units of lux.

$$\text{Room index} = \frac{\text{length} \times \text{width}}{\text{mounting height} \times (\text{length} + \text{width})}$$

The room index is needed to know the minimum number of measuring positions from which average illuminance may be calculated.

Room index	Minimum no. of measuring positions
Less than 1	4
1 to below 2	9
2 to below 3	16
3 or greater	25

These must be arranged on a grid basis. For example,

+	+	+
+	+	+
+	+	+

(*Reference:* Appendix 2 of 1977 IES Code.)

Results

1. Room 300 mm × 300 mm × 300 mm high (approximate dimensions)

    Room index = _____

    = _____

    ∴ no. of measuring points = _____

Measuring point	Illuminance (lux)

    ∴ Average illuminance = _____

2. Room 600 mm × 300 mm × 300 mm high

   Room index = _____

   = ===========

   ∴ no. of measuring points = _____

Measuring point	Illuminance (lux)

   ∴ Average illuminance = ===========

3   Room 600 mm × 600 mm × 300 mm high

Room index = _____

= $\underline{\underline{\qquad\qquad}}$

∴ no. of measuring points = _____

Measuring point	Illuminance (lux)

∴ Average illuminance = $\underline{\underline{\qquad\qquad}}$

Experiment 8.4(d)   To study the effect of spacing/mounting height ratios on illuminance

Method

The apparatus is set up as described in the previous experiments with all six lamps again controlled at 6 V. One partition is removed from the box, producing a room with dimensions approximately 600 mm × 300 mm × 300 mm with two lamps. The photocell is placed on the floor at the centre of one end and the illuminance measured. It is then moved 25 mm along the centre line and the illuminance measured again. This is repeated at 25 mm intervals along the centre line.

When this is completed, a false floor is put in at a height of 100 mm to produce a room approximately 200 mm high. (This may be supported on rods through small holes in the side of the box.) The measurements are then repeated.

This is repeated for the floor at 200 mm (room height 100 mm). The ratio of spacing (S) to mounting height (H_m) is calculated for each position of floor, and graphs of lux vertically to position horizontally are plotted. These graphs are then compared to illustrate the influence of the ratio S/H_m on the uniformity of illuminance.

Theory

$$\text{Uniformity ratio} = \frac{\text{minimum illuminance}}{\text{average illuminance}}$$

both at working plane height and should not be less then 0.8 (for lumen design method).

354

Results

1. *Room height approx. 300 mm*

Distance between lamp centres, $S =$ _____ mm

Height of lamps above photocell, $H_m =$ _____ mm

$$\therefore \frac{S}{H_m} = \underline{\qquad\qquad}$$

Position	Illuminance (lux)

2. *Room height approx.* 200 mm

Distance between lamp centres, $S =$ _____ mm

Height of lamps above photocell, $H_m =$ _____ mm

$$\therefore \quad \frac{S}{H_m} = _____$$

Position	Illuminance (lux)

356

3. *Room height approx.* 100 mm

Distance between lamp centres, $S =$ _____ mm

Height of lamps above photocell, $H_m =$ _____ mm

$$\therefore \quad \frac{S}{H_m} = \underline{\underline{\qquad\qquad}}$$

Position	Illuminance (lux)

Questions

1. The vertically downward intensity from a luminaire is 500 cd. Calculate the illuminance 2.8 m beneath it.
2. Calculate the direct illuminance at two positions A and B 1.5 m apart on a horizontal working plane if A is 2 m vertically beneath a luminaire whose distribution is shown in Fig. 8.25 and is fitted with a 1200 lumen source.
3. Four uniform sources are arranged on a 2.5 m square grid 2 m above a working plane. Calculate the illuminance due to all the sources at a point on the working plane below one of the sources and also at a point on the working plane at the centre of the group.
   The lumen output of each source is 1885 lumens.
4. Define the term utilisation factor.
5. List the points which influence the value of the utilisation factor.

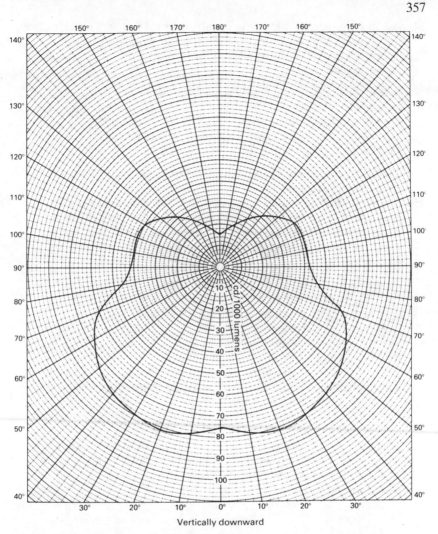

Fig. 8.25 Polar curve for Question 2

6. How do room reflectances affect illuminance value?
7. Explain what is meant by spacing/mounting height ratio.
8. What is a light output ratio as applied to a luminaire?
9. Sketch the intensity distribution curves for a BZ 8 and a BZ 1 luminaire.
10. Give brief details of the British Zonal classification.
11. State the lumen design formula.
12. Describe the objectives of a PSALI scheme.
13. State a method of defining the PSALI zone and a formula for the illuminance required in it.

14. An illuminance of 300 lux is required in a $15\,\text{m} \times 7\,\text{m}$ room from BZ 3 luminaires mounted 3 m above the working plane. The lamp flux is 4000 lighting design lumens and the maintenance factor is 0.85.

Lower flux utilance, LFU   = 0.9
Upper flux utilance, UFU   = 0.45
Downward light output ratio, DLOR = 70 per cent
Upward light output ratio, ULOR   = 0 per cent
Maximum S/H_m for BZ 3   = 1 : 1

Calculate room index, utilisation factor, number of lamps and check the S/H_m ratio.

15. An average illuminance level of 750 lux is specified for a deep plan office. Use the equipment and data listed in Table 8.4 to design a lighting installation, assuming discomfort glare will be within acceptable limits for either orientation of the linear fittings.

Room data: Length 25 m, width 20 m
       Height 3.6 m to false ceiling
       Working plane height 0.85 m
       Reflection factors:
        Ceiling   70 per cent
        Walls   30 per cent
        Working plane 20 per cent
Luminaires: Triple tube air handling luminaires recessed into ceiling system
       Maximum $S/H_m = 1.42 : 1$
       Dimensions: $1800\,\text{mm} \times 300\,\text{mm}$
Lamps: 1800 mm 85 W Plus White fluorescent tubes
       Lighting design lumens – 5850 lumens per tube
       Three tubes per luminaire
Maintenance: 0.875
  factor

Table 8.4 Luminaire photometric data sheet for Question 15

Utilisation factors Room reflectances				Room index									
Working plane	Ceiling	Wall	H_s/H_m	0.6	0.8	1.0	1.25	1.5	2.0	2.5	3.0	4.0	5.0
20%	70%	50%	0	0.22	0.25	0.27	0.29	0.31	0.33	0.34	0.35	0.36	0.37
		30%	0	0.20	0.23	0.25	0.27	0.29	0.31	0.33	0.34	0.35	0.36
		10%	0	0.19	0.21	0.23	0.26	0.27	0.30	0.32	0.33	0.34	0.35
	50%	50%	0	0.22	0.24	0.26	0.29	0.30	0.32	0.33	0.34	0.35	0.36
		30%	0	0.20	0.22	0.25	0.27	0.28	0.30	0.32	0.33	0.34	0.35
		10%	0	0.19	0.21	0.23	0.25	0.27	0.29	0.31	0.32	0.33	0.34
	30%	30%	0	0.20	0.22	0.24	0.26	0.28	0.30	0.31	0.32	0.33	0.34
		10%	0	0.19	0.21	0.23	0.25	0.26	0.29	0.30	0.31	0.32	0.33
0–100%	0%	0%	0	0.18	0.20	0.22	0.24	0.25	0.27	0.29	0.29	0.31	0.31
BZ Classification				—	2	2	2	2	2	2	2	3	3

ULOR 0% DLOR 35% LOR 35%

Chapter 9

Daylighting

Introduction

Natural lighting or daylighting is provided by direct sunlight, light from a clear sky and daylight from an overcast sky. The total hours of availability of daylight at a given latitude are a function of the relative positions of the sun and earth. The variation in daylight hours throughout the year is shown for two places in the UK in Figs. 9.1 and 9.2.

The daylight illuminance varies both during the day and through the year. Figure 9.3 shows the outdoor illuminances that occur during the year based on the averages of past observations. At any particular time there will be variations in illuminance due to local weather conditions. For calculation purposes an overcast sky is taken as representing the worst type of daylighting situation. Other sky conditions would produce better illuminance levels but will not be considered here. An overcast sky is one in which the outline of the sun cannot be seen through the cloud cover. Two mathematical models have been used for the luminance distribution of an overcast sky: uniform overcast sky; and CIE standard overcast sky.

Uniform overcast sky (Fig. 9.4)

The luminance is the same for all parts of the hemispherical sky. If the uniform or average luminance is B cd/m^2 then the horizontal illuminance can be shown to be πB lux. A uniformly overcast sky condition is used in legal arguments about 'rights of light'.

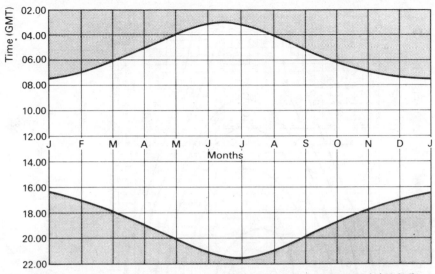

Fig. 9.1 Approximate hours of daylight in southern England (52 °N)

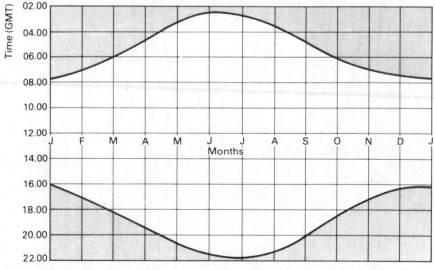

Fig. 9.2 Approximate hours of daylight at the Clyde Valley (56 °N)

CIE standard overcast sky (Fig. 9.5)

The real overcast sky is not of uniform luminance. The luminance
varies with angle above the horizon. The CIE standard overcast sky is
a model which bears close comparison to the luminance distribution
of a real overcast sky.

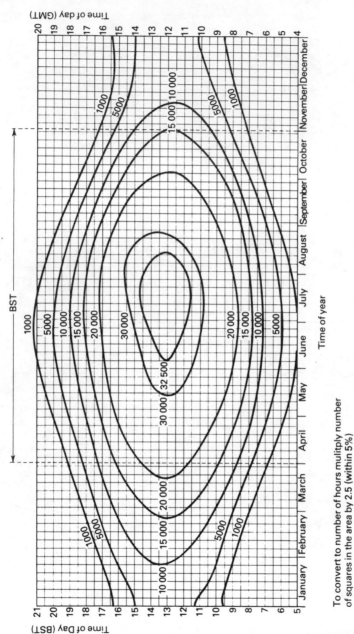

To convert to number of hours mulitply number
of squares in the area by 2.5 (within 5%)

Fig. 9.3 Sky illuminance (lux) at different times of day throughout the year

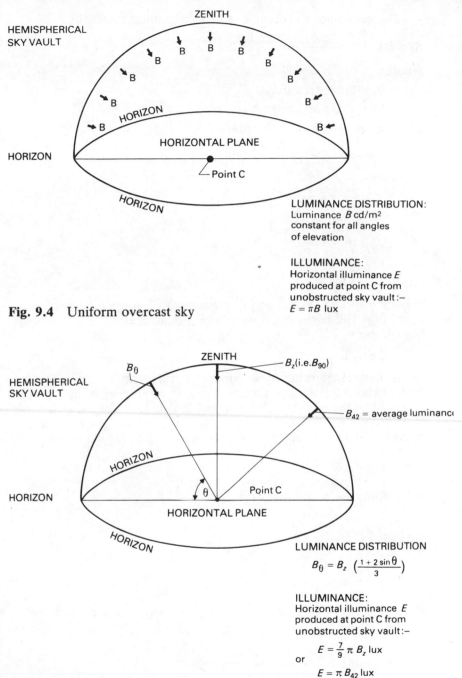

Fig. 9.4 Uniform overcast sky

LUMINANCE DISTRIBUTION:
Luminance B cd/m² constant for all angles of elevation

ILLUMINANCE:
Horizontal illuminance E produced at point C from unobstructed sky vault :–
$E = \pi B$ lux

LUMINANCE DISTRIBUTION

$$B_\theta = B_z \left(\frac{1 + 2\sin\theta}{3}\right)$$

ILLUMINANCE:
Horizontal illuminance E produced at point C from unobstructed sky vault :–

$$E = \frac{7}{9}\pi B_z \text{ lux}$$
or
$$E = \pi B_{42} \text{ lux}$$

Fig. 9.5 CIE standard overcast sky

The luminance distribution is defined mathematically by:

$$B_\theta = B_Z \frac{(1 + 2\sin\theta)}{3}$$

Where:

B_θ = luminance in cd/m^2

θ = angle above the horizontal

B_Z = luminance at the zenith

If $\theta = 90°$

$\sin\theta = 1.0$

and $B_{90} = B_Z \left[\dfrac{1+2}{3} = 1 \right]$

If $\theta = 0°$

$\sin\theta = 0$

and $B_\theta = \frac{1}{3}B_Z$

The luminance at the horizon is only one-third of the luminance at the zenith. The horizontal illuminance produced at the centre of the hemisphere is

$$= \tfrac{7}{9}\pi B_Z \text{ lux}$$

If the average value of brightness for a CIE sky occurs at some angle θ above the horizon then the average brightness B_θ multiplied by π will be the illuminance (cf. a uniform sky).

We can find out at what angle above the horizon the average brightness occurs by equating the two expressions for illuminance.

$$\pi B_\theta = \frac{7}{9}\pi B_Z$$

Substituting for B_θ from $B_\theta = B_Z \left(\dfrac{1 + 2\sin\theta}{3} \right)$

$$\pi B_Z \frac{1 + 2\sin\theta}{3} = \frac{7}{9}\pi B_Z$$

$$\frac{1 + 2\sin\theta}{3} = \frac{7}{9}$$

$$2\sin\theta = \frac{21}{9} - 1$$

$$\sin\theta = \frac{6}{9}$$

$$\therefore \quad \theta \simeq 42°$$

Therefore if we measure the luminance of the real sky at an angle of 42° above the horizon and multiply it by π we will obtain the illuminance provided by the unobstructed overcast sky.

Daylight enters the interior of a building either through windows in the walls or by roof lights. The windows provide a source of daylight, allow a view out and can be used for natural ventilation. Unfortunately they are also a weakness in the thermal and acoustic envelope of the building.

The illuminance due to daylight inside the building will vary with position inside the room and will vary during the day and throughout the year. The illuminance is not constant at any one point in a room due to the variability of daylight itself.

In order to assess the influence of window size, shape and position on the penetration of daylight into a building, the daylight at a point in a room is quantified by means of a daylight factor.

Daylight factor

The CIE definition of daylight factor is the ratio of the daylight illuminance at a point on a given plane due to the light received directly and indirectly from a sky of assumed or known luminance distribution, to the illuminance on a horizontal plane due to an unobstructed hemisphere of this sky. Direct sunlight is excluded for both values of illuminance.

The daylight factor is usually expressed as a percentage. When measuring daylight factors in an existing room, the internal illuminance at a point is measured against the simultaneously occurring external illuminance of the unobstructed real overcast sky, but for calculation purposes either a uniform sky or more usually a CIE overcast sky condition is assumed. The expression for daylight factor can then be written:

The daylight factor at a point in a room

$$= \frac{\text{internal illuminance at that point}}{\substack{\text{simultaneously occurring external} \\ \text{illuminance from an unobstructed} \\ \text{CIE standard overcast sky condition}}} \times 100 \text{ per cent}$$

or more briefly:

$$\text{Daylight factor} = \frac{\text{internal illuminance}}{\text{external illuminance}} \times 100 \text{ per cent}$$

An increase in the external illuminance will give rise to an increase in the internal illuminance but the daylight factor at a point will remain constant. The point is usually taken on the horizontal working plane although other planes are sometimes of interest such as the vertical plane of a blackboard.

Example 9.1 Calculate the illuminance at a point in a room where there is a daylight factor of 5 per cent if the external illuminance is 8000 lux.

$$\text{Daylight factor} = \frac{\text{internal illuminance}}{\text{external illuminance}} \times 100$$

$$\text{Internal illuminance} = \frac{\text{DF} \times \text{external illuminance}}{100}$$

$$= \frac{5 \times 8000}{100}$$

$$= 400 \, \text{lux}$$

If the external illuminance falls to 5000 lux the internal illuminance at the point falls to 250 lux.

A picture of the distribution of daylight on a working plane may be obtained by producing daylight factors at grid intersection points in the room and by treating them in a similar way to spot heights. On a level survey daylight factor contours can be produced (Fig. 9.6).

Ideally the interpolation for contours between grid point values should be logarithmic in accordance with psychophysical laws, but a linear interpolation will not produce too great an error in most cases. (See pages 48–49 *Architectural Physics: Lighting*, by R. G. Hopkinson, HMSO.)

Components of daylight factor

For calculation purposes the daylight factor at a point in a room is divided into three components, each expressed as a percentage of the external illuminance.

Sky component (SC)

The sky component of the daylight factor is the ratio of that part of the daylight illuminance at a point on a given plane which is received directly from a sky of assumed or known luminance distribution to the illuminance on a horizontal plane due to an unobstructed hemisphere of this sky. Direct sunlight is excluded for both values of illuminance.

This is the component of daylight factor coming directly from the area of sky as seen at the measurement point (Fig. 9.7). (Sky factor is

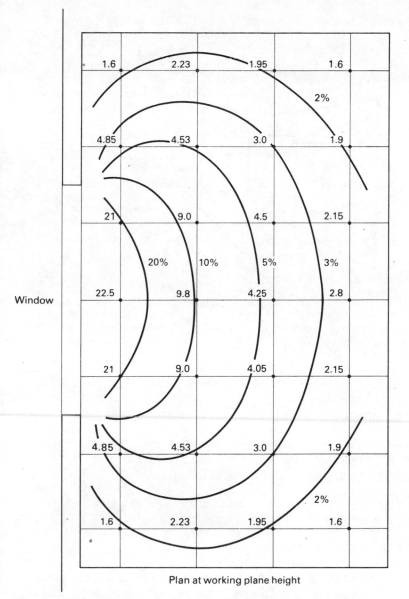

Plan at working plane height

Fig. 9.6 Daylight factor contours drawn from daylight factors measured at grid intersection points in a room

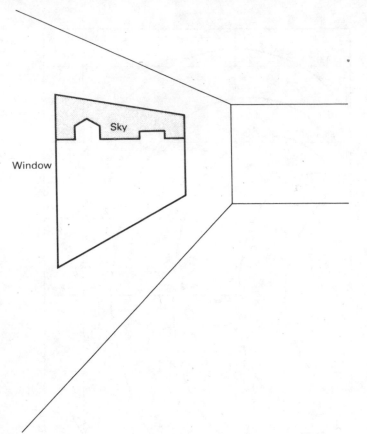

The area of sky as seen from the measuring point is the source of light
for this component

Fig. 9.7 Sky component of daylight factor

identical to sky component for an unglazed opening and uniform
sky – it is seldom used except in legal cases.)

Externally reflected component (ERC)

The externally reflected component of the daylight factor is the ratio
of that part of the daylight illuminance at a point on a given plane
which is received directly from external reflecting surfaces illuminated
directly or indirectly by a sky of assumed or known luminance
distribution to the horizontal illuminance on a horizontal plane due
to an unobstructed hemisphere of this sky. Contributions of direct
sunlight to the luminances of external reflecting surfaces and to the
illumination of the comparison plane are excluded.

This is the component which is due to light being reflected off
external obstructions to the measurement point (Fig. 9.8).

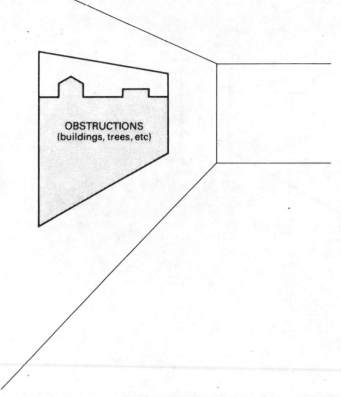

The area of obstruction seen through the window reflects light to the measuring position

Fig. 9.8 Externally reflected component of daylight factor

Internally reflected component (IRC)

The internally reflected component of the daylight factor is the ratio of that part of the daylight illuminance at a point on a given plane which is received from internal reflecting surfaces, the sky being of assumed or known luminance distribution, to the illuminance on a horizontal plane due to an unobstructed hemisphere of this sky. The effects of direct sunlight are excluded.

This component is due to light which has entered the room and is now reflected on to the measurement point by the room surfaces (Fig. 9.9).

The daylight factor is then the sum of the three components.

DF = SC + ERC + IRC

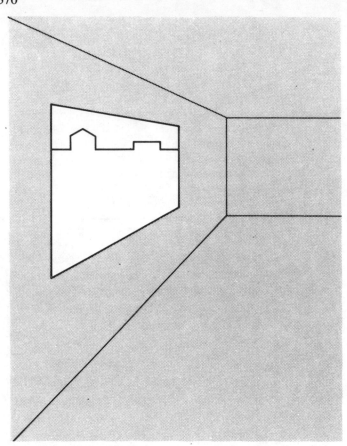

Light reflected from internal surfaces to the measuring point

Fig. 9.9 Internally reflected component of daylight factor

The sky component and internally reflected component are the more important components. The externally reflected component makes only a small contribution to the daylight factor in side-lit rooms. The sky component is the most important component for areas near to the windows. Both the sky component and the externally reflected component decrease with depth into the room and near the rear of a side-lit room the internally reflected component can be of major importance to the daylighting. The internally reflected component is fairly uniform over much of the room interior and an average value can be calculated which is appropriate to most room positions. In the furthest positions from the windows and in dark corners a minimum value of internally reflected component is used. The higher the reflectances of the room surfaces the higher the internally reflected component for a given window size.

Influence of window size, shape and position on daylight factors

Maximum daylight would be obtained if the window wall were entirely glazed (100 per cent fenestration). Limiting the fenestration and arranging it either as one window or several smaller windows alters the distribution pattern and character of the daylight admitted. However, people prefer the windows to be greater than 20 per cent of the window wall in side-lit rooms. A side-lit room can only look at a quarter sphere of sky at best. At the window, therefore, the maximum daylight factor cannot be more than 50 per cent and will be less than this allowing for glass losses and a CIE sky distribution.

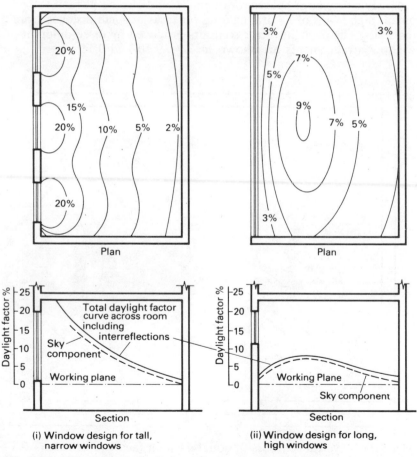

Fig. 9.10   The effect of window design on daylight factor

Daylight factors away from the window reduce rapidly to quite small percentage values.

In a side-lit room daylight penetration is deepest in line with windows. The taller the window for a given width the larger the sky component and the deeper the daylight factor contours are pushed back. Wide windows give a better distribution across the width of the room but do not produce such high daylight factors in depth if their height is limited. The only contribution to daylight factor on the horizontal working plane from parts of windows below working plane height will be to enhance the internally reflected component.

A high level clerestory window will give better daylighting further from the wall in which it is set than closer to the wall. Fall-off of daylight factors with depth is very rapid. Windows in more than one wall will produce better penetration of daylight. They will also reduce sky glare by brightening the surroundings and add interest to the room. For very large spaces roof lights would also be needed to obtain high levels of natural lighting for tasks at some distance from the walls. The daylight factor contours produced by three different window arrangements are shown in Figs. 9.10 and 9.11.

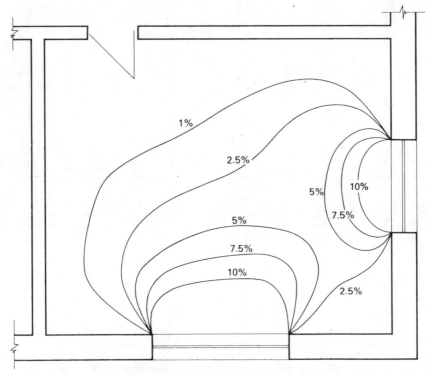

Fig. 9.11 Typical contours of equal daylight factor for a room with windows in adjacent walls

Daylight levels

Data on the percentage of working hours when daylight illuminance exceeds specified values at various daylight factors are given in Appendix 1 of IES Technical Report No. 4.

Prediction methods

Prediction methods for daylight factor components and total daylight factors include:

(a) BRS daylight protractors and IRC formula – components;
(b) BRS simplified daylight tables – components;
(c) Model room analysis – total daylight factor;
(d) Pilkington sky dots – components;
(e) Waldram diagrams – components;
(f) Computer programs – components and total daylight factor;
(g) Pleijel diagrams – components.

Only methods (a), (b) and (c) will be considered further here.

BRS daylight protractors

The sky component is a function of the angular size of the window at the point in a room, the angle which the incident light makes with the working plane and the angle of elevation of the patch of sky visible from the point. The sky component can be evaluated in terms of these angles by means of specially designed protractors. The protractors can also be used to measure an externally reflected component if there are external obstructions.

The ten protractors comprise two sets. Five for a uniform overcast sky and five for a standard CIE overcast sky. Each set of five consists of protractors for:

(i) vertical glazing;
(ii) horizontal glazing;
(iii) glazing sloping at 30° to the horizontal;
(iv) glazing sloping at 60° to the horizontal;
(v) unglazed openings.

This enables both conventional windows and roof lights to be measured.

Full details for the use of these protractors are available in *BRS Daylight Protractors*, J. Longmore, HMSO. Reference should also be made to *BRS Digest*, 41 and 42.

The use of protractor No. 2 – vertical glazing, CIE sky
The protractor (Fig. 9.12) is used in conjunction with an elevation and plan of a room to calculate the sky component (SC) and the externally reflected component (ERC) at a point in the room. The

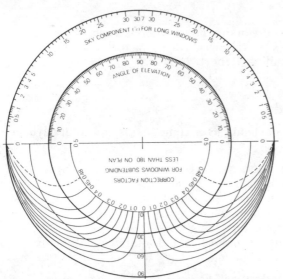

Fig. 9.12   BRS sky component protractor for vertical glazing and CIE overcast sky

internally reflected component (IRC) is calculated by another method. The point is taken to be on the horizontal working plane.

The protractor consists of two halves. The sky component and angle of elevation scales for infinitely long windows are used on the elevation (Fig. 9.13(a)). The auxiliary scale to correct for a window of finite length is used on the room plan (Fig. 9.13(b)).

A computation sheet helps in systematically recording the information (Table 9.1(a), (b), (c) and (d)).

Procedure
Room elevation

1. Draw a site line from the measurement point to the outside top edge of the window opening.

2. Draw a site line from the measurement point to the top of any obstruction. If the obstruction is irregular it may be replaced by an equivalent obstruction line (Fig. 9.14).

3. Draw a site line from the measurement point to the inside edge of the window sill. If the window sill is lower than the measurement point a horizontal line is drawn as light cannot reach the point through a desk. It will be noticed that if there are no obstructions then the lines drawn in (2) and (3) would be the same.

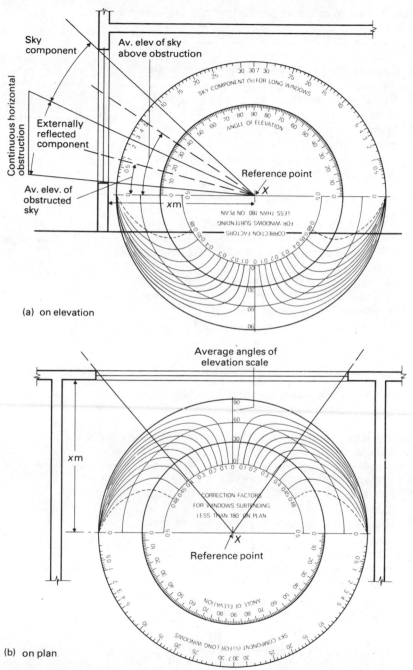

Fig. 9.13 BRS sky component protractor for vertical glazing and CIE overcast sky

Table 9.1(a) Worksheet for finding the sky component of daylight factor

---

a. Sky component
  1. Reading on primary sky component scale
     (upper)   =
  2. Reading on primary sky component scale
     (lower)   = _____
  3. Primary sky component for window of
     infinite length $(1-2)$ = _____%
  4. Angle of elevation of visible sky (upper)   =
  5. Angle of elevation of visible sky (lower)   = _____
                                                                            °
  6. Average angle of elevation $\dfrac{4+5}{2}$ = _____
  7. Reading on auxiliary scale (left)   =
  8. Reading on auxiliary scale (right)   = _____
  9. Correction factor to be applied to primary sky
     component (7 ± 8) = _____
 10. Sky component for window of finite length
     (3×9) = _____%
 11. Allowance for glazing material   =
 12. Allowance for window frame and glazing bars   =
 13. Allowance for dirt on glazing   = _____
 14. Total allowance for effects of glazing
     $(11 \times 12 \times 13)$ = _____
 15. Corrected sky component (10×14) = _____%

Table 9.1(b) Worksheet for finding the externally reflected component

---

b. Externally reflected component
 16. Reading on primary sky component scale
     (upper)   =
 17. Reading on primary sky component scale
     (lower)   = _____
 18. Equivalent primary sky component for
     window of infinite length (16–17)   = _____%
 19. Angle of elevation of visible obstruction
     (upper)   =
 20. Angle of elevation of visible obstruction
     (lower)   = _____
                                                                            °
 21. Average angle of elevation $\dfrac{(19+20)}{2}$ = _____
 22. Reading on auxiliary scale (left)   =
 23. Reading on auxiliary scale (right)   = _____

24. Correction factor to be applied to equivalent
    primary sky component (22 ± 23)   = _____
25. Equivalent sky component for window of
    finite length (18 × 24)   = _____%
26. Correction for relative luminance of obstruction
    (× 0.1 for uniform sky protractors or ×
    0.2 for CIE sky)   =
27. Externally reflected component (25 × 26)   = _____
28. Allowance for glazing material   =
29. Allowance for window frame and glazing bars   =
30. Allowance for dirt on glazing   = _____
31. Total allowance for effects of glazing
    (28 × 29 × 30)   = _____
32. Corrected externally reflected component
    (27 × 31)   = _____%

Table 9.1(c) Tabulation of the internally reflected component

---

c. *Internally reflected component*
33. IRC by calculation or from BRS nomogram   =   %
34. Allowance for glazing material   =
35. Allowance for window frame and glazing bars   =
36. Allowance for dirt on glazing   =
37. Allowance for dirt on internal surfaces   =
38. Total allowances for effects of glazing
    and dirt on internal surfaces
    (34 × 35 × 36 × 37)   =
39. Corrected internally reflected component
    (33 × 38)   =   %

Table 9.1(d) Worksheet for finding the total daylight factor

---

d. *Daylight factor*
15. Corrected sky component   =   %
32. Corrected externally reflected component   =   %
39. Corrected internally reflected component   = _____%
34. Total daylight factor (15 + 32 + 39)   = _____%

---

4. The upper and lower primary sky component figures are obtained by placing the protractor with its base horizontal at the measurement point. The difference between lines (1) and (2) gives the value for a window of infinite length.
5. The angles of elevation for lines (1) and (2) are read from the second scale.
    Similar measurements are made to obtain the externally reflected component, which at this stage is called an equivalent sky component.

Room plan The lower half of the protractor is then used to obtain correction factors for windows of finite length. Site lines defining the window size are drawn from the measurement point to the window on a plan of the room. The protractor is aligned parallel to the window and correction factors read off appropriate to the average angle of elevation of the sky component and externally reflected component. (The intersection of site line and angle of elevation is traced back to the correction scale parallel to one of the curved lines.) Correction factors are added if the measurement point lies on a line drawn back normal to the window, and the difference taken if the measurement point lies off the line of the window.

The primary scale readings for windows of infinite length are now multiplied by the net correction factors to give SC and ERC values for the finite window length.

Corrections to SC and ERC
SC corrections The SC has been measured for a clear glazed aperture (allowance for glazing material = 1.0). Allowance must now

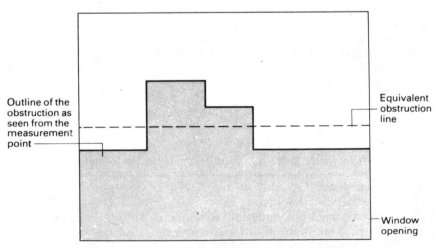

Fig. 9.14 An irregular obstruction replaced by an equivalent obstruction line

be made for:

(a) obstruction due to glazing bars (Table 9.2); and
(b) average light loss due to dirt on the glazing (glazing mainte-
     nance factor) (Table 9.3) (See also IES Technical Report
     No. 9.)

Table 9.2 Light losses due to glazing bars

For large-paned windows	Reduction factor
All metal windows	0.8–0.85
Metal windows in wood frames	0.75
All wood windows	0.65–0.70

Table 9.3 Maintenance factors to be applied to the total daylight factor or to each of its three components to allow for dirt on glazing

Location of building	Inclination of glazing	Maintenance factor	
		Non-industrial or clean industrial work	Dirty industrial work
Non-industrial or clean industrial area	Vertical	0.9	0.8
	Sloping	0.8	0.7
	Horizontal	0.7	0.6
Dirty industrial area	Vertical	0.8	0.7
	Sloping	0.7	0.6
	Horizontal	0.6	0.5

ERC corrections So far the ERC has been measured as an equivalent sky component. The ERC involves light reflected off an external
obstruction. In the case of a CIE sky the average obstruction is taken
as providing only 0.2 of the light that would be provided by the sky it
covers up. The 'ERC' value (equivalent sky component) obtained
above is multiplied by 0.2 to give the ERC. The allowance for glazing
material is still 1.0.

   Further corrections are needed to this value for:

(a) glazing bars; and
(b) dirt on the glazing;

by using the same multipliers as for the sky component.

Internally reflected component (IRC)

The internally reflected component of daylight factor can be calculated from the BRS split-flux formula or read off from the BRS nomogram.

BRS split-flux formula for average internally reflected component

The formula combines the effects of light entering the room above and below the mid-plane through the window and being reflected from the internal surfaces to the measurement point (Figs. 9.15 and 9.16).

$$\text{Average IRC} = \frac{0.85W}{A(1 - R)} \left[CR_{\text{fw}} + 5R_{\text{cw}} \right] \text{ per cent}$$

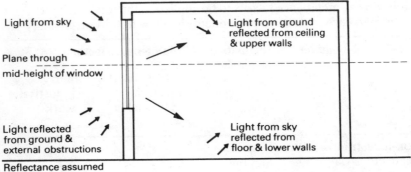

Light from sky

Plane through mid-height of window

Light reflected from ground & external obstructions

Light from ground reflected from ceiling & upper walls

Light from sky reflected from floor & lower walls

Reflectance assumed as 5%

Fig. 9.15 BRS split-flux principle for internally reflected component

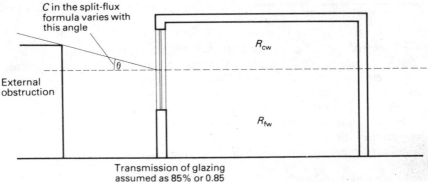

C in the split-flux formula varies with this angle

R_{cw}

θ

External obstruction

R_{fw}

Transmission of glazing assumed as 85% or 0.85 for clear flat glass

Fig. 9.16 BRS split-flux method for internally reflected component with an external obstruction at an angle θ

where: A = total internal surface area

W = window area

R = average reflectance of internal surfaces

C = a coefficient to take into account any external obstruction. The variation of C with the angle of obstruction is given in Table 9.4 where θ is shown in Fig. 9.16.

Table 9.4   Effect of external obstruction or internally reflected component

Angle of obstruction measured from centre of window (degrees above horizontal)	Coefficient (C)
no obstruction	39
10°	35
20°	31
30°	25
40°	20
50°	14
60°	10
70°	7
80°	5

R_{fw} = average reflectance of floor and lower walls (below mid-plane of window) but not including window wall.

R_{cw} = average reflectance of ceiling and upper walls (above mid-plane of window) but not including window walls.

0.85 represents the transmission factor of a single sheet of clear glass (diffuse transmittance) and 5 represents the assumed reflection factor of the ground and obstruction of 5 per cent

BRS nomogram method for internally reflected component

The value of window area/total surface area ratio is found on A and average internal reflectance is found on B (Fig. 9.17). A line joining these two points intersects C to give a value of IRC without external obstruction. The value of the angle of obstruction θ (as shown in Fig. 9.16) is found on D. A line from this point on D through the point already found on C projected to cut E will give the value of the IRC with obstruction. (Note for $\theta = 0°$ the value on E and C should be the same. Swivelling about the point on C, as θ increases the value on E decreases.)

382

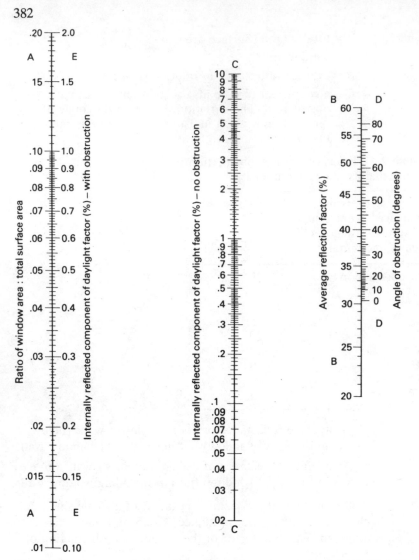

Fig. 9.17 Average internally reflected component of daylight factor: Nomogram I

Minimum internally reflected component (IRC)

The average internally reflected component applies to most room positions but for checking points in dark corners or remote from the window a minimum IRC is used instead. This may be obtained either from a BRS nomogram II (Fig. 9.18) or from the conversion table (Table 9.5).

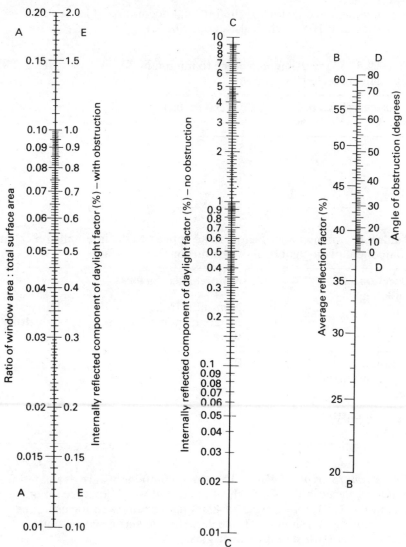

Fig. 9.18 Nomogram II: minimum internally reflected component of daylight factor

Corrections to internally reflected component

The internally reflected component must be corrected for:

(a) glazing bars;
(b) dirt on glazing;

and an additional correction is required for:

(c) deterioration of the internal surface reflectance (i.e. a

maintenance factor for the internal decoration) (Table 9.6).
(See also IES Technical Report No. 9.)

Table 9.5 Conversion of average to minimum
internally reflected component

Average reflectance $R(\%)$	Conversion factor
30	0.54
40	0.67
50	0.78
60	0.85

Table 9.6 Maintenance factors to be applied to the internally reflected
component of daylight to allow for dirt on room surfaces

Location of building	Maintenance factors	
	Non-industrial or clean industrial work	Dirty industrial work
Non-industrial or clean industrial area	0.9	0.7
Dirty industrial area	0.9	0.6

Correction for type of glass The three components are evaluated for
windows glazed with a single sheet of clear grass (Allowance for grazing
material = 1.0.). Other types of glass may be allowed for using the
correction factors shown in Table 9.7. Double glazing correction
factor ≃ 0.9 (two sheets of clear glass).

Example 9.2 Assess the daylight factor at the position A on the desk
shown in Fig. 9.19 using BRS Daylight Protractor No. 2 plus calculations for the internally reflected component, using dimensions
shown in Fig. 9.20.

Room dimensions:	4.55 m × 3.33 m × 3 m (high)
Window dimensions:	1.6 m × 1.25 m (single sheet of clear glass)
Midpoint of window:	1.75 m above floor

The surface reflection factors are:

Ceiling 70 per cent
Walls 50 per cent
Floor 15 per cent
Glass 15 per cent

Table 9.7 Conversion factors to allow for the reduced light transmission of some typical glazing materials

Glazing material	Conversion factor to be applied to sky component
Transparent glasses	
Flat drawn sheet or float	1.0
6 mm polished plate	1.0
6 mm polished wired	0.95
Patterned and diffusing glasses	
3 mm rolled	0.95
6 mm rough cast	0.95
6 mm wired cast	0.9
Cathedral	1.0
Hammered	1.0
Arctic	0.95
Reeded	0.95
Small Morocco	0.9
Special glasses	
Heat absorbing tinted plate	0.9
Heat absorbing tinted cast	0.6–0.75
Laminated insulating glass	0.6–0.7
Plastic sheets	
Corrugated resin-bonded glass-fibre reinforced roofing sheets:	
Moderately diffusing	0.9
Heavily diffusing	0.75–0.9
Very heavily diffusing	0.65–0.8
3 mm diffusing plain opal acrylic plastic sheets (depending on grade)	0.65–0.9

Note: Conversion factors multiplied by 0.85 will give approximate diffuse transmittance of these glazing materials.

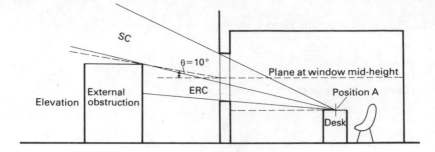

Fig. 9.19 Worked example 9.2 using BRS daylight factor protractor No. 2

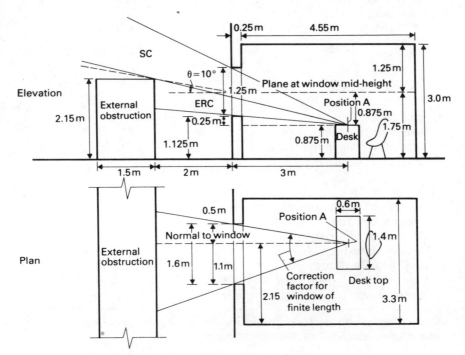

Fig. 9.20 Worked example 9.2 dimensions for calculations.

Obstruction angle 10° from midpoint of window

Use the following correction factors:

Allowance for window frame and glazing bars = 0.8
Allowance for dirt on the glazing = 0.9
Allowance for deterioration of the = 0.9
  internal decoration
Note: clear glass correction factor = 1.0

The answer is calculated as shown in Table 9.8(a), (b), (c) and (d).

Table 9.8(a)

a. Sky component
  1. Reading on primary sky component scale (upper)   =   2.95
  2. Reading on primary sky component scale (lower)   =   0.70

  3. Primary sky component for window of infinite
     length (1 − 2)   =   2.25%

  4. Angle of elevation of visible sky (upper)   =   27°
  5. Angle of elevation of visible sky (lower)   =   14°

  6. Average angle of elevation $\dfrac{4+5}{2}$ =   20.5°

  7. Reading on auxiliary scale (left)   =   0.21
  8. Reading on auxiliary scale (right)   =   0.1
  9. Correction factor to be applied to
     primary sky component (7 + 8)   =   0.31

10. Sky component for window of finite length (3 × 9)   =   0.7%

11. Allowance for glazing material   =   1.0
12. Allowance for window frame and glazing bars   =   0.8
13. Allowance for dirt on glazing   =   0.9

14. Total allowance for effects of
     glazing (11 × 12 × 13)   =   0.72
15. Corrected sky component (10 × 14)   =   0.5%

Table 9.8(b)

b. Externally reflected component
16. Reading on primary sky component scale (upper)   =   0.7
17. Reading on primary sky component scale (lower)   =   0.09

18. Equivalent primary sky component for window of
     infinite length (16–17)   =   0.61%

Table 9.8(b) (*Cont.*)

19. Angle of elevation of visible obstruction (upper)	=	14°
20. Angle of elevation of visible obstruction (lower)	=	5°

21. Average angle of elevation $\dfrac{(19 + 20)}{2}$ = 9.5°

22. Reading on auxiliary scale (left)	=	0.22
23. Reading on auxiliary scale (right)	=	0.1
24. Correction factor to be applied to equivalent primary sky component (22 + 23)	=	0.32

25. Equivalent sky component for window of finite length (18 × 24)   =   0.2%

26. Correction for relative luminance of obstruction (× 0.1 for uniform sky protractors or × 0.2 for CIE sky)   =   0.2

27. Externally reflected component (25 × 26)   =   0.04%

28. Allowance for glazing material	=	1.0
29. Allowance for window frame and glazing bars	=	0.8
30. Allowance for dirt on glazing	=	0.9
31. Total allowance for effects of glazing (28 × 29 × 30)	=	0.72
32. Corrected externally reflected component (27 × 31)	=	0.03%

Table 9.8(c)

---

c. Internally reflected component

33. Average IRC by calculation	=	0.63%
34. Allowance for glazing material	=	1.0
35. Allowance for window frame and glazing bars	=	0.8
36. Allowance for dirt on glazing	=	0.9
37. Allowance for dirt on internal surfaces	=	0.9
38. Total allowances for effects of glazing and dirt on internal surfaces (34 × 35 × 36 × 37)	=	0.648
39. Corrected internally reflected component (33 × 38)	=	0.41%

Table 9.8(d)

---

d. Daylight factor

15. Corrected sky component	=	0.5%
32. Corrected externally reflected component	=	0.03%
39. Corrected internally reflected component	=	0.41%
34. Total daylight factor (15 + 32 + 39)	=	0.94%

Calculation of average internally reflected component

Window area $= 1.6 \times 1.25$
$$= 2\,\text{m}^2$$

Total surface area $= (2 \times 4.55 \times 3.33) + (2 \times 4.55 \times 3)$
$$+ (2 \times 3.33 \times 3)$$
$$= 77.583\,\text{m}^2$$

Average reflectance by area weighted calculation:

$R_{av} \times A = R_1 A_1 + R_2 A_2 + R_3 A_3 + \text{etc.}$

$R_{av} = \{(1.6 \times 1.25)0.15 + (3.33 \times 3 \times 0.5)$
$$+ [(3.33 \times 3) - (1.6 \times 1.25)]0.5$$
$$+ (2 \times 4.55 \times 3 \times 0.5)$$
$$+ (4.55 \times 3.33 \times 0.7)$$
$$+ (4.55 \times 3.33 \times 0.15)\} \div 77.583\,\text{m}^2$$

$\therefore$ $R_{av} = 0.46$ or 46 per cent.

For R_{fw}:

Area of lower walls and floor not including window wall
$$= (2 \times 1.25 \times 4.55) + (3.33 \times 1.25) + (4.55 \times 3.33)$$
$$= 36.904\,\text{m}^2$$

Areas $\times$ reflectances
$$= (2 \times 1.25 \times 4.55 \times 0.5) + (3.33$$
$$\times 1.25 \times 0.5) + (4.55 \times 3.33 \times 0.7)$$

$$= 13.149$$

$$R_{fw} = \frac{13.149}{36.904}$$

$$= 0.356 \text{ or } 35.6 \text{ per cent.}$$

For R_{cw}:

Area of upper walls and ceiling not including window wall
$$= (2 \times 1.25 \times 4.55) + (3.33 \times 1.25) + (4.55 \times 3.33)$$
$$= 30.69\,\text{m}^2$$

Area $\times$ reflectances $= (2 \times 1.25 \times 4.55 \times 0.5) + (3.33$
$$\times 1.25 \times 0.5) + (4.55 + 3.33 \times 0.7)$$
$$= 18.37$$

$$R_{cw} = \frac{18.37}{30.69}$$
$$= 0.599 \text{ or } 59.9 \text{ per cent.}$$

C for $10°$ from Table 9.4 = 35

Average IRC $= \dfrac{0.85W}{A(1 - R)} [CR_{fw} + 5R_{cw}]$ per cent

$\qquad = \dfrac{0.85 \times 2}{77.583(1 - 0.46)} [35 \times 0.356 + 5 \times 0.599]$ per cent

$\qquad = 0.63$ per cent.

Corrected for glazing bars, dirt on glazing and deterioration of interior decoration

Corrected average IRC $= 0.63 \times 0.8 \times 0.9 \times 0.9$

$\qquad\qquad\qquad\qquad\quad = 0.41$ per cent.

Alternatively:
Average IRC using Nomogram 1 (Fig. 9.17)

$\qquad = 0.62$ per cent

$\qquad = 0.4$ per cent (corrected).

Minimum internally reflected component
The desk position is fairly far back into the room and a worst daylighting condition could be considered using the minimum internally reflected component value.

Using the conversion from Table 9.5 of 0.736 for an average reflectance of 46 per cent the minimum internally reflected component

$\qquad = 0.63 \times 0.736$ or 0.62×0.736

$\qquad = 0.46$ per cent.

Minimum internally reflected component from Nomogram II (Fig. 9.18)

$\qquad = 0.44$ per cent.

Corrected minimum internally reflected component (using either value)

$\qquad = 0.29$ per cent (to 2 decimal places).

If the minimum value of the internally reflected component is used instead of the average internally reflected component

Daylight factor $= 0.5 + 0.03 + 0.29$

$\qquad\qquad\qquad\quad = 0.82$ per cent.

In either case the desk position has a daylight factor less than 1 per cent and supplementary lighting will be needed.

BRS simplified daylight tables

The simplified daylight tables were produced to enable a quick estimation to be made of sky component, externally reflected component and minimum internally reflected component. The full set of tables appeared in National Building Studies, Special Report No. 26, *Simplified Daylight Tables*, Hopkinson, Longmore and Murray Graham (HMSO 1958).

Table 9.9 is for calculating sky component and externally reflected component for vertical rectangular windows with clean clear glass and CIE overcast sky conditions.

Table 9.10 is for calculating minimum internally reflected component.

Referring to Fig. 9.21 the window is described in terms of two ratios: height(H) of window head above working plane to distance (D) from window wall to measurement point (H/D) and effective window width(W) to one side of a normal from measurement point to window wall to distance(D)(W/D). In the case of the window shown in Fig. 9.21 values would be read off the table appropriate to H/D, W_1/D ratios and H/D, W_2/D ratios and the values added to give a sky component.

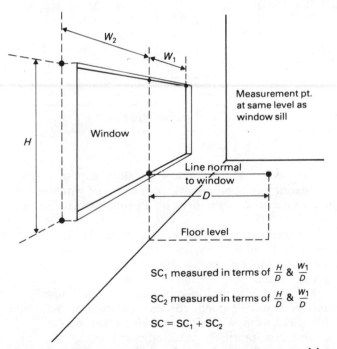

SC$_1$ measured in terms of $\frac{H}{D}$ & $\frac{W_1}{D}$

SC$_2$ measured in terms of $\frac{H}{D}$ & $\frac{W_1}{D}$

SC = SC$_1$ + SC$_2$

Fig. 9.21 Sky component for a measurement position on the line of the window and at sill level

Table 9.9 Sky components (CIE standard overcast sky) for vertical rectangular windows with clean clear glass

	Ratio H/D = height of window head above													
0	*0.1*	*0.2*	*0.3*	*0.4*	*0.5*	*0.6*	*0.7*	*0.8*	*0.9*	*1.0*	*1.1*	*1.2*	*1.3*	*1.4*
0.1	0	0	0.1	0.1	0.2	0.2	0.3	0.4	0.5	0.6	0.6	0.7	0.8	0.8
0.2	0	0.1	0.1	0.2	0.4	0.5	0.7	0.8	1.0	1.1	1.3	1.4	1.5	1.6
0.3	0	0.1	0.2	0.3	0.5	0.7	1.0	1.2	1.5	1.7	1.9	2.1	2.3	2.4
0.4	0	0.1	0.3	0.4	0.7	1.0	1.3	1.6	1.9	2.2	2.5	2.7	2.9	3.2
0.5	0	0.1	0.3	0.5	0.8	1.2	1.5	1.9	2.2	2.6	3.0	3.3	3.6	3.8
0.6	0	0.1	0.3	0.6	1.0	1.3	1.7	2.2	2.6	3.0	3.4	3.8	4.1	4.4
0.7	0	0.2	0.4	0.7	1.0	1.5	1.9	2.4	2.8	3.3	3.8	4.2	4.5	4.8
0.8	0.1	0.2	0.4	0.7	1.1	1.6	2.1	2.6	3.1	3.6	4.1	4.5	4.9	5.2
0.9	0.1	0.2	0.4	0.8	1.2	1.7	2.2	2.7	3.3	3.8	4.3	4.8	5.2	5.6
1.0	0.1	0.2	0.4	0.8	1.3	1.8	2.3	2.9	3.4	4.0	4.6	5.0	5.5	5.9
1.2	0.1	0.2	0.5	0.9	1.4	1.9	2.5	3.1	3.7	4.3	4.9	5.4	5.9	6.4
1.4	0.1	0.2	0.5	0.9	1.4	1.9	2.5	3.2	3.8	4.5	5.1	5.7	6.2	6.7
1.6	0.1	0.2	0.5	0.9	1.4	2.0	2.6	3.3	3.9	4.6	5.3	5.9	6.4	7.0
1.8	0.1	0.2	0.5	1.0	1.4	2.0	2.6	3.3	4.0	4.7	5.4	6.0	6.6	7.2
2.0	0.1	0.2	0.5	1.0	1.5	2.0	2.6	3.3	4.0	4.7	5.4	6.1	6.7	7.3
2.5	0.1	0.2	0.5	1.0	1.5	2.1	2.6	3.3	4.0	4.8	5.5	6.2	6.8	7.4
3.0	0.1	0.2	0.5	1.0	1.5	2.1	2.7	3.4	4.1	4.8	5.6	6.2	6.9	7.5
4.0	0.1	0.2	0.5	1.0	1.5	2.1	2.7	3.4	4.1	4.9	5.6	6.3	6.9	7.5
6.0	0.1	0.2	0.5	1.0	1.5	2.1	2.8	3.4	4.2	5.0	5.7	6.3	6.9	7.6
∞	0.1	0.2	0.5	1.0	1.5	2.1	2.8	3.4	4.2	5.0	5.7	6.3	7.0	7.6
0	6°	11°	17°	22°	27°	31°	35°	39°	42°	45°	48°	50°	52°	54°

Ratio W/D = Width of window to one side of normal: distance from window

Angle of obstruction

For a position off the line of the window the sky component can be obtained by evaluating sky components for a series of windows and obtaining the required sky component by a process of subtraction and addition (Fig. 9.22).

Externally reflected components can be determined by finding the equivalent sky component and multiplying it by 0.2. Alternatively the combined sky compònent and externally reflected component can be found by reading a sky component off the table against the scales at the bottom and left of the table, multiplying this value by 0.8 and subtracting it from the sky component for the full window, although this is only possible if the window sill is at working plane height.

The table for minimum internally reflected component relates to a room roughly $6\,\text{m} \times 6\,\text{m} \times 3\,\text{m}$ high with a window extending from working plane height to ceiling. Conversion factors are available (Table 9.11) to extend its use to floor areas from $10\,\text{m}^2$ to $100\,\text{m}^2$.

Table 9.9 (*cont.*)

working plane: distance from window														
1.5	*1.6*	*1.7*	*1.8*	*1.9*	*2.0*	*2.2*	*2.4*	*2.6*	*2.8*	*3.0*	*3.5*	*4.0*	*5.0*	*∞*
0.9	0.9	0.9	1.0	1.0	1.0	1.1	1.1	1.1	1.1	1.2	1.2	1.2	1.2	1.3
1.7	1.8	1.9	1.9	2.0	2.0	2.1	2.2	2.2	2.3	2.3	2.4	2.4	2.4	2.5
2.6	2.7	2.8	2.9	3.0	3.1	3.2	3.3	3.4	3.4	3.5	3.6	3.6	3.7	3.7
3.3	3.5	3.6	3.8	3.9	4.0	4.1	4.3	4.4	4.5	4.5	4.6	4.7	4.8	4.9
4.0	4.2	4.4	4.6	4.7	4.8	5.0	5.2	5.3	5.4	5.5	5.7	5.8	5.9	5.9
4.6	4.9	5.1	5.3	5.4	5.6	5.8	6.0	6.2	6.3	6.4	6.6	6.7	6.8	6.9
5.1	5.4	5.6	5.8	6.0	6.2	6.4	6.6	6.8	7.0	7.1	7.3	7.4	7.6	7.7
5.6	5.8	6.1	6.3	6.5	6.7	7.0	7.3	7.5	7.6	7.8	8.0	8.2	8.3	8.4
5.9	6.2	6.5	6.7	6.9	7.1	7.4	7.7	7.9	8.1	8.2	8.5	8.7	8.8	9.0
6.2	6.5	6.8	7.1	7.3	7.5	7.9	8.1	8.4	8.6	8.7	9.0	9.2	9.4	9.6
6.8	7.2	7.5	7.8	8.1	8.3	8.7	9.1	9.3	9.6	9.8	10.1	10.3	10.5	10.7
7.1	7.5	7.8	8.2	8.5	8.7	9.1	9.5	9.8	10.0	10.2	10.6	10.9	11.1	11.6
7.4	7.8	8.2	8.5	8.8	9.1	9.6	10.0	10.2	10.5	10.7	11.1	11.4	11.7	12.2
7.6	8.1	8.5	8.8	9.2	9.5	10.0	10.4	10.8	11.1	11.3	11.8	12.0	12.3	12.6
7.8	8.2	8.6	9.0	9.4	9.7	10.2	10.7	11.1	11.4	11.7	12.2	12.4	12.7	13.0
7.9	8.4	8.8	9.2	9.6	9.9	10.5	11.0	11.4	11.7	12.0	12.6	12.9	13.3	13.7
8.0	8.5	8.9	9.3	9.7	10.0	10.7	11.2	11.7	12.0	12.4	12.9	13.3	13.7	14.2
8.0	8.6	9.0	9.4	9.8	10.1	10.8	11.3	11.8	12.2	12.5	13.2	13.5	14.0	14.6
8.1	8.6	9.1	9.5	9.9	10.2	10.9	11.4	11.9	12.3	12.6	13.2	13.6	14.1	14.9
8.1	8.6	9.1	9.5	9.9	10.3	10.9	11.5	11.9	12.3	12.7	13.3	13.7	14.2	15.0
56°	58°	60°	61°	62°	63°	66°	67°	69°	70°	72°	74°	76°	79°	90°

Angle of obstruction

The table assumes a ceiling reflectance of 70 per cent and allowance
for other ceiling reflection factors may be made using Table 9.12. An
external obstruction angle of 20° is assumed. Conversion minimum to
average IRC can be made using Table 9.13.

Example 9.3 Find the sky component (SC) at point A in Example 9.2 using Table 9.9.
The heights and widths are shown in Fig. 9.23(a) and (b). The sky
component is obtained by considering four windows as in
Fig. 9.23(b).
Window (1):

$$H_1/D = \frac{1.5}{3.0}$$

$$= 0.5$$

Table 9.10 The minimum internally reflected component of daylight factor (%).
Assuming ceiling reflection factor = 70 per cent, angle of external obstruction = 20 degrees

Ratio of window area: floor area	Window area as percentage of floor area	Floor reflection factor											
		10%				20%				40%			
		Wall reflection factor											
		20%	40%	60%	80%	20%	40%	60%	80%	20%	40%	60%	80%
1:50	2	—	—	0.1	0.2	—	0.1	0.1	0.2	—	0.1	0.2	0.2
1:20	5	0.1	0.1	0.2	0.4	0.1	0.2	0.3	0.5	0.1	0.2	0.4	0.6
1:14	7	0.1	0.2	0.3	0.5	0.1	0.2	0.4	0.6	0.2	0.3	0.6	0.8
1:10	10	0.1	0.2	0.4	0.7	0.2	0.3	0.6	0.9	0.3	0.5	0.8	1.2
1:6.7	15	0.2	0.4	0.6	1.0	0.2	0.5	0.8	1.3	0.4	0.7	1.1	1.7
1:5	20	0.2	0.5	0.8	1.4	0.3	0.6	1.1	1.7	0.5	0.9	1.5	2.3
1:4	25	0.3	0.6	1.0	1.7	0.4	0.8	1.3	2.0	0.6	1.1	1.8	2.8
1:3.3	30	0.3	0.7	1.2	2.0	0.5	0.9	1.5	2.4	0.8	1.3	2.1	3.3
1:2.9	35	0.4	0.8	1.4	2.3	0.5	1.0	1.8	2.8	0.9	1.5	2.4	3.8
1:2.5	40	0.5	0.9	1.6	2.6	0.6	1.2	2.0	3.1	1.0	1.7	2.7	4.2
1:2.2	45	0.5	1.0	1.8	2.9	0.7	1.3	2.2	3.4	1.2	1.9	3.0	4.6
1:2	50	0.6	1.1	1.9	3.1	0.8	1.4	2.3	3.7	1.3	2.1	3.2	4.9

$$W_1/D = \frac{0.5}{3.0}$$
$$= 0.167$$

Interpolation from Table 9.9 gives

$$SC_1 = 0.334$$

Window (2):

$$H_2/D = \frac{0.75}{3.0} \qquad W_1/D = \frac{0.5}{3.0}$$
$$= 0.25 \qquad\qquad = 0.167$$
$$SC_2 = 0.084$$

Window (3):

$$H_1/D = \frac{1.5}{3.0} \qquad W_2/D = \frac{1.1}{3.0}$$
$$= 0.5 \qquad\qquad = 0.367$$
$$SC_3 = 0.634$$

Window (4):

$$H_2/D = \frac{0.75}{3.0}$$
$$= 0.25$$
$$W_2/D = \frac{1.1}{3.0}$$
$$= 0.367$$
$$SC_4 = 0.184$$

Required $SC = (SC_1 - SC_2) + (SC_3 - SC_4)$
$$= (0.334 - 0.084) + (0.634 - 0.184)$$
$$= 0.7 \text{ per cent.}$$

This value would have to be corrected for glazing bars and dirt on the glazing.

$\therefore$ Corrected sky component $= 0.5$ per cent.

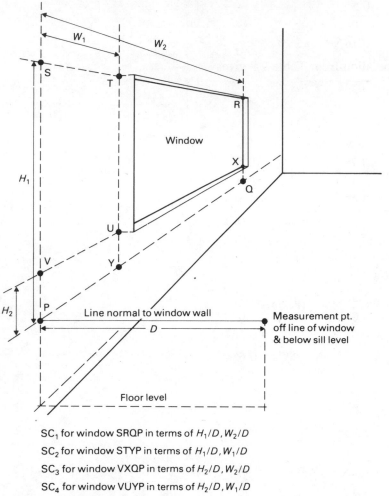

SC$_1$ for window SRQP in terms of H_1/D, W_2/D
SC$_2$ for window STYP in terms of H_1/D, W_1/D
SC$_3$ for window VXQP in terms of H_2/D, W_2/D
SC$_4$ for window VUYP in terms of H_2/D, W_1/D
SC = SC$_1$ − SC$_2$ − SC$_3$ + SC$_4$

Fig. 9.22 Sky component for a measurement point that is off the line of the window and below sill level

Table 9.11 Conversion factors for rooms whose floor area corresponds to 10 m^2 and 100 m^2
(Note: 36m^2 = 1.0 for all wall reflection factors)

Floor area	Wall reflection factor			
	20%	40%	60%	80%
10 m^2	0.6	0.7	0.8	0.9
100 m^2	1.4	1.2	1.0	0.9

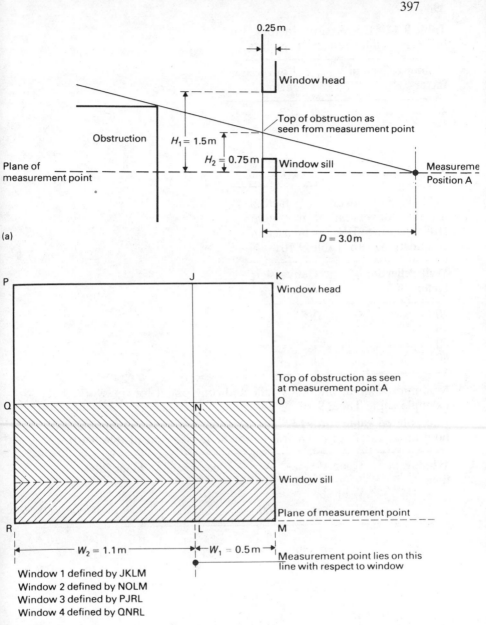

(a)

(b)

Window 1 defined by JKLM
Window 2 defined by NOLM
Window 3 defined by PJRL
Window 4 defined by QNRL

Fig. 9.23(a) Dimensions for worked example
(b) 'Windows' for worked examples

Table 9.12 Conversion factors for different ceiling reflection factors

Ceiling reflection factor %	Conversion factor
40	0.7
50	0.8
60	0.9
70	1.0
80	1.1

Table 9.13 Conversion factors for the conversion of minimum IRC to average IRC (to be used in conjunction with Table 9.10)

Wall reflection factor %	Conversion factor
20	1.8
40	1.4
60	1.3
80	1.2

Minimum value of internally reflected component for the worked example using Table 9.10.

Interpolated value for a 13.2 per cent window area as percentage of floor area, and floor reflection factor of 15 per cent (Table 9.10).

Window to floor	Floor 10%			Floor 15%	Floor 20%		
	Wall %				Wall %		
	40	50	60		40	50	60
10%	0.2	0.3	0.4		0.3	0.45	0.6
13.2%		0.428		0.503		0.578	
15%	0.4	0.5	0.6		0.5	0.65	0.8

Minimum internally reflected component

= 0.503 per cent

Adjustment for floor area from Table 9.11

$10\,\mathrm{m}^2 = 0.75$

$15.1515\,\mathrm{m}^2 = 0.8$

$36\,\mathrm{m}^2 = 1.0$

$0.503 \times 0.8 = 0.4$ per cent.

Adjustment for obstruction angle 10° instead of 20°, from Table 9.4

C = 35 for 10°

and   C = 31 for 20°

$$\therefore \quad 0.4 \times \frac{35}{31} = 0.45 \text{ per cent}$$

Minimum internally reflected component

= 0.45 per cent.

This value would then need further corrections for glazing bars, dirt on glazing and deterioration of decoration.

Corrected minimum internally reflected component

= 0.29 per cent.

(*Note*: average IRC = minimum IRC × 1.35 from Table 9.13.)

Daylight factor measurement in model rooms

Daylight factors can be measured directly in scale models of proposed or existing rooms. Models also allow a visual inspection of the effect of daylight in a room interior.

Three basic requirements for model studies are:

1. A scale model room.
2. A light source.
3. An instrument to measure daylight factors.

The model

A model of a proposed or existing room can be made to any suitable scale (for example, 1:10 or 1:20). The daylight factor measurements do not need to be corrected for the particular scale chosen (the ratios of window size and room proportions are the same in the model and full size room whatever scale is used.

The model can be fully detailed. The same surface finishes can be used in the model as in the real room. This means that reflection factors of walls, ceiling and floor can be accurately simulated. Model furniture, equipment and fittings can also be introduced and their effect assessed. Window glass of the same type as the real room can be used.

Provision for visual inspection can be made by having small peep holes in the walls or a hole in the floor large enough to allow a head through. The holes are covered during measurement.

If the effects of varying a number of parameters are to be investigated a flexible model is required where changes of room size, window size and finishes can be made quickly. Finishes can be changed by means of different coloured card inserts, and windows screened with card to change their size.

Access to the model can be made either by having a removable wall or ceiling.

The light source

Two approaches are adopted:

(a) The real sky.
(b) An artificial sky.

The real sky

This approach has the virtue that the model will be lit by the same source as the real room.

The model is placed on a table and located so that it has an unobstructed view of the sky vault. This requires either a flat rooftop or a large field. The model is orientated in the same azimuth as the real room. Scale models of obstructions can be introduced outside the model at scale distances away from it. The model can have windows on more than one wall and roof lights can also be investigated.

Measurements should only be undertaken on days when the sky condition is overcast unless the shading and sunlight patterns are also of interest. Limitations on measurement in terms of day and weather conditions restrict the use of this source.

Artificial skies An artificial sky enables measurements to be made under a standard set of conditions at any time.

There are various forms of artificial sky ranging from very large hemispherical domes (such as planetarium domes) to large, intermediate and small rectilinear mirror versions (Figs. 9.24–9.27). The

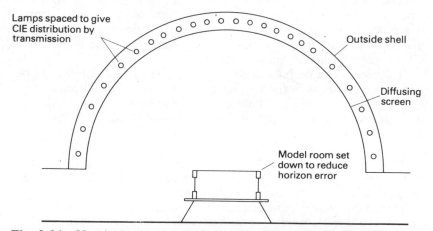

Fig. 9.24 Hemispherical transmission sky – a large sky with the model room inside – windows on more than one side and roof lights may be assessed

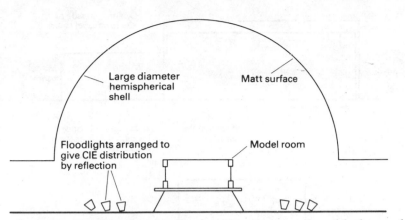

Fig. 9.25 Hemispherical reflection sky – a large sky with the model room inside – windows on more than one side and roof lights may be assessed

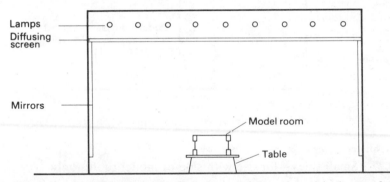

Fig. 9.26 Rectilinear mirrored sky – a large sky with the model room inside – windows on more than one side and roof lights may be assessed

larger artificial skies enable models to be fully immersed in them so that windows on all four walls and roof lights can be studied. Small versions might only be able to light models with windows in one wall.

The feature that all these skies must have in common is that the luminance distribution produced in them should approximate to a standard CIE overcast sky condition (or a uniform overcast sky). The actual illuminance can be varied by dimmers.

Usually the worst source of error in hemispherical skies is due to light originating near the horizon being thrown directly on to the ceiling of the model, a condition which never occurs in a real room. Too much light also gets on to the rear wall of the model. Setting the

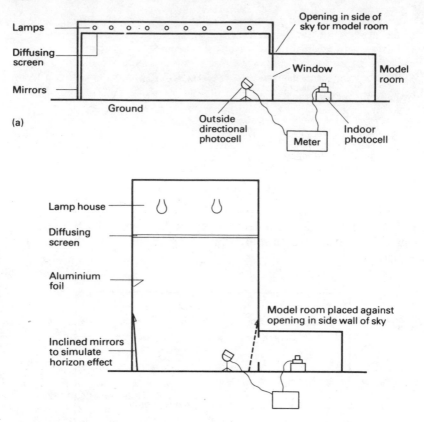

(b) Two patterns of small rectilinear mirrored sky

Fig. 9.27 Small skies – for use with side-lit model rooms only

model down with respect to the sky vault and screening the horizon helps. The horizon error is reduced in rectilinear mirror skies by reproducing an artificial horizon, apparently at infinity, by inter-reflection.

Instrumentation for model studies

The same types of daylight factor meters that are used in real rooms can be used in models.

Any daylight factor meter must be able to measure the external illuminance from the sky (real or artificial) and as near simultaneously as possible the internal illuminance at a point in the model.

The external photocell can be one of two forms:

(a) A colour and cosine corrected cell – this cell needs to 'view' a completely unobstructed sky vault.

(b) A colour corrected photocell at the end of a tube which can be
     directed at an unobstructed portion of sky at 42° above the
     horizon. (The luminance at 42° multiplied by π is the external
     illuminance for a CIE sky.)
     (Colour correction – the photocell responds to the spectrum in a
     similar way as the human eye instead of having a flat response.
     Cosine correction – the cell can accept light at any angle from
     the sky vault. Without this correction light at high angles of
     incidence might be totally reflected from the cell surface and not
     contribute to the measurement.)
     The internal photocell is a colour and cosine corrected cell.

When measuring using a real sky it is essential that an external
illuminance measurement is obtained each time an internal illuminance measurement is made at a new point in the model because
the sky condition is varying continuously. Artificial skies provide
a more stable value of external illuminance and one external reading
might be sufficient for several internal measurement positions.

The ratio internal to external illuminance expressed as a percentage is then calculated to give the daylight factor.

Some instruments such as the Megatron BRS daylight photometer
have been developed to give a direct scale reading of daylight factor.
The reading from the external cell is adjusted to a calibration value
and a button is depressed to connect the internal cell and read out
the daylight factor on a meter dial.

Another version for use specifically with models has 12 small
photocells, one of which is used to monitor the sky, the others are
deployed at grid points in the model. A selector switch enables the
cells to be read very quickly in turn with minimum disturbance to the
model. If the artificial sky can be adjusted to a suitable illuminance
value, the expression of the internal illuminances as percentages of
the external illuminance can be simplified.

Model studies enable rapid measurements to be made of daylight
factors at many points in the room. Investigations can then be carried
out in which room conditions are altered and new sets of data can be
obtained quickly.

Examples of investigations that could be carried out in models:

1. Variation of daylight factor patterns with window size.
2. Influence of window shape and position on daylight distribution
    for a given window size (Fig. 9.28(a) and (b)).
3. Influence of internal reflectances on daylight factors (Fig. 9.29).
4. Measurement of sky component only by using a black velvet
    lined model.
5. Effect of different types of glass on daylight factors (Fig. 9.29).
6. Effect of solar gain controls on daylight penetration.

404

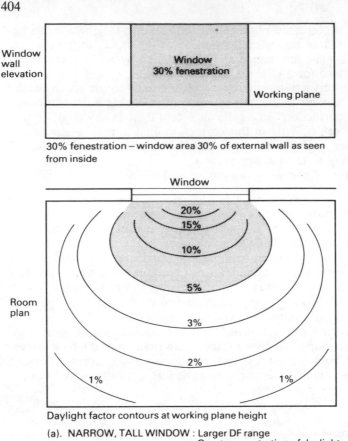

30% fenestration – window area 30% of external wall as seen from inside

Daylight factor contours at working plane height

(a). NARROW, TALL WINDOW : Larger DF range
Greater penetration of daylight
in line with window but not such
good distribution to the sides of
the room

Fig. 9.28 Model study of daylight distribution: comparison of the daylight distribution from two different arrangements of the same area of glazing

Daylighting recommendations

Recommendations on daylighting are to be found in British Standard Code of Basic Data for the Design of Buildings CP 3, Chapter 1, Part 1, 1964 (being revised) and the IES Code 1977. Some recommendations from this Code are given in Tables 9.14 and 9.15.

The IES daylighting schedule gives recommendations in terms of average daylight factor in a room, minimum daylight factor at a point and limiting daylight glare index (see Chapter 10).

For task lighting a minimum daylight factor is specified usually on the working plane. For areas with less then 1 per cent daylight factor supplementary electric lighting would be required.

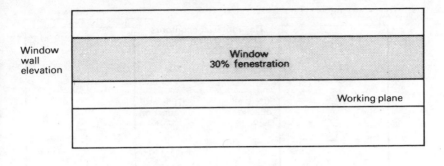

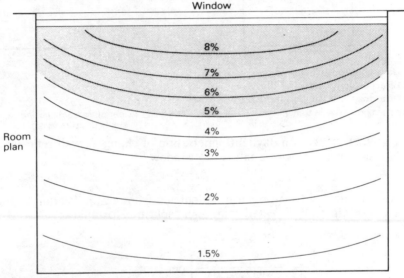

Daylight factor contours at working plane height

In this comparison the wider window lights a greater area of the working plane to more than 5% DF than the narrower window

(b). WIDE, SHORT WINDOW : Smaller D.F. range
Penetration smaller but better distribution across width of room

Fig. 9.28 (*Cont.*)

Where amenity natural lighting is required an average daylight factor is given.

Average daylight factor

$$= \frac{\text{total incident light flux on working plane}}{\text{outdoor illuminance} \times \text{area of working plane}}$$

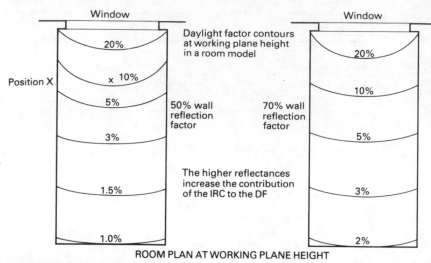

Quick check on influence of glass :-
DF at position X with glass : 10.3%
DF at position X without glass : 12.0%
Approximate glass transmission $= \frac{10.3}{12.0} \times 100 = 86\%$ (Clear glass transmission usually taken as 85%)

Fig. 9.29 Effect on daylight penetration of changing wall reflectance in a side-lit room with a tall, wide window

Average daylight factor can be calculated from a modification of the average internally reflected component split-flux formula.

Average daylight factor

$$= 0.85\, W \left[\frac{C}{A_{\text{fw}}} + \frac{CR_{\text{fw}} + 5R_{\text{cw}}}{A(1-R)} \right] \text{ per cent}$$

Table 9.14 Some recommendations from the IES Code Daylighting Schedule

	Average daylight factor (%)	Minimum daylight factor (%)	Position of measurement	Limiting daylight glare index
General offices	5	2	Desks	23
Bank-counters	5	2	Desks	23
Library–reading and reference room	5	1.5	On tables	23
College classrooms	5	2	Desks	21

Table 9.15 Daylight recommendations for dwellings

	Recommended daylight factor (%)	
Kitchen	2	Over at least 50% of the floor area (minimum 4.5 m²)
Living Room	1	Over at least 50% of the floor area (minimum 7 m²)
Bedroom	0.5	Over at least 75% of (minimum 5.5 m²)

Where symbols have the meaning given on page 381 plus A_{fw} = area of floor and lower walls not including window wall. Rearrangement of the formula will give the window area needed to provide a certain average daylight factor.

A uniformity ratio defined as the ratio of minimum daylight factor to average daylight factor might also be used in future recommendations.

Experiments

Experiment 9.1 Determination of the daylight factor and equal-illuminance contours in a room

Apparatus
Megatron daylight photometer, (Fig. 9.30) tape measure, graph paper.

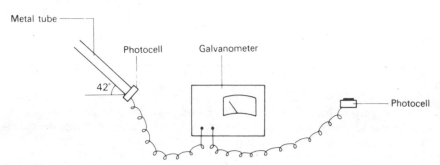

Fig. 9.30 Daylight photometer

Method
The room is divided into a grid of squares of 1 m side and the graph paper is used to prepare a scale plan of the room.

The outside directional photocell is set up outdoors, levelled and set to 42° above the horizontal so that it points to an unobstructed part of the overcast sky. The meter is adjusted to its calibration figure. The indoor photocell is placed at the intersection of two grid lines at working plane level. The daylight factor percentage may be obtained directly from the meter by depressing the range knob. The indoor photocell is then moved to each of the other grid intersection points and readings taken. The calibration is checked before each measurement in case the sky condition has altered. The results are recorded on the graph paper at the appropriate points. Daylight factor contours are then drawn on the plan.

Theory
The daylight factor at a point in a room is the ratio of the illuminance indoors to the simultaneously occurring outdoor illuminance expressed as a percentage. When an outdoor directional photocell is used the outside illuminance (E) is related to the luminance (B) of the sky at 42° above the horizontal such that $E = \pi B_{42}$. The meter automatically takes this relationship into account.

Questions

1. State the luminance distribution and illuminance formulae for a CIE standard overcast sky.
2. Define the term daylight factor.
3. Why are daylight factors used to measure daylight?
4. State three components of daylight factor and describe how they originate.
5. State the further corrections that must be made to daylight factor components after they have been obtained by protractors, nomograms or tables.
6. Give a brief outline of how a BRS daylight protractor is used to measure a sky component.
7. Compare three methods for predicting daylight factors.
8. Explain how to make daylight factor measurements in models of rooms.
9. Calculate the daylight factor in the room in the worked example (Example 9.2) on page 384 if the measurement point is on the same normal to the window but only 0.5 m from the inside face of the window wall (0.75 m from outside face):
   (a) using a BRS daylight protractor and nomogram;
   (b) using the BRS tables.
10. Calculate the average daylight factor for the worked example (Example 9.2) on page 384 and comment on your result.
11. Calculate the window area needed to give a corrected average daylight factor of 5 per cent for the worked example (Example 9.2) on page 384.

Chapter 10

Illumination for human comfort

The lighting in a room produces comfort conditions when it achieves its objective of lighting the task adequately without being distracting and glaring or causing unnecessary visual fatigue. The emphasis should be on the task to be lit and not on the source of light.

Lighting can be used to reveal the shape, form, texture and colour of both the task and the interior but it can also be the source of glare and uncomfortable flicker, cause veiling reflections on glossy surfaces and induce eyestrain.

To produce comfort conditions the lighting must be designed to suit the requirements of our visual mechanism (the combination of eyes and brain).

Light and the eye

Light is that part of the electromagnetic spectrum which occurs between the infra-red and the ultra-violet. The velocity of light is approximately 300 Mm/s in a vacuum. The range of wavelengths to which the eye is sensitive is from approximately 350 to 700 nm. The mixture of all visible wavelengths is called white light (cf. white noise in sound). Unlike the ear, the eye and the brain cannot recognise the individual wavelengths making up the light and can only accept the overall effect of the mixture. Newton demonstrated that white light could be dispersed through a prism and that colour sensations are associated with different parts of the spectrum. The eye is not equally sensitive to different wavelengths (Fig. 10.1).

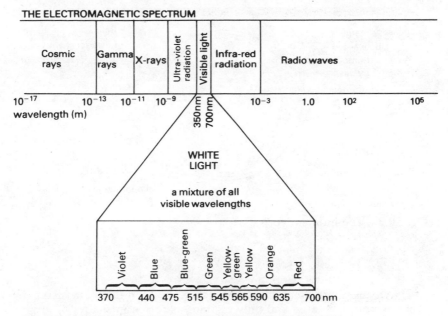

COLOURS ASSOCIATED WITH THE VISIBLE PART OF THE SPECTRUM

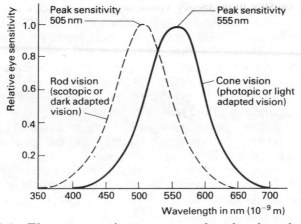

Fig. 10.1 Electromagnetic spectrum, colour bands and eye sensitivity

Light enters the eye and is focused by a lens on to the retina (Fig. 10.2). The light receptors of the retina send electrical messages, via the optic nerve, to the brain for interpretation. The light receptors are of two kinds: cones which are used when the eye is light adapted as in most lighting design situations, and rods which automatically function when the eye is dark adapted. The cones are

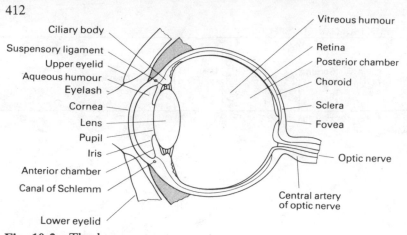

Fig. 10.2 The human eye

tri-colour pigmented and enable objects to be seen in colour. Rod vision is not pigmented and only enables us to see in shades of grey. The fovea is the central part of the retina on the optic axis of the eye and consists entirely of cones. The end of the optic nerve is deficient in rods and cones and forms the blind spot in each eye.

Objects at different distances from the eye are focused on to the retina by adjusting the curvature of the convex eye lens, a process called accommodation. The adjustment is controlled by the ciliary muscles which are relaxed when the eye is viewing scenes from 6 m to infinity and contract causing the lens to thicken and increase its curvature when focusing down to the near point. The near point recedes from about 100 to 750 mm during the course of a working lifetime due to hardening of the lens. Prolonged viewing of near objects tends to produce fatigue because not only is muscular energy being used to produce accommodation but muscular energy is also used to maintain convergence of the two eyes.

The iris exercises control over the quantity of light entering the eye and is capable of an 8:1 reduction. As the iris closes the depth of focus increases and detail can be recognised over a wider range of distances as illuminance increases.

As the eye ages the muscles controlling the iris lose their ability to open it as wide and this coupled with an increase in absorption and scattering of light in the eye means that older people require higher illuminance levels than younger people for equal ease of seeing.

The horizontal field of view when looking straight ahead is about 190° wide, although binocular vision only exists over 120° of this (Fig. 10.3). The vertical field of view is from approximately 45° above the horizontal to 50° below it.

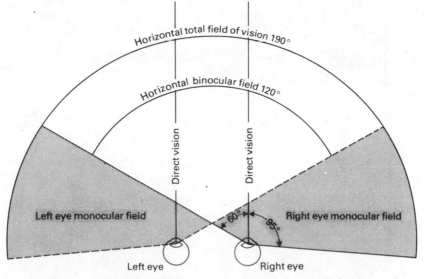

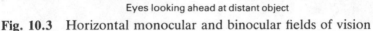

Eyes looking ahead at distant object

Fig. 10.3   Horizontal monocular and binocular fields of vision

Adaptation

Adaptation is the process by which the eye adjusts itself to different luminances. The full range of luminances over which the eye can adjust is from a visual threshold of 10^{-6} cd/m^2 to an upper limit of 10^6 cd/m^2, above which retinal damage would occur due to excessive absorption of radiation. The eye and brain cannot cope with this full range simultaneously but can only extract complete information from the visual field within a luminance range which is less than 10 000:1. The eye therefore acts as a self-optimising device which automatically adjusts itself up and down the full range in order to extract the maximum available visual information.

The actual adaptation level at any time is conditioned by the average brightness of the field of view. At low levels of adaptation a luminance which is only $\frac{1}{10}$ of the adaptation level may be too low for detail to be seen clearly while at the same time a luminance 100 times the adaptation level may not be too bright for comfort. At high levels of adaptation however, the eye can see detail which is only $\frac{1}{100}$ of the adaptation level but luminances only 10 times greater may cause discomfort. For this reason the danger of glare occurring increases as the illuminance level rises. At intermediate adaptation levels the useful luminance range has a more even spread either side of the adaptation level (Fig. 10.4).

The adaptation level governs not only the particular range of

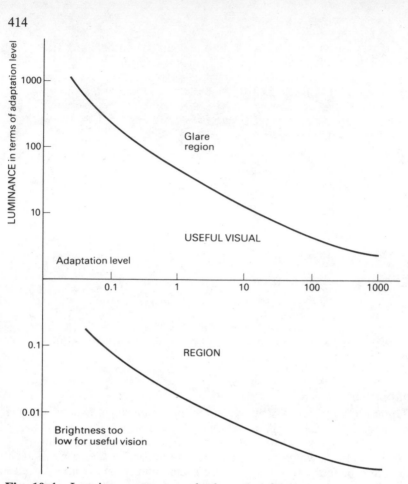

Fig. 10.4 Luminance range and adaptation level

luminances over which we can see but also the ability to appreciate differences in luminance (i.e. contrast), colour and detail. We do not see equally well at all adaptation levels. Under moonlight it is only possible to detect 10 per cent differences in luminance while in daylight luminance differences as small as 1–2 per cent are noticeable. We can never adapt to moonlight so that the surroundings appear as bright as or have the same apparent brightness relationships as under daylight. Our appreciation of colours and small differences in colour increases with adaptation level. At very low levels we cannot appreciate colour at all because the eye is functioning on rod vision. Small differences in detail can be picked out more easily or may even be noticed for the first time as the level is raised. This ability to discriminate contrast, colour and detail more easily at higher adaptation levels must be taken into account when deciding the illuminance level suitable for a particular task.

The eye takes a finite time to alter its adaptation level. This is particularly noticeable when the eye becomes dark adapted and involves the transition from cone to rod vision. When there is a sudden change in luminance vision is impaired while adaptation takes place. The adaptation process has three components:

1. A rapid phase during which cell or nerve behaviour readjusts.
2. A longer phase while there is readjustment of pupil size.
3. A slow phase which controls the overall adaptation time during which there is stabilisation of the ratio between bleached and unbleached photopigment in the light receptors.

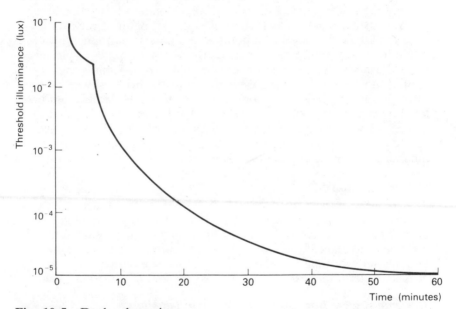

Fig. 10.5 Dark adaptation

We have all experienced walking into a darkened room and gradually being able to see more detail as time goes on. As can be seen from Fig. 10.5 dark adaptation takes as much as 60 minutes. Adapting back to a high level of luminance does not take as much time, perhaps 3–6 seconds. Fortunately adaptation between different luminances as we scan around a visual scene is a fairly rapid process at the luminance levels produced by the levels of illuminance and range of reflectances recommended for task lighting. However, the visual mechanism still has to cope with the problems of excessive brightness.

Glare and its control

Glare occurs when luminaires, windows or other sources, seen either directly or by reflection, are too bright compared with the general brightness of the interior. Glare is always present in a lit environment but it is only the more severe cases which cause real problems. Two categories of glare are recognised according to whether it is more difficult to carry out the task in hand or more uncomfortable to do so – these are called disability and discomfort glare respectively. They may be present simultaneously or separately.

Disability glare

This is glare which severely interferes with the ability to see detail without necessarily causing visual discomfort. Disability glare has a physiological basis related to the shift in adaptation level. Bright lights or their images and windows close to the point of regard may raise the adaptation level to a point where objects at a much lower brightness level cannot be seen properly. A bright spotlight placed to shine down a staircase may prevent the stair treads from being seen. A portable chalkboard viewed against the bright sky seen through a large window may be impossible to read. Care should be taken in the location of visual tasks so that bright lights, or their images, or windows, are not close to the line of sight. Failing this, remedies include screening the source from view, reducing its luminance or increasing the task illuminance.

Discomfort glare

This is glare which causes visual discomfort without necessarily reducing the ability to see the task. It is a psychological effect producing annoyance and distress rather than a direct reduction in visual performance. However, it is much more likely to occur in a building interior than disability glare and its cumulative effect can cause people to feel excessively tired at the end of a working day.

Discomfort glare from artificial lighting installations

When looking at a task the eye is aware of bright lights within the visual field ahead; i.e. within a vertical angle of 45° above the horizontal (Fig. 10.6). It is known that the magnitude of the glare sensation is related directly to the luminance of the glare source and its apparent size as seen by the observer. The discomfort is reduced if the source is seen in surroundings of high luminance and is off the line of sight.

The degree of discomfort produced by a lighting scheme can be reduced by limiting the luminance of the sources in the direction of the eye, screening the sources from view, raising the background luminance against which they are seen or repositioning the work station. The luminance may be limited by restricting the distribution

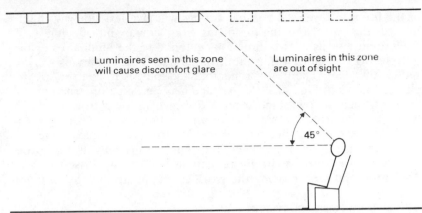

Luminaires seen in this zone will cause discomfort glare

Luminaires in this zone are out of sight

45°

Fig. 10.6 Discomfort glare – visual field

Trough

Bowl silvered lamp

'Black hole' reflector with gloss black finish

Symmetric louvre

White or matt aluminium: medium brightness Grey or black: low brightness

Cut-off

Specular aluminium dark appearance

Specular aluminium low brightness

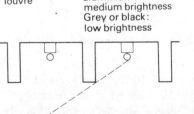

Coffered ceiling

Cornice

Pelmet

Fig. 10.7 Use of reflectors and louvres to control direct glare

of the fitting – lower BZ number fittings will give less sideways light towards the eye. Screening the fitting from view by introducing downstand screens in two directions external to the luminaires or by introducing screening within the fitting in the form of louvres will reduce the glare. The use of reflectors and louvres to control direct glare is shown in Fig. 10.7. Advantage may be taken of structural features, such as downstand beams, to conceal fittings from view. The ceiling can be made brighter by choosing a fitting with more upward flux or specifying a higher reflectance floor to reflect more of the downward flux back on to the ceiling. This is especially useful in the case of recessed fittings. Bright reflections will also give rise to discomfort and repositioning the work can help improve the situation.

The IES glare index system

The IES glare index system enables the acceptability of discomfort glare to be assessed for proposed or existing lighting schemes.

Experimental work at the Building Research Station established that discomfort glare from a light source was a function of the source luminance, the background luminance against which it is seen, the angular size of the source and its position relative to the observer.

A glare constant (g) can be calculated for each light fitting using the BRS glare formula:

$$g = \frac{B_s^{1.6}}{B_b} \times \frac{\omega^{0.8}}{p^{1.6}} \times 0.478$$

where: B_s = the source luminance in cd/m^2

B_b = the average background luminance in cd/m^2 against which the source is seen

ω = the angular size of the source in steradians as seen at the eye

p = a position index which indicates the effect of the position of the source on its capacity to produce discomfort glare.

The IES glare index for the lighting scheme is then given by:

Glare index = $10 \log_{10} (0.5 \times \Sigma g)$

where Σg = the sum of the glare constants produced by the individual fittings

Glare index values range from less than 10 for low levels of discomfort glare to 28 for high levels of discomfort glare. A change of 3 units on the scale represents a significant change in the level of discomfort. However, instead of relating the glare index directly to sensations of glare the IES carried out field studies from which it was able to set different limiting values of glare index for rooms in such a way that experience indicated that complaints from glare discomfort would be few or entirely absent. This would always be the case if the limiting glare index were set at a very low value for each room type, but it was recognised that in some situations tolerance of glare would be much greater than in others. Higher levels of discomfort can be tolerated for instance in a warehouse where somebody spends only 10 minutes collecting items from a shelf and then alters his position or goes elsewhere, than in a room such as an office where the direction of view may be fixed, the visual task more demanding, and the time spent at a work position may extend for hours.

Appropriate limiting glare index values from the series 10, 13, 16, 19, 22, 25 and 28 are given in the IES Code for each room type. Note however, that luminous ceilings should not be used in situations where the limiting glare index is less than 19 and elsewhere their luminance should be limited to less than $500 \, cd/m^2$.

Although the above formulae could be used to calculate the glare index for a lighting scheme it is more convenient to use sets of tables for BZ classified luminaires.

A proposed scheme can be checked by the calculation technique given in IES Technical Report No. 10, 'The Evaluation of Discomfort Glare', as long as the room is rectangular and has a regular array of fittings. The technique applies to a direction of view which is horizontal, straight across the room and 1.2 m above the floor (the eye level of a seated person). It also assumes that glare is additive, so that the glare from all the luminaires would be the same as for a single luminaire of the same luminance, but whose angular size (ω) is the total of all the fittings.

An initial glare index is read off Standard Tables in Technical Report No. 10 or alternatively from a manufacturer's photometric data sheets, and then is corrected for the actual conditions of downward flux, luminous area of the luminaires, and the height above a 1.2 m eye level. Further corrections can be made for lighter floors and also the orientation of linear fittings. (A manufacturer's data sheet may incorporate this last correction and the luminous area correction in the initial glare index.)

The scheme needs to be revised if the calculated glare index exceeds the recommended limiting glare index in the Code, even if this is only by a small amount. Modifications to the scheme include replacing the luminaire by one of lower BZ classification, increasing the flux fraction ratio so that more light is thrown on to ceiling, reorientating linear fittings, changing the mounting height and increasing the reflection factors of the room surfaces.

Glare index calculations

(For a fuller account and the complete sets of tables and graphs referred to below see IES Technical Report No. 10.)

Outline procedure

1. Transfer all information to computation sheet.
2. Calculate:
   (a) the height (H) of the luminaires above a 1.2 m eye level;
   (b) the dimensions, X and Y expressed in terms of H; i.e. $3H$, $2\frac{1}{2}H$, etc. (see small diagram on page 425) for definition of X and Y relevant to line of sight);
   (c) the flux fraction ratio for the luminaire.

   Flux fraction ratio

   $$= \frac{\text{upward light output ratio per cent}}{\text{downward light output ratio per cent}}$$

   $$= \frac{\text{ULOR}}{\text{DLOR}}$$

   (d) the downward flux from the luminaire.

   Downward flux

   $$= \text{lighting design lumens per luminaire} \times \text{downward light output ratio}$$

3. Initial glare index
   (a) Refer to sheets of initial glare indices (Tables 1.1–1.10 in IES Technical Report No. 10) and select table appropriate to BZ number of luminaire to be used; i.e. 1.5 for BZ 5.
   (b) Select appropriate values for X and Y in terms of H at left-hand side of table.
   (c) Select appropriate flux fraction ratio and reflection factor group (for ceiling, walls and floor).
   (d) Read off initial glare index.
   Note: linear interpolation can be used for awkward values.
4. Corrections
   Corrections to the initial glare index are made as follows:
   (a) Downward flux – use graph Fig. 2 (IES Technical Report No. 10) to obtain a correction for the downward flux calculated in 2(d) above (Fig. 10.8)
   (b) Luminous area – the area of the orthogonal projection of the luminaire on to a plane vertically below it (plus some area of side prisms if fitted). Use 'Table A1' (IES Technical Report No. 10) to apply a correction for luminous area (Table 10.1).
   (c) Height (H) above a 1.2 m eye level. Use 'Table A1' (IES Technical Report No. 10) to correct for the actual value of H in metres (Table 10.1).
   (d) Correct for floor reflection factor and endwise or crosswise viewing of linear luminaires (using Tables 3.1–3.10 in IES Technical Report No. 10).

5. Glare index
    Sum positive and negative columns and then obtain algebraic
    difference. This is the glare index.
6. Compare glare index obtained with the value of the IES Limiting
    'glare' index for that type of room. If the value is below the limit,
    the discomfort glare should be acceptable for that type of room
    usage.

Worksheet for glare index evaluation

A typical worksheet for calculating glare index is shown below. It is
split into three parts: details of lighting installation; details of lighting
fittings and lamps; and final glare index. This type of worksheet is
used to calculate a glare index in Example 10.1.

Details of lighting installations

Description of room ....

Room height (ceiling to floor) .... m

Mounting height above 1.2 m eye level (H) .... m

Room dimension X (at right angles to line .... m = ... H
    of sight)

Room dimension Y (parallel to line of sight) .... m = ... H

Reflection factors ⎰ceiling .... per cent
                   ⎱walls .... per cent
                    floor .... per cent

Orientation of linear fittings (relative   endwise/crosswise
    to line of sight)

Details of lighting fittings and lamps

Fitting type and catalogue number ....

Lamp type and number per fitting ....

Light output of lamps (lighting design lumens) .... lumens

BZ classification ....

Flux fraction ratio $\left\{\dfrac{\text{ULOR}}{\text{DLOR}} = \dfrac{\text{per cent}}{\text{per cent}}\right\}$ ....

Downward flux (DLOR per cent) .... lumens

Luminous area .... m^2

| *Final glare index* | + | − |
| Initial glare index | .... | .... |

Conversion terms:

Downward flux	....	....
Luminous area	....	....
Height (H)	....	....
30 per cent floor reflection factor	....	....
Orientation of linear fittings	....	....
Sub-totals	....	....
Final glare index (net total)	....	

Note: If there is no definite line of sight, it should in general be taken parallel to the larger of the two dimensions to obtain the worst conditions of discomfort glare. However, if the fittings are linear, a check should be made for the direction at right angles to this, taking note of the orientation of the fittings.

Table 10.1 'Table A1' from IES Technical Report No. 10

Luminous area(A) (cm²)	(m²)	Conversion term	Height(H) above 1.2 m eye level (m)	Conversion term
50		+ 8.9	1.0	− 1.2
75		+ 7.5	1.5	− 0.8
100		+ 6.5	2.0	− 0.5
150		+ 5.1	2.5	− 0.3
200		+ 4.1	3.0	0.0
300		+ 2.7	3.5	+ 0.2
500	0.05	+ 0.9	4.0	+ 0.4
750	0.075	− 0.5	5.0	+ 0.7
1000	0.1	− 1.5	6.0	+ 1.0
1500	0.15	− 2.9	8.0	+ 1.4
2000	0.2	− 3.9	10.0	+ 1.8
3000	0.3	− 5.3	12.0	+ 2.1
5000	0.5	− 7.1		
7500	0.75	− 8.5		
10 000	1.0	− 9.5		
	1.5	− 10.9		
	2.0	− 11.9		
	3.0	− 13.3		

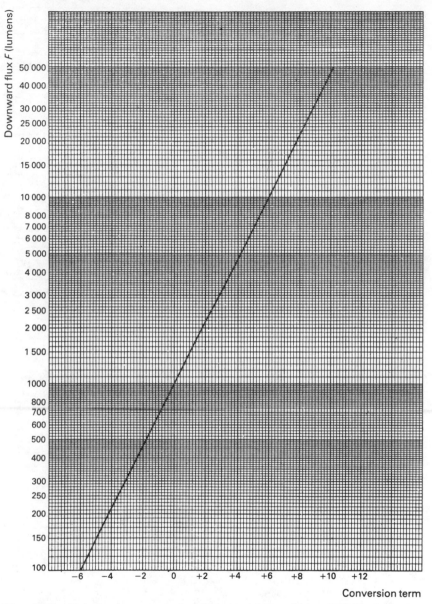

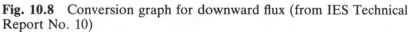

Fig. 10.8 Conversion graph for downward flux (from IES Technical Report No. 10)

Table 10.2 Initial glare index: Light distribution classification BZ 8 (Table 1.8 from IES Technical Report No. 10)

Flux fraction ratios of lighting fittings

Reflection factors of room surfaces (%) — Ceiling / Walls / Floor

Room dimension		0 $\left(\dfrac{\text{UFF}=0\%}{\text{LFF}=100\%}\right)$					0.33 $\left(\dfrac{\text{UFF}=25\%}{\text{LFF}=75\%}\right)$					1.0 $\left(\dfrac{\text{UFF}=50\%}{\text{LFF}=50\%}\right)$					3.0 $\left(\dfrac{\text{UFF}=75\%}{\text{LFF}=25\%}\right)$				
X	Y	70/50/14	70/30/14	50/50/14	50/30/14	30/30/14	70/50/14	70/30/14	50/50/14	50/30/14	30/30/14	70/50/14	70/30/14	50/50/14	50/30/14	30/30/14	70/50/14	70/30/14	50/50/14	50/30/14	30/30/14
2H	2H	14.4	17.3	14.9	17.6	18.1	12.3	14.7	13.2	15.5	16.5	10.1	11.2	11.3	12.9	14.4	6.5	7.9	7.9	9.3	11.3
	3H	17.8	20.5	18.3	20.7	21.2	15.8	17.8	16.6	18.6	19.6	13.3	14.8	14.5	16.0	17.5	9.8	11.0	11.2	12.5	14.4
	4H	19.4	22.0	19.9	22.3	22.6	17.3	19.1	18.2	20.0	21.1	14.9	16.4	16.1	17.6	19.1	11.3	12.4	12.7	13.9	15.8
	6H	21.1	23.7	21.7	24.0	24.4	19.1	20.9	20.0	21.8	22.8	16.6	17.9	17.8	19.1	20.6	12.8	13.9	14.2	15.3	17.3
	8H	22.1	24.5	22.7	25.1	25.3	20.0	21.6	20.9	22.6	23.7	17.5	18.7	18.7	20.0	21.6	13.9	14.7	15.3	16.3	18.3
	12H	23.1	25.4	23.6	25.9	26.3	21.0	22.5	21.9	23.6	24.6	18.4	19.6	19.6	20.9	22.5	14.8	15.6	16.2	17.2	19.2
4H	2H	15.7	18.2	16.2	18.5	18.9	13.6	15.4	14.5	16.3	17.4	11.2	12.7	12.3	13.8	15.4	7.6	9.0	8.7	10.2	12.1
	3H	19.5	21.8	20.0	22.2	22.6	17.3	18.9	18.2	19.9	21.0	14.7	15.9	15.9	17.2	18.9	11.2	12.0	12.6	13.5	15.5
	4H	21.5	23.8	22.1	24.2	24.7	19.3	20.7	20.3	21.8	22.9	16.8	17.8	18.0	19.2	20.8	13.0	13.7	14.5	15.4	17.3
	6H	23.2	25.4	23.7	25.8	26.3	21.0	22.4	21.9	23.4	24.6	18.4	19.2	19.7	20.7	22.3	14.6	15.3	16.1	16.9	18.8
	8H	24.4	26.4	25.0	26.9	27.4	22.1	23.4	23.0	24.5	25.6	19.6	20.4	20.8	21.8	23.3	15.8	16.4	17.3	18.0	19.8
	12H	25.5	27.4	26.0	27.9	28.5	23.2	24.4	24.1	25.5	26.6	20.6	21.3	21.8	22.7	24.2	16.8	17.3	18.3	18.9	20.7

Initial glare indices

8H	4H	13.9	14.4	15.3	16.1	17.9	17.6	18.4	18.9	19.8	21.4	20.2	21.5	21.1	22.5	23.7	22.5	24.5	23.0	25.0	25.4
	6H	16.0	16.6	17.4	18.2	20.2	19.8	20.6	21.0	21.9	23.7	22.5	23.7	23.5	24.7	26.0	24.9	26.8	25.4	27.3	27.9
	8H	17.2	17.8	18.7	19.4	21.3	21.0	21.7	22.2	23.2	24.9	23.7	24.8	24.7	25.9	27.2	26.1	27.9	26.8	28.4	29.1
	12H	18.4	18.9	19.9	20.5	22.4	22.1	22.9	23.4	24.3	25.9	25.0	25.9	26.0	27.1	28.3	27.3	29.0	28.0	29.6	30.2
12H	4H	14.1	14.6	15.6	16.3	18.1	17.9	18.6	19.1	20.0	21.5	20.5	21.7	21.4	22.8	23.9	22.8	24.7	23.3	25.2	25.8
	6H	16.4	17.0	17.9	18.7	20.5	20.2	21.0	21.5	22.5	24.0	23.0	24.0	24.0	25.2	26.3	25.3	27.1	26.0	27.7	28.4
	8H	17.7	18.2	19.2	19.9	21.7	21.5	22.1	22.7	23.6	25.2	24.2	25.2	25.2	26.4	27.6	26.6	28.2	27.2	28.8	29.5
	12H	19.1	19.5	20.5	21.0	23.1	22.8	23.5	24.0	24.8	26.7	25.6	26.5	26.6	27.7	29.0	28.1	29.6	28.7	30.2	30.9

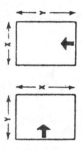

H Height of fittings above 1.2 m eye level.

X Room dimension at right angles to the line of sight in terms of the height(H).

Y Room dimension parallel to the line of sight in terms of the height(H).

Example 10.1 The 1977 IES Code suggests a limiting glare index of 19 for college seminar rooms. Calculate the glare index for the following seminar room and comment on your result:

Room dimensions $\qquad$ $9\,m \times 6\,m \times 2.7\,m$ high

Preferred viewing direction is down the long axis of the room.

Luminaire type	spherical opal diffuser mounted close to ceiling
Lamp type	200 watt GLS pearl
Lighting design lumens	2400 lumens
BZ classification	BZ 8
Downward light output ratio (DLOR)	33 per cent
Upward light output ratio (ULOR)	33 per cent
Luminous area	$0.05\,m^2$
Reflection factors	ceiling 70 per cent walls 30 per cent floors 14 per cent

(Refer to Tables 10.1 and 10.2 and Fig. 10.8.)

Details of lighting installation

Description of room	College seminar room
Room height (ceiling to floor)	2.7 m
Mounting height above 1.2 m eye level (H)	1.5 m
Room dimension X (at right angles to line of sight)	$6\,m = 4H$
Room dimension Y (parallel to line of sight)	$9\,m = 6H$
Reflection factors: ceiling	70 per cent
walls	30 per cent
floor	14 per cent

Details of lighting fittings and lamps

Fitting type and catalogue number	Spherical opal diffuser
Lamp type and number per fitting	200 W GLS Pearl 1 per fitting
Light output of lamps (lighting design lumens)	2400 lumens
BZ classification	8
Flux fraction ratio $\dfrac{ULOR}{DLOR} = \dfrac{33 \text{ per cent}}{33 \text{ per cent}}$	1.0

Downward flux (DLOR 33 per cent)		800 lumens
Luminous area		$0.05\,\mathrm{m}^2$

Final glare index	+	−
Initial glare index	19.2	
Conversion terms:		
Downward flux		0.6
Luminous area	0.9	
Height (H)		0.8
Sub-totals	20.1	1.4
Final glare index (net total)		18.7

The glare index is below the limiting glare index for this room and discomfort glare should be at an acceptable level.

Example 10.2 (using a manufacturer's data sheet)   Calculate a glare index in the direction of the long axis of the college room with the data below. Comment on your result if the IES limiting glare index is 16 in that direction towards a chalkboard.

The manufacturer's data sheet, Table 10.3, gives initial glare indices for both crosswise and endwise orientation of the luminaires and these also incorporate the correction for luminous area. Reference must still be made to 'Fig. 2' (Fig. 10.8) and 'Table A1' (Table 10.1) from IES Technical Report No. 10 for the conversion terms for downward flux, and height (H) above 1.2 m eye level.

Data:

Room dimensions	12 m long × 8 m wide × 3.2 m high
	Working plane at 0.850 m
Reflection factors	Ceiling 70 per cent
	Walls 50 per cent
	Floor 14 per cent
Maintenance factor	0.74
Luminaires	1800 mm twin tube with opal diffuser
	(photometric data sheet Table 10.3)
	Ceiling mounted
	Downward light output ratio 44 per cent
	Maximum S/H_m 1.73:1
	Dimensions: 1812 mm long × 200 mm wide
Lamps	1800 mm 75 W Plus White; 5500 lighting
	design lumens per lamp.
	Two lamps per luminaire

Details of lighting installation

Description of room	College room
Room height (ceiling to floor)	3.2 m
Mounting height above 1.2 m eye level (H)	2.0 m

Room dimension X (at right angles to line of sight)	$8\,m = 4\,H$
Room dimension Y (parallel to line of sight)	$12\,m = 6\,H$
Reflection factors: ceiling	70 per cent
walls	50 per cent
floor	14 per cent
Orientation of linear fittings (relative to line of sight)	crosswise

Details of lighting fittings and lamps

Fitting type and catalogue number	Twin tube with Opal Diffuser
Lamp type and number per fitting	MCFE 75 W 1800 mm Plus White 2 per fitting
Light output of lamps (lighting design lumens)	$11\,000(2 \times 5500)$
BZ classification	6
Flux fraction ratio $\dfrac{\text{ULOR}}{\text{DLOR}} = \dfrac{40 \text{ per cent}}{44 \text{ per cent}}$	0.91
Downward flux (DLOR 44 per cent)	4840 lumens
Luminous area	$0.35\,m^2$

Final glare index $+$ $-$

Initial glare index	13.5	from Table 10.3 (viewed crosswise)

Conversion terms:

Downward flux	4.1	
Luminous area	Already allowed for in initial glare index	
Height (H)	0.5	
Orientation of linear fittings	Already allowed for in initial glare index	
Sub-totals	$\underline{17.6}$	0.5
Final glare index (net total)	$\underline{17.1}$	

The calculated glare index of 17.1 for the scheme is above the recommended limiting glare index of 16 and discomfort glare will not be acceptable to people particularly at the rear of the room looking at the chalkboard.

    Try orientating the luminaires endwise to the direction of view.

Change in initial glare index = 13.5 − 11.8

$$= 1.7$$

Final glare index $\quad\quad = 17.1 - 1.7$

$$= 15.4$$

This is below the IES limiting glare index of 16 and a reorientation of the luminaires would produce an acceptable condition.

Discomfort glare from windows – sky glare

The discomfort glare experienced from the sky seen through windows is a function of the sky luminance and the general background luminance against which it is seen.

The reader is referred to IES Technical Report No. 4, 'Daytime Lighting In Buildings' and its supplement, for details of the glare constant and glare index formulae used in this situation.

It has been found that for practical conditions glare index values do not vary much with window size and are nearly constant above a window/floor area greater than 1–2 per cent for any set of conditions. The major controlling factor is the sky luminance and not the window size. The surround luminance will depend on the sky luminance and the room reflectances.

Table 10.4 gives glare indices for a sky condition (average luminance 8900 cd/m², horizontal illuminance 28000 lux) that is only exceeded for 15 per cent of the year. The average reflectances of 40 per cent and 60 per cent are representative of the extremes found in normal practice.

Table 10.5 gives glare indices for various sky luminances. If the ceiling height is greater than 4 m the glare index may be 1–2 units above these values. The glare indices apply to worst conditions – a single window directly ahead along the line of sight.

It is not possible at the moment to combine or compare the glare indices for artificial lighting and sky glare for assessing a permanent supplementary artificial lighting installation (PSALI) scheme. However, if the glare index for each is lower than the appropriate IES limiting glare index in each case then the overall effect is likely to be acceptable.

Methods for controlling glare from windows

1. Temporary measures for use only on bright days – adjustable curtains or blinds which may be controlled by the user:
   (a) Translucent curtains or blinds fitted internally or externally to the window.
   (b) Vertical or horizontal louvred or slatted blinds or fins fitted internally or externally.
2. Permanent measures in operation all the time:
   (a) Overhangs, canopies, awnings and deeply recessed windows – these restrict the view of the sky but daylight

430

Table 10.3 (For use with Example 10.2)

Glare indices

Room dimension		Viewed crosswise					Viewed endwise				
Ceiling RF%		70	70	50	50	30	70	70	50	50	30
Wall RF%		50	30	50	30	30	50	30	50	30	30
Floor RF%		14	14	14	14	14	14	14	14	14	14
X	Y										
2H	2H	6.2	7.5	7.3	8.6	10.2	6.0	7.3	7.1	8.4	9.9
	3H	9.0	10.2	10.1	11.3	12.9	8.3	9.5	9.4	10.6	12.2
	4H	10.4	11.5	11.5	12.7	14.2	9.3	10.4	10.4	11.6	13.1
	6H	11.8	12.8	12.9	14.0	15.6	10.1	11.1	11.2	12.3	13.9
	8H	12.5	13.5	13.7	14.7	16.3	10.4	11.4	11.5	12.6	14.1
	12H	13.3	14.3	14.5	15.5	17.1	10.6	11.6	11.7	12.7	14.3
4H	2H	7.2	8.3	8.4	9.5	11.1	7.1	8.2	8.2	9.4	10.9
	3H	10.4	11.3	11.5	12.5	14.1	9.7	10.7	10.9	11.9	13.5
	4H	11.9	12.8	13.1	14.0	15.6	10.9	11.7	12.0	12.9	14.6
	6H	13.5	14.3	14.7	15.6	17.2	11.8	12.6	13.0	13.9	13.5
	8H	14.4	15.1	15.6	16.4	18.0	12.3	13.0	13.5	14.2	15.9
	12H	15.3	16.0	16.5	17.2	18.9	12.6	13.3	13.8	14.5	16.2
8H	4H	12.6	13.4	13.8	14.6	16.2	11.8	12.5	13.0	13.7	15.4
	6H	14.5	15.2	15.8	16.4	18.1	13.0	13.7	14.3	14.9	16.6
	8H	15.6	16.2	16.8	17.4	19.1	13.6	14.2	14.9	15.3	17.1
	12H	16.7	17.3	18.0	18.5	20.2	14.1	14.7	15.4	15.9	17.6

12H

	4H									
4H	12.7	13.4	14.0	14.7	16.3	12.0	12.7	13.3	14.0	15.6
6H	14.8	15.4	16.0	16.6	18.3	13.5	14.1	14.7	15.3	17.0
8H	15.9	16.5	17.2	17.7	19.4	14.2	14.7	15.4	16.0	17.7
12H	17.3	17.7	18.5	19.0	20.7	14.9	15.3	16.1	16.6	18.3

ULOR – 40%   DLOR – 40%   LOR – 84%   Flux fraction ratio 0.91

Table 10.4 Glare indices for average sky luminance of 8900 cd/m²

Average reflectances of room surfaces	Glare index
0.4	26
0.6	24

Table 10.5 Glare indices for various sky luminances

Horizontal illuminance from whole sky (lux)	Average sky luminance (cd/m²)	Proportion of annual working hours exceeded (%)	Reduction in glare index
28 000	8900	15	0
20 000	6400	38	0.5
15 000	4800	54	1.5
10 000	3200	68	2.5
5 000	1600	87	4.0

      penetration is also reduced although some advantage can be gained by using light-coloured pavings and high reflection factor ceilings to reflect more light into the room.
(b) Low transmission glass – sky luminance is reduced but daylight levels are also permanently lowered.
3. Miscellaneous measures:
    (a) Increase internal reflectances.
    (b) Splay window reveals to reduce the contrast between the window and window wall.
    (c) Splay or taper window bars, transoms and mullions and keep light coloured to avoid excessive contrast.
    (d) Make sill as low as possible to increase illumination of the interior and raise adaptation of the eye.
    (e) Raise the luminance of the surround to the windows by using wall washing luminaires.

Figure 10.9 shows sketches illustrating some window detailing designs which have advantages for the control of glare.

Indirect or reflected glare caused by veiling reflections
This type of glare occurs when the images of bright lights or the sky are seen in surfaces which are not matt. The contrast in the task is reduced and makes the task more difficult to see. The more glossy the task the greater the reduction in contrast.

Printing inks themselves are glossy and many information sheets and brochures are printed on glossy paper. Meters and visual display unit screens are difficult to read if veiling reflections occur in the glass front. Veiling reflections not only cause interference with vision but can be distracting and cause visual discomfort.

The problem is largely one of the geometry of the source, task and observer. Repositioning luminaires or the observer, or re-positioning or angling the task, can minimise the effects of veiling reflections (Fig. 10.10).

Other remedial measures include using luminaires with special distributions or incorporating polarising diffusers in the luminaires

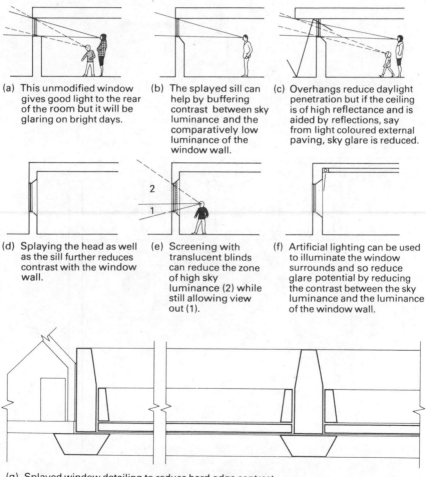

(a) This unmodified window gives good light to the rear of the room but it will be glaring on bright days.

(b) The splayed sill can help by buffering contrast between sky luminance and the comparatively low luminance of the window wall.

(c) Overhangs reduce daylight penetration but if the ceiling is of high reflectance and is aided by reflections, say from light coloured external paving, sky glare is reduced.

(d) Splaying the head as well as the sill further reduces contrast with the window wall.

(e) Screening with translucent blinds can reduce the zone of high sky luminance (2) while still allowing view out (1).

(f) Artificial lighting can be used to illuminate the window surrounds and so reduce glare potential by reducing the contrast between the sky luminance and the luminance of the window wall.

(g) Splayed window detailing to reduce hard edge contrast

Fig. 10.9 Control of glare by window design

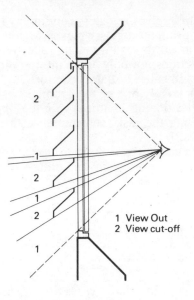

1 View Out
2 View cut-off

(h)

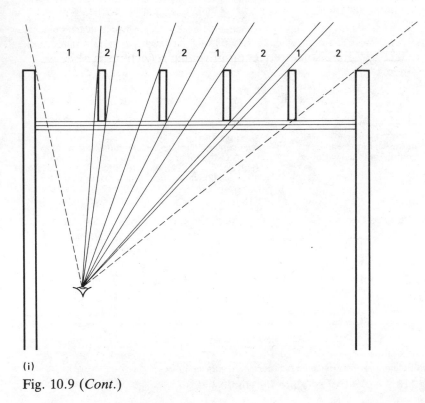

(i)

Fig. 10.9 (*Cont.*)

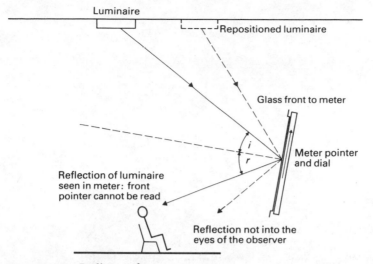

Fig. 10.10 Indirect glare

which produce more vertically polarised light than opal or prismatic diffusers. Further polarisation by reflection occurs at the glossy surface and the image intensity is much reduced. Some windows have also been fitted with polarizing films and although primarily intended to reduce sky glare some reduction in veiling reflections is also obtained.

Visual acuity

Visual acuity is the ability to see small detail. It is measured as the reciprocal of the angle subtended at the eye in minutes of arc by the smallest detail which can be picked out (Fig. 10.11).

The highest known visual acuity is 2.5 (measured for the resolution of two points subtending an angle of 24 seconds at the eye). Clinically normal acuity is 1.0 – the ability to see detail subtending an angle of 1 minute at the eye, under ideal conditions, although the visual acuity of a proportion of the population is as low as 0.5.

The ability of the healthy eye to see detail is a function of the illuminance on the detail, the contrast between the detail and its background, the size of the detail and the time spent viewing it (although many tasks are self-paced).

Contrast is used subjectively to describe the difference in appearance of two parts of a visual field seen simultaneously or successively. The difference may be one of brightness or colour or both. Objectively the term expresses the luminance difference ratio

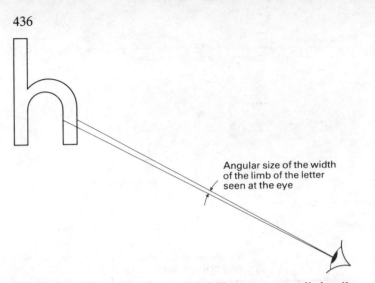

Angular size of the width
of the limb of the letter
seen at the eye

Fig. 10.11 Visual acuity – the ability to see small detail

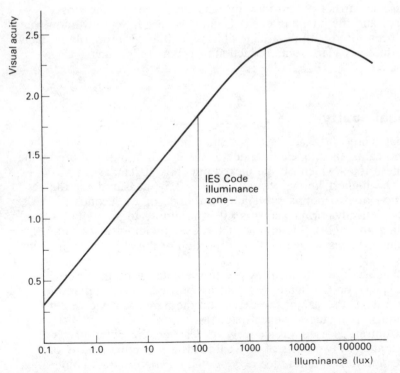

Fig. 10.12 The relationship between visual acuity and illuminance

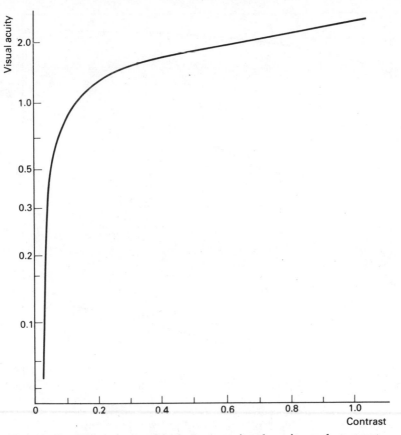

Fig. 10.13 The relationship between visual acuity and contrast

numerically by such relationships as:

$$\text{Contrast} = \frac{L_d - L_b}{L_d}$$

where L_d = luminance of detail in cd/m^2
L_b = luminance of background in cd/m^2

The variation of visual acuity with illuminance and contrast is shown in Figs. 10.12 and 10.13. The higher the illuminance up to some maximum level, and the greater the contrast the better small detail can be seen. These two factors affect visual performance (measured in terms of speed and accuracy of doing a task. Figure 10.14 indicates how visual performance improves for a given size and contrast task as the illuminance increases. It should be noted however, that very small low contrast tasks will never be seen as well

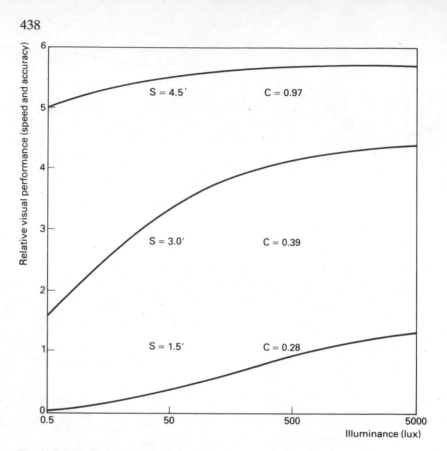

Fig. 10.14 Relationship between visual performance and illuminance for different values of apparent size (S) and contrast (C)

as large good contrast tasks no matter how much the illuminance is increased or how much time is spent viewing them.

In certain cases visual acuity can be improved by viewing small detail in monochromatic light thus avoiding chromatic aberration in the eye.

Visual fatigue

The causes of visual fatigue or eyestrain fall into three broad groups:

1. ocular;
2. constitutional;
3. environmental.

The ocular causes are concerned with bad eyesight and the effects of ageing changes. The constitutional causes are to do with the general state of health of the individual. However, it is the environmental causes which might produce headaches and eye strain with which we are most concerned here.

The environmental causes of visual fatigue can be divided into those which arise out of the visual task itself and those which arise out of the visual environment in which the task is performed.

The visual task

The brain tries to interpret the visual information sent to it. When this information is of poor quality the brain attempts to improve it by feedback to the eye. Strain results when this is an almost continuous process because of inadequacies in the task. These inadequacies in the visual task are when:

(a) the size of detail is too small causing focusing and convergence problems;
(b) the contrast between parts of the detail is too low;
(c) the visual task moves;
(d) the surface texture of the task makes seeing difficult;
(e) the pattern of the task may be disturbing.

The visual environment

There are a number of factors within the working environment which singly or in combination can give rise to visual fatigue.

(a) Inadequate illuminance – not having sufficient light on the task commensurate with its size, contrast and degree of difficulty.
(b) Too great a contrast between the task and its background. (For example, white paper on a black desk top or working on a drawing board using a local light while the rest of the room is in darkness.)
(c) Disability and discomfort glare.
(d) Veiling reflections from the task due to incorrectly positioned overhead luminaires.
(e) Flicker from fluorescent lamps.
(f) A psychological factor concerned with the individual's satisfaction with the environment as a whole. This is affected by such things as the appearance of the environment, the presence or absence of windows, the colour appearance and the colour rendering of the lamps and the degree of modelling produced.

Conditions necessary for good illumination

When we say a room has good lighting we mentally balance a number of visual conditions existing within the room. Good task lighting requires good overall room lighting in order to avoid discomfort. Think of the visual fatigue which results when working by a desk lamp in a darkened room or when watching a television set without any background lighting.

In task lighting situations it is necessary to consider:

(a) the way the task is lit;
(b) the way the interior as a whole is lit and relates to the task.

The effectiveness of task lighting is judged by the criterion of visibility; i.e. the ease with which important details of the task can be recognised. Visibility is influenced by the size of the task, the contrast between details of the task and their immediate background, as well as the amount, direction and colour rendering of the incident light. The room interior is judged by the appearance of the environment as a whole. The appearance should be appropriate to the function of the interior and be visually pleasing and free from discomfort caused by excessive glare. The appearance is affected by the general brightness of the interior, the patterns of light, shade and colour throughout the space, the degree of glare and the modelling of people, objects and structural features.

Specific conditions

Illuminance and daylight factors

The illuminance needed for the task depends on:

(a) the visual difficulty and complexity of the task;
(b) the average standard of eyesight;
(c) the level of visual performance required.

Some large, well-contrasted tasks could be performed at quite low illuminance values but the appearance of an interior looks gloomy below 200 lux and this should be regarded as the lowest value of illuminance for good illumination in working environments. Visual performance increases with illuminance but follows a law of diminishing returns (Fig. 10.15).

The illuminance recommendations in the IES Code are related to the visual requirements of the task weighed against the need for cost effective use of energy and the lessons of practical experience. Average and minimum daylight factor recommendations are also given in the IES Code.

Brightness

The interior appearance is affected by its general brightness which depends on the distribution of light in the room and the 'lightness' of room surfaces. (Luminance of surfaces is a function of both illuminance and reflectance.) The impression of general brightness of a room with diffusing luminaires and light decoration corresponds to the horizontal illuminance but this breaks down if the horizontal illuminance is high and the walls and ceiling are dark (either pro-

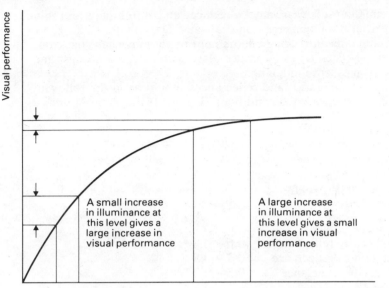

Fig. 10.15 Visual performance and illuminance

duced by low wall and ceiling reflection factors or downlighters with little sideways distribution).

Ceilings appear more prominent as room size increases. Small rooms can tolerate low reflectance ceilings but this is not the case with large rooms where a dark ceiling would create a gloomy appearance. There is more freedom with the reflectance of walls. Areas of high chroma can add interest to the scene. In large rooms

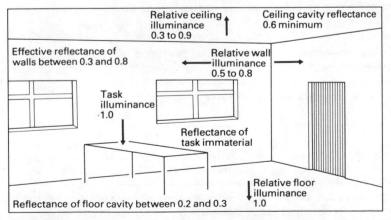

Fig. 10.16 Recommended ranges of reflectance and relative illuminance for room surfaces

light emission sideways may be reduced to control glare and this can restrict the illuminance on the walls. Dark floors make good modelling difficult. Lighter floors help to get more light back on the ceiling which helps to reduce glare with recessed fittings, and produces pleasant modelling.

Illuminance ratios and reflectances should be in the following ranges to produce good conditions (Fig. 10.16):

Illuminance ratios

Ceiling/task in the range 0.3 – 0.9	Below 0.3 the ceiling looks dark even if of high reflectance.
Wall/task in the range 0.5 – 0.8	Wall illuminance should not exceed 750 lux.

Reflectance ranges

Ceiling cavity reflectance	0.6 – 0.8
Effective wall reflectance	0.3 – 0.8
Floor cavity reflectance	0.2 – 0.3

(Refer to IES Technical Report No. 2 for calculation of cavity reflectances.)

Flow of light

The directional qualities of the lighting will affect:

(a) modelling – the ability of the light to reveal solid form;
(b) the emphasis of surface texture;
(c) the presence or absence of veiling reflections produced in glossy finish tasks.

Good directional qualities are obviously desirable. Further information on this subject is available in the IES Code.

Colour

1. *Colour appearance:* The colour appearance of lamps is classified into three groups:

Lamp colour appearance	Colour temperature (K)
cool	6500
intermediate	3500 – 4300
warm	2100 – 3000

The larger the room the cooler the lamp appearance can be but intermediate or warm types are needed for small rooms or for less than 300 lux. Intermediate types are suitable for daylight/artificial light combinations.

The colour temperature indicates that the colour appearance of the lamp is the same as that of a Planckian black body radiator (a full radiator) at that temperature. The colour temperature of incandescent lamps is also a guide to the filament temperature. Discharge lamps do not necessarily operate at such high temperatures and therefore the term correlated colour temperature is used for these sources.

2.   *Colour rendering:*   The spectral output of a lamp will affect how the colour of a particular surface is rendered. For example, a red rose will only appear red if the appropriate wavelengths of light are reflected from it. If these wavelengths are partially absent in the incident light the colour would be rendered differently. In the extreme case of a complete absence of the necessary wavelengths the surface would appear black. Taking the range of possible surface colours as a whole, the power balance and presence or absence of certain wavelengths in the incident light causes differences in colour rendering under different sources.

Daylight has a continuous spectrum and is usually thought to render colours 'naturally'. However, the emphasis within the spectrum alters during the course of a day (e.g. bluer at noon and redder in late afternoon), and affects the rendering of colours.

Incandescent lamps produce continuous spectra. A tungsten filament radiates more powerfully at the red end of the spectrum and produces more emphasis there, the yellows and greens are less emphasised and the blues are strongly subdued Fig. 10.17.

Discharge lamps produce spectral outputs which exhibit marked variations of power output across the spectrum. The discharge itself produces characteristic spectral lines, with more wide band radiation from phosphor coatings. Complex combinations of phosphor are used to improve the colour balance although at some loss in efficacy. A new generation of phosphors promises good colour rendering coupled with high efficacy. Halides introduced into high pressure mercury lamps also improve their colour rendering.

The IES Code gives recommendations on the suitability of lamps for different types of rooms, both as regards colour appearance and colour rendering.

Glare

This should be controlled within the IES limiting glare index and limiting daylight glare index. Disability glare should not be allowed to occur. Veiling reflectances in the task should be avoided by a careful consideration of the flow of light on to the task.

Flicker and stroboscopic effects

These undesirable effects are particularly produced by discharge lamps working off the 50 Hz AC mains.

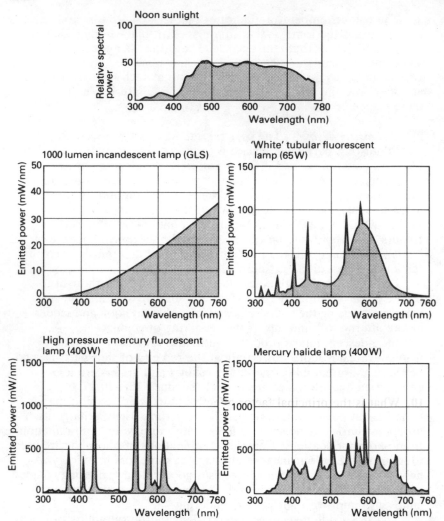

Fig. 10.17 Examples of typical spectral outputs of some light sources

Flicker can be reduced by shielding the electrodes or screening the ends of fluorescent tubes. Mixing tungsten lamps with discharge lamps helps.

Stroboscopic effects are reduced by:

(a) dividing lamps across three phases;
(b) using lead-lag circuits;
(c) using a high frequency mains supply.

A more detailed discussion of some of the conditions for good illumination referred to above and the techniques involved to bring them about is beyond the scope of this book, but at least the reader should be aware that there is more to good lighting than just producing a certain illuminance level. Quality of lighting is as important as quantity of light.

Questions

1. State the range of wavelengths to which the eye is sensitive.
2. Sketch a graph of the variation of eye sensitivity with wavelength.
3. Explain why the eye adapts for different illumination levels.
4. What influence does adaptation level have on our ability to see?
5. How long does the eye take:
   (a) to become dark adapted?
   (b) to adapt to 'normal' luminances?
6. State two types of glare and discuss ways in which each may be reduced.
7. Describe briefly the IES glare index system.
8. Calculate the glare index in the direction of the short axis of the room for Example 10.1.
9. Calculate the glare index for the college room in Example 10.2:
   (a) down the long axis of the room but with the luminaires turned through 90° to give an endwise condition of viewing;
   (b) across the short axis of the room, luminaires viewed endwise;
   (c) across the short axis of the room, luminaires viewed cross-wise.
10. What is the principal factor affecting sky glare?
11. State methods for reducing the glare from windows.
12. Define visual acuity.
13. State environmental factors which affect visual acuity.
14. Discuss factors which affect visual fatigue.
15. What are the conditions necessary for good lighting?

Section IV

Chapter 11

Fire

Introduction

The aim of this chapter is to consider the factors that affect the development and spread of fire in buildings. This includes the properties of materials in so far as they affect the fire loading or possible spread of flame.

There are three requirements for a fire to ignite and grow. These are heat, fuel and air. The relative availability of each of these factors controls the behaviour of the fire. Combustible materials are materials which burn, whereas inflammable materials burn with a flame and tend to spread a fire more quickly. A flame is caused by a region of burning gas released by the action of heat on the matter concerned. If the rate of burning is very rapid then an explosion may result.

The fact that a material is combustible does not necessarily make it unsafe in a fire. It may even be safer than a non-combustible material. Large sections of timber, particularly hardwood such as oak, are much safer than light steel sections. The timber ignites initially but quickly chars on the outside. The charcoal glows but does not ignite readily and consequently does not have a flame and does not spread fire easily. The charcoal, because of its thermal insulation, also prevents the inner part of the timber from burning. Hence the wood retains most of its strength. Steel loses much of its strength at high temperatures and as a result may buckle and collapse.

There are a number of other aspects which affect safety in a fire. One related to materials is the possible smoke emission. Many materials produce smoke on ignition but some produce gases which are toxic and others fumes which affect the eyes. Such effects can be just as important to safety as inflammability.

Calorific values of common building materials

All the combustible material in a building represents the available fuel. A measure of the amount of fuel may be obtained from a knowledge of the materials used and their calorific value. The calorific value for a material is the amount of heat given out per unit mass of the material when completely burnt. It is measured in kilojoules per kilogram. A few examples are given in Table 11.1.

While the calorific value can enable a figure for the available fuel to be obtained it does little to indicate the fire hazard. Large concentrated masses of materials are less likely to ignite and spread fire than thin sections. Thin vertical sheets of combustible material are far the worst in that they encourage flames to travel upwards and thus spread rapidly. There will be a difference if the sheet material is fixed to a heavy non-combustible substance, such as brickwork, compared to the situation with plenty of air (oxygen) on both sides. Furnishings such as curtains represent one of the worst situations when they hang vertically and away from a window or wall in thin sections with plenty of air available. The temperature at which ignition can arise is also very important.

Table 11.1 Approximate calorific values of materials commonly found in buildings

Material	kJ/kg
Natural gas	54 000
Petrol and fuel oil	46 000
Asphalt	39 800
Linseed oil	39 400
Butter	33 000
Coal	29 000
Wood	17 000
Paper	16 300
Sugar	16 300
Rubber	14 000
Straw	13 800
Bread	11 300

Fire tests on building materials and structures

Non-combustibility test for materials

The aim of this test is to determine which materials can be described as combustible or non-combustible. Originally described as the combustibility test it is now called the non-combustibility test in line with international thinking on the subject as the aim must be to use the least combustible materials in building construction.

The test involves three blocks of material 40 mm × 40 mm × 50 mm each heated in an electrical tubular furnace (Fig. 11.1). Where a material of suitable thickness is not normally available the blocks may be composed of layers to produce the required

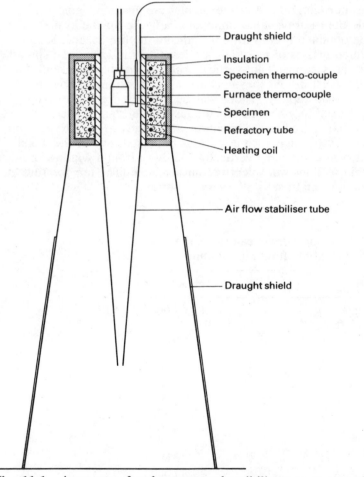

Draught shield

Insulation

Specimen thermo-couple

Furnace thermo-couple

Specimen

Refractory tube

Heating coil

Air flow stabiliser tube

Draught shield

Fig. 11.1   Apparatus for the non-combustibility test on materials

dimensions. The furnace is heated electrically to 750 °C for 10 minutes. One specimen is then lowered centrally into the tube and the two thermocouples measure the temperature in the sample and in the tube respectively. A non-combustible material is defined by means of this test and is one which shows no continuous flame for more than 10 s and where neither thermocouple rises by more than 50 °C above the initial temperature. All three samples must comply.

It is perhaps worth noticing that this test does not measure how combustible a material is but merely aims to find non-combustible ones. These should add nothing to the fire loading within a building.

Surface spread of flame tests for materials

This is a most important test. Its aim is to determine the tendency of materials to spread flames horizontally across their surface. The test is intended for the classification of exposed surfaces of walls and ceilings according to the rate and distance of spread of flame across them.

Four classes of materials are defined according to their behaviour in this test as shown in Table 11.2.

Table 11.2 Classification of materials for surface spread of flame

Classification	Flame spread after 90 s		Final spread of flame	
	Limit (mm)	+ tolerance for only 1 specimen (mm)	Limit (mm)	+ tolerance for 1 sample (mm)
Class 1	165	25	165	25
Class 2	215	25	455	45
Class 3	265	25	710	75
Class 4	>265 or >25	or	>710 or >75	
	(i.e. worse than Class 3)			

It can be seen that Class 1 materials are the best with the least flame spread. The materials must always be tested in the condition in which they are used. Hence if they are normally painted then that is how they should be tested. Some paints and some impregnating materials are intended to retard the spread of flames and may up-grade materials.

The test itself consists of a large (1 m × 1 m approx) vertical gas fired radiating surface maintained at 800 °C. The material sample (900 mm × 225 mm) is mounted lengthways at right angles with an igniting flame near to the hot surface (Fig. 11.2). The aim is to simulate the fire situation where materials may be heated by other

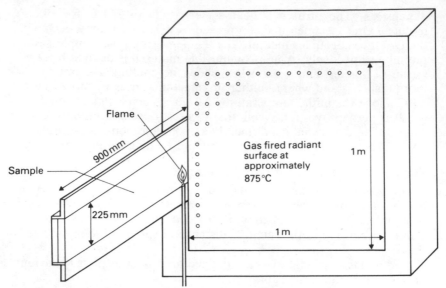

Fig. 11.2 Surface spread of flame test apparatus

sources and ignited by a flame. Six samples should be chosen so that the results may reasonably represent the material under test. It can be seen that one of the six samples may be allowed to fall slightly short of that normally required for the class.

In order to check the radiant heat, radiometers are mounted at 75 mm intervals on a 9 mm thick asbestos board in the place of the specimens. The radiation intensity for the apparatus may be checked. In practice it will be realised that the amount of radiant heat absorbed is dependent upon the colour of the material surface, dark surfaces absorbing more than light ones.

Table 11.3 shows the typical classification of several construction materials. They can vary with the surface finish used.

Table 11.3 Commonly accepted classification of some building materials

Class	Material
1	Plasterboard
2	Decorative plastic laminates
3	Chipboard, hardboard, plywood and timber of density $> 400 \, kg/m^3$
4	Wood fibre insulating board

Fire propagation test

The aim of this test is to assess the heat contribution which a material makes to a fire. Heat is supplied to the apparatus by means of gas and electric heaters. The contribution by the material is found from the temperature of the flue gases.

The apparatus consists of an asbestos box, internally 190 mm × 190 mm × 90 mm. One side 190 mm square contains the sample (Fig. 11.3). The sample itself is 228 mm × 228 mm and therefore completely covers this side. It is fixed into a holder which clamps

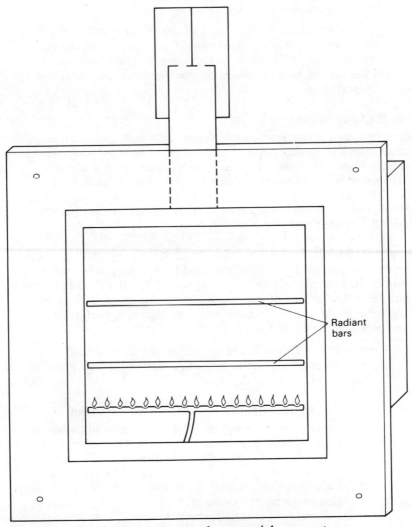

Fig. 11.3 Fire propagation test for materials apparatus

on to the side of the heated box. Three samples should be tested. Temperatures are taken for each sample of the flue gases; at $\frac{1}{2}$ minute intervals up to 3 minutes, then at 1 minute intervals up to 10 minutes, and 2 minute intervals up to 20 minutes. Then for each sample its index of performance is calculated from:

$$I = \sum_{1/2}^{3} \frac{\theta_m - \theta_c}{10t} + \sum_{4}^{10} \frac{\theta_m - \theta_c}{10t} + \sum_{12}^{20} \frac{\theta_m - \theta_c}{10t}$$

where θ_m = temperature in °C for the material at time t minutes

θ_c = temperature in °C of the calibration curve for the apparatus at time t minutes

θ_m is best obtained by drawing a graph of time and temperature and taking the value from it. θ_c is obtained by repeating the experiment with asbestos in place of the sample. The lower the index the less the rate of heat evolution by the specimen and therefore the better its performance.

Test for ignitability

Some materials ignite easily while others although combustible are far more difficult to ignite. A piece of newspaper held vertically ignites easily as does a thin piece of wood the thickness of a matchstick. Large sections of wood are difficult to light. Some liquids such as ether and petrol ignite very easily because their boiling points are only just above room temperature. Other liquids such as paraffin cannot be ignited easily in mass but burn if in small droplets.

The aim of the test is to determine the ignition characteristics of the exposed surfaces of flat building materials in the vertical position. The sample should be 225 m square and of normal product thickness mounted in a vertical frame. A gas burner is set at 45° to the vertical 3 mm from the centre of the specimen. The burner is fired in a standard manner for 10 s and moved away. Ignition performance is met if none of the three specimens:

(a)   flames for more than 10 s after burner removal; and if
(b)   burning does not extend to any edge during the test flame or within 10 s of its removal.

Fire resistance test of elements of building construction

The tests described previously apply to small samples of materials used in construction. Needless to say they are not representative of the whole construction and the loading which applies. The aim of the fire resistance test is to study complete sections of a construction to see how they behave under controlled fire conditions. The test varies slightly according to the type of element: wall, floor, column, beam, etc. There are three main criteria to be considered: stability, integrity and insulation.

Stability is important to ensure that the construction does not collapse too soon in case of fire. Thus a stability of 45 minutes would indicate that collapse took place 45 minutes after the start of the standard heating cycle.

Integrity is said to have failed when openings appear through which flames or hot gases can pass which would ignite material on the other side of the building element. Cotton wool is used for that material and its ignition indicates loss of integrity.

Insulation is important to prevent fires spreading from one compartment in a building to another. Insulation failure is said to occur in the test when the side of the element away from the fire reaches a temperature of 140 °C above the initial value.

For each of these three criteria the measure is one of time to failure. The Building Regulations regard fire resistance in terms of half-hour or hour steps. For example, $\frac{1}{2}$h, 1h, 2h, 3h, 4h, etc.

External fire exposure roof test

The aim of this test is to study the ability of a roof or such components as roof lights to resist penetration by external fire from radiation and flame and to see the extent of external surface ignition. The test applies to both sloping and flat roofs.

It is clear that some types of roof construction, such as thatch are very vulnerable to being set light by external sources. While the other tests study materials or internal building elements this test is concerned with fire spreading from one building to another. Roofs are graded from the best at A grade to the worst performance at D grade.

Growth and decay cycle of fires

A typical growth and decay cycle for a fire is shown in Fig. 11.4. It may conveniently be divided into three phases: ignition and growth; fully developed burning; and the decay period.

Ignition and growth

This phase is shown on the graph (Fig. 11.4) by the first three events:

(a) The ignition period.
(b) Fire growth with smouldering conditions producing smoke and gases. The time for this period may vary from a few minutes to many hours.
(c) The period of rapid temperature rise to the fully developed fire. During this period the fire has spread from burning of some materials to the burning of nearly all combustible substances.

It is usually possible for people to escape during the first two sections before the rapid rise in temperature commences, provided

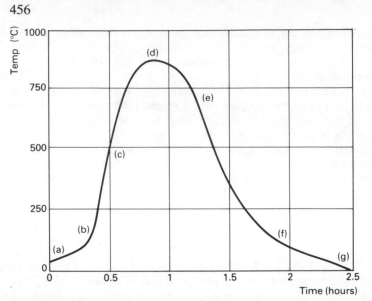

Fig. 11.4 Typical temperature–time graph for the growth and decay of a fire

they know the exit routes. This makes it important that fires are detected very quickly as this period may be fairly brief.

Fully developed burning

(d) After the surface materials are all alight other solid combustibles will commence to burn with slowly rising temperature for a long period. This time may be much longer than for the first phase depending upon the materials available and the air supply.

(e) There is a small total increase in temperature after which it starts to drop.

Decay period

After the peak temperature has been reached the fire begins to diminish until eventually flames cease. The total period may be very long until the debris can be handled. There are two sections to this phase:

(f) The smouldering condition when most of the combustible material is being used up.

(g) The burn-out period.

Fire load and fire density

The fire load and fire density are interrelated.

Fire load is the calorific value of the combustible material per square metre of floor area (MJ/m^2).

Fire density is the mass of combustible material per square metre of floor area. Table 11.4 gives some indication of typical fire densities.

Table 11.4 Typical fire loads for different spaces

Type of room	Fire load (kg/m^2)
Houses and flats	40–50
Classrooms	35
Library	180
Offices	10–225
Toilet areas	10–50

The duration of a fire is related to the fire load as shown in Fig. 11.5. In this figure the fire load is the equivalent mass of wood per square metre of floor area. Thus for a typical house or flat the fire duration could be expected to be about 1 hour. This may not be correct in practice as ventilation also controls the rate of burning and hence the duration.

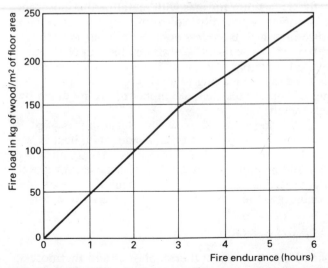

Fig. 11.5 Relationship between the fire load in timber and the fire endurance

Fire grading

This is normally defined in terms of time to failure in the standard fire test on elements. It may fail on integrity, stability or insulation. In practice this time is given in hours or fractions of an hour.

Properties of building materials at high temperatures

As already discussed materials can be divided into two distinct types, combustible and non-combustible. The effects of high temperatures on either may be disastrous but for quite different reasons. The properties of the materials will determine behaviour in fire. The method of classification of fires agreed with other European countries indicates the general failure by fire of some different materials.

Class A   Fires involving solid materials, usually organic, in which combustion generally involves the formation of glowing embers.

Class B   Fires of flammable liquids, or solids which melt to form flammable liquids.

Class C   Fires involving gases.

Class D   Metal fires.

Wood based materials

Timber is a very common form of fuel for fires. Thin sections used for wall finishes or furniture will burn very fast, particularly if they are placed vertically. Those thin finishes with metal or thermosetting plastic facings do not ignite so readily. Softwoods tend to ignite more quickly than hardwoods which have less volatile resinous material. Large sections of timber, particularly hardwoods, tend to char and be self-protective by the formation of a layer of charcoal. A temperature of about 250 °C is needed for charring to start.

The properties of composites such as chipboard are modified by the other materials included.

Paper is used as an internal finish on walls. Because it is fixed rigidly to the plaster air is only present on one side while heat is conducted away by the other, so it does not represent a significantly increased hazard. Building paper or cardboard boxes used for storage can represent an appreciable hazard. This is particularly true of the warehouse situation where items are stored vertically to a considerable height.

Plastics

These can be divided into two types: thermoplastics and thermosets. In general the thermosets do *not* burn or only with difficulty. Thermoplastics melt often at low temperatures and many burn, some very easily. A problem with many plastics is the smoke and fumes which are produced.

PVC

PVC melts because it is thermoplastic and burns only while a flame is applied. The fumes of hydrogen chloride which are produced extinguish the flames. The fumes are particularly unpleasant and thus PVC can be dangerous.

Polyesters

These are thermosetting plastics and although they burn they are self-extinguishing.

Polythene and polypropylene

These have a molecular structure which is similar to paraffin wax and when burning smell rather like a candle, producing smoke.

Nylon

This melts very easily and burns producing fumes with a characteristic odour. It tends to be self-extinguishing.

Polystyrene

This melts and burns very easily producing large quantities of dense black smoke. Expanded polystyrene containing air will burn extremely fast. In a fire ceiling tiles are particularly dangerous as molten drops of burning polystyrene can ignite clothes at the same time as producing black smoke which makes escape more difficult.

Non-combustible materials

It is often thought that non-combustible materials are safe in the event of a fire. This is not necessarily the case. Combustible materials are almost inevitably present as furniture if not as part of the building. Thus some fuel is present for a fire. The burning of this fuel may change the properties of non-combustible materials with a resultant change in their structural properties. Expansion of certain components may produce large strains causing cracking or distortion.

Steel

Unfortunately the strength of steel is greatly reduced at even moderately high temperatures. The result for an unprotected steel frame to a building in a fire would be deformation followed possibly by collapse.

Where steel framed buildings are constructed the steel may be protected in some manner to ensure it does not reach these high temperatures. This is done by increasing thermal inertia and insulation. Encasing the steel in concrete increases both of these. If lightweight aggregates are used then the insulation is even greater and there is less danger of the aggregates changing. Siliceous aggregates tend to spall at high temperatures and reinforcement is needed to ensure continuity of cover to steel. Limestone aggregates will remain intact to about 1000 °C.

Reinforced concrete
When a fire develops rapidly temperatures of 1200 °C are quite likely at a concrete surface. However such temperatures would very seldom be reached at the centre particularly as the steel reinforcement conducts heat from one part to another. Temperatures from 400 °C to 600 °C would cause a very large loss (up to 80 per cent) in compressive strength of concrete and an even larger loss in tensile strength. Combined with thermal expansion large deflections can result. The strength losses can be permanent particularly if temperatures are high enough to remove water of combination from the cement. In general concrete behaves well in fire but is dependent upon the aggregate used and temperatures reached.

Factors affecting the spread of fire

Heat travels in three different ways: by conduction, convection or radiation. Each of these factors influences the way a fire spreads. The problem can be considered under three different headings:
Fire spread:

(a)   within a room or compartment;
(b)   from one room to another within the same building;
(c)   to another building.

The factors affecting the first situation have been largely considered already. The properties of the materials in the room together with the supply of oxygen are the main ones. The spread of fire will be affected by the thermal insulation of the compartment, the higher the insulation the hotter the fire after a given time.

Spread of fire from one compartment to another depends upon the properties of the walls and whether there are any openings. In practice it should be the aim to contain any fire within a room. Closed doors reduce oxygen supply and convection. Windows may allow radiant heat to reach other inflammable materials. The thermal insulation of partitions needs to be good enough to reduce conduction and avoid materials on the opposite side reaching ignition temperatures.

Fire spread to other buildings is controlled by radiation and convection. In particular convection currents can carry burning material downwind for a considerable distance. Materials used on the exterior of buildings are clearly important as are materials near to the outside. Thus dry thatch and trees can contribute to fire spread.

Compartmentation

As already discussed the aim of compartmentation is to:

(a) prevent spread of fire out of the compartment;
(b) prevent entry of fire from nearby rooms or buildings;
(c) contain smoke which would make escape from other parts of the building more difficult and make fire fighting more hazardous.

Doors and window openings must be closed in an emergency. How resistant a partition needs to be depends upon its situation. The classification used in the Building Regulations for types of use indicates the standards needed (Table 11.5). Typically 225 mm of solid brickwork or 275 mm cavity brickwork will give over 4 hours' fire resistance. Similarly 25 mm of plasterboard on joists will provide 30 minutes' resistance.

Table 11.5 Minimum periods of fire resistance. In each case the Building Regulations give details of the sizes of spaces in relation to the fire resistance time

Class	Use	Min. fire resistance, excluding basements (hours)
1	Small residential up to 3 storeys	$\frac{1}{2}$
2	Institutional such as a school	$1-1\frac{1}{2}$
3	Other residential such as hotels	$\frac{1}{2}-1\frac{1}{2}$
4	Offices	$0-1\frac{1}{2}$
5	Shops	$0-2$
6	Factories	$0-2$
7	Assembly halls	$0-1\frac{1}{2}$
8	Storage and general	$0-4$

Behaviour of smoke

Unvented spaces

The products of combustion will usually include finely divided solid particles and often other components which may irritate or poison. In an unventilated room the smoke will quickly fill the space obscuring the source of the fire (Fig. 11.6). In large buildings the smoke may appear to originate from the opposite direction to its real source. The

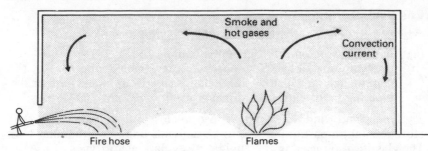

Fig. 11.6 Unventilated buildings in the event of a fire fill with smoke making escape and fire fighting difficult

result is that in trying to escape people may actually go towards the fire. A large unvented factory may be filled with smoke in only a few minutes.

Vented spaces

The aim of venting is to ensure that in the event of a fire smoke is removed from the building to ensure easy escape and ability to fight the fire (Fig. 11.7). The fire will burn more readily due to the greater supply of oxygen. However it is important that the oxygen content does not become too low while people are present. An oxygen content below about 15 per cent affects the brain so that incorrect judgements may be made. In large buildings such as factories and warehouses it is also important that some compartmentation is included from roof level to avoid smoke spread along the roof.

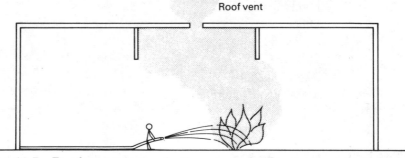

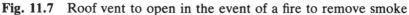

Fig. 11.7 Roof vent to open in the event of a fire to remove smoke

Experiments

Experiment 11.1 Demonstration of the effect of material thickness and size of ignition source on fire development

Apparatus

Oven dried 300 mm long hardwood dowel rods of 20 mm, 15 mm, 10 mm, and 5 mm diameter; retort stands and clamps; ignition sources; matches, taper, methylated spirits, candle, bunsen burner.

Method

The dowel rods are clamped horizontally. Using a match each is ignited and the result noted. This is repeated for the other sources of ignition. In the case of the methylated spirits it is poured over the rods and ignited with a match.

Theory

Ignition is the process whereby the volatile products of pyrolysis, formed when the material is heated, are ignited by an adjacent flame or are heated to a sufficiently high temperature that the material flames of its own accord. The first situation is known as pilot ignition and the latter spontaneous ignition. The likelihood of ignition is dependent upon the amount of heat available and the rate of dissipation by the material. Typical results are shown in Table 11.6 on page 464.

Results

Ignition source	Diameter of dowel in mm			
	20	15	10	5
Match				
Taper				
Meths				
Candle				
Bunsen burner				

Table 11.6 Typical results of ignition experiment with different sources

Ignition source	Diameter of dowel (mm)			
	20	15	10	5
Match	Just ignites but extinguishes when match is removed	Extinguishes after about 10 s	Continues to burn for about 30 s	Continues to burn
Taper	Ignites after about 35 s but extinguished when taper is removed	Burns for about 30 s	Ignites after about 20 s and continues to burn for similar period	Continues to burn
Meths	Extinguishes when meths is burnt	Extinguishes about 30 s after meths burnt	Extinguishes about 45 s after meths burnt	Extinguishes about 60 s after meths burnt
Candle	Ignites after 20 s and continues to burn for about 30 s after candle removed	Ignites after 15–20 s and continues to burn for about 30 s after candle removed	Extinguishes about 45 s after removal of candle	Continues to burn
Bunsen burner	Ignites after some 20 s but extinguishes about 35 s after heat removal	Continues to burn for about 60 s after burner removal	Continues to burn	Continues to burn

Experiment 11.2   Effect of specimen angle on the rate of spread of fire

Apparatus

300 mm × 25 mm × 3 mm strips of hardboard, similar strips of corrugated cardboard, methylated spirits, retort stands, stopwatch, ruler.

Method

The hardboard strips are mounted in the retort stands – horizontal, at 45°, and vertical. The end of each is soaked in methylated spirits and ignited with a match. The rate of flame propagation is measured in millimetres per second. The experiment is repeated for the corrugated cardboard (which may be rolled into a tube for rigidity) both with the corrugations parallel to the direction of propagation and perpendicular to it.

Theory

Fire spreads by heating the material ahead of the flame by convection, radiation and sometimes conduction from the flame. Convection and radiation are enhanced if the unburnt fuel is above the burning material. Many materials are anisotropic in that the rate of burning may be different in different planes. Corrugated cardboard is one example but wood is also significantly anisotropic though it is less easy to demonstrate.

Results

Angle	Flame propagation rates (mm/s)		
	Material		
	Hardboard strip	Corrugated cardboard (corrugated parallel to direction of propagation)	Corrugated cardboard (corrugated perpendicular to direction of propagation)
Horizontal			
45°			
Vertical			

Experiment 11.3 To investigate the effect of material thickness on the rate of flame spread

Apparatus
Filter papers, stand, stopwatch.

Method
Different thicknesses of filter papers are produced from 1 to 10 and fixed at their edges. A small hole is punched at the centre. Two concentric rings are drawn on the sets of papers. The single paper is ignited at the centre and the time for the flame to spread between the two rings is noted. This is repeated for each of the sets of filter papers.

Theory
Most thin homogeneous materials spread fire at a rate which is inversely proportional to the thickness, or

$$R = \frac{k}{m}$$

where $R =$ linear rate of spread
  $m =$ mass/unit area
  $k =$ constant

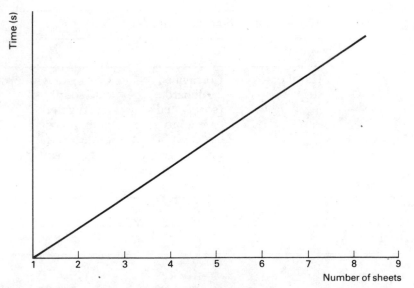

Fig. 11.8

But $\quad R = \dfrac{\text{distance}}{\text{time}} = \dfrac{\text{constant}}{t}$

$\therefore \quad \dfrac{c}{t} = \dfrac{k}{m}$

But $\quad m\ \alpha$ number of layers (n)

$\therefore \quad \dfrac{c}{t} = \dfrac{k_2}{n}$

If the burning is in radial form then edge effects are eliminated and a graph of time against number of layers of filter paper should be plotted. A straight line verifies the expectation (Fig. 11.8).

Results

No. of sheets	Time (s)
1	
2	
3	
4	
5	
6	
7	
8	
9	
10	

Questions

1. State the three requirements for a fire.
2. With the aid of a graph explain the phases in the progress of a fire in a building.
3. Explain what is meant by non-combustibility and describe with a sketch of the apparatus how a sample of material may be tested.
4. The degree of fire resistance to be provided in a building is influenced by the 'fire loading' for that building. Discuss this statement.
5. Describe the following fire tests and discuss the relevance of each test when dealing with building materials and elements:
   (a) combustibility;
   (b) ignitability;
   (c) fire propagation;
   (d) surface spread of fire;
   (e) fire resistance.

Section V

Chapter 12

Climatic effects

Wind speed

Not all parts of the earth's surface are heated equally by the sun.
Variation in latitude and the fact that land masses experience more
rapid and more extreme changes in temperature than the sea, cause
air temperature gradients which lead to the atmospheric density and
pressure differences which result in global air movement or wind. The
revolution of the earth about the sun and the rotation of the earth
about its own axis cause the seasonal and diurnal weather patterns
respectively.

The speed of the air in contact with the earth's surface is zero.
Above the ground wind speed increases with height until the fric-
tional drag of the earth's surface becomes negligible and the wind
attains the wind speed of the free atmosphere. This boundary layer
region in which both the wind speed and the direction of the wind are
influenced by the nature of the earth's surface varies between a
height of about 300 m over flat open country and a height of about
500 m over large towns. The rate of increase of the wind speed with
height within this boundary layer is known as the velocity gradient
and depends upon the roughness of the earth's surface. An estimate
of the wind speed (U) at a height (z) above the ground may be
obtained from the equation

$$\frac{U}{U_{\mathrm{m}}} = Kz^{a}$$

where K and a depend upon the terrain and U_m, the wind speed at an equivalent height of 10 m in open countryside, may be obtained from Meteorological Office measurements.

In Fig. 12.1 the height (z) is plotted against wind speed ratio (U/U_m) for the four types of terrain:

(a) Open level country.
(b) Flat country with scattered windbreaks (such as trees and occasional buildings).
(c) Urban.
(d) City (the centres of large towns and cities where the buildings are fairly closely spaced with a general roof height of about 25 m or more).

Wind speed increases with height at a slower rate over urban and city areas than over open level country.

Wind speed profiles are also influenced by the local topography. Wind speeds may be increased by exposed hills which rise well above the level of the surrounding land. If a hill is very steep the wind breaks away from the ground in the vicinity of the hilltop and considerable wind turbulence follows. Wind blowing along a valley which is diminishing in width may also be accelerated but reduced wind speeds may occur in steep-sided enclosed valleys. As the wind moves over the ground, obstructions such as trees and buildings disturb the ordered free flow of the air, creating disorder in the air movement or turbulence. The gusts and lulls in the wind depend upon the size of the obstruction, for example tall buildings cause more interference than small buildings.

Air flow around buildings

Air flow around a building is determined by the shape of the building and adjacent buildings and the direction, speed and gustiness of the oncoming wind.

When the direction of the wind is perpendicular to the face of a building, the windstream is retarded by the windward wall, but deflected and accelerated around the wall edges and over the roof (Fig. 12.2). The windward wall experiences a pressure increase, but the rapid air movement over the end walls and roof of the building leads to a pressure reduction (or suction). The leeward wall is also a low pressure region in which vortices (whirling air masses) may form. For other wind directions turbulence, whirls of air, occur around the wall and roof edges and any projecting feature of the building.

In the suburban and traditional town environment the wind is channelled between low rise buildings. Sheltered areas may be found and adequate protection from high winds at street level is usually provided. The construction of tall buildings which project over the urban landscape has resulted in high wind speeds at street level. Wind incident above the top one-third to one-quarter height of a tall

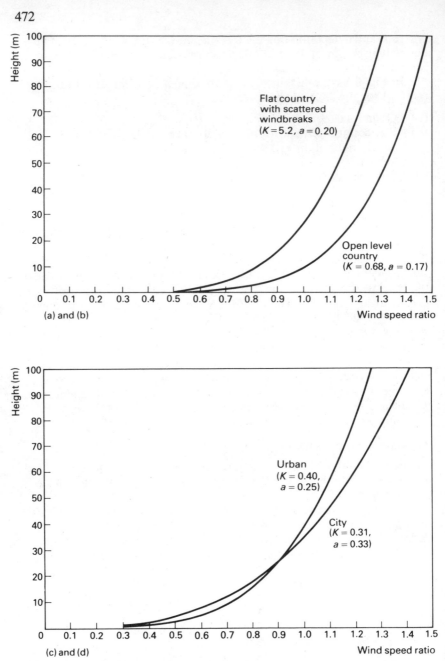

Fig. 12.1 Variation in wind speed ratio (U/U_m) with height (z). (The values for factors K and a are as recommended in *BRE Digest*, 210.)

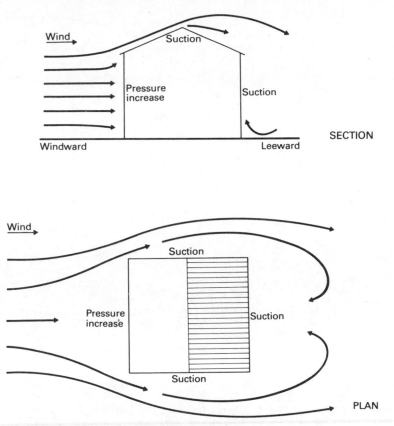

Fig. 12.2 Wind flow around a building

building flows up the building surface and over the roof. Below this level wind is deflected downwards and

(a) forms a horizontal vortex in front of the tall building;
(b) sweeps around the windward corners to form corner streams; and
(c) flows from the high pressure region on the windward side of the building through any opening at ground level to the low pressure region on the leeward side of the building (Fig. 12.3).

These regions of increased wind speed depend especially upon the building layout and dimensions but are less likely to occur when the building height is below 25 m.

For the comfort and safety of pedestrians wind speeds should be kept below 5 m/s for as much of the time as possible. A wind speed of 5 m/s raises dust, dry soil and loose paper. At a wind speed of 10 m/s umbrellas can be used only with difficulty and at 20 m/s people may be blown over by gusts of wind.

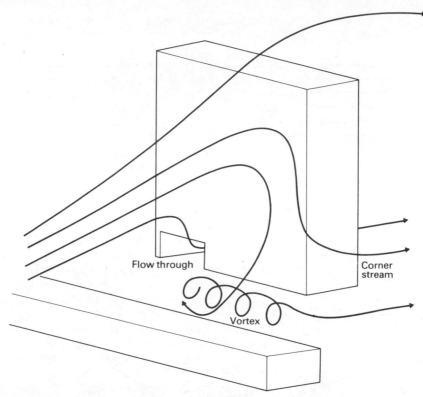

Fig. 12.3 Typical air flow pattern around a tall building

Air pollution

The likelihood of air pollution effects in a particular district is associated with the speed, direction and turbulence of the prevailing wind. The amount of pollution is a balance between the rate at which pollutants are released into the air and the rate at which they are swept away and diluted by the wind. Pollutants are released into the atmosphere by the burning of fossil fuels (coal, oil and natural gas) for heating, industrial processes and the combustion of petrol and diesel fuels in motor vehicles.

Common pollution forms are (a) grit and dust, (b) smoke and (c) gases. The solid particles of grit and dust are heavy enough to settle out of the air and usually fall close to and downwind of the industrial chimneys which emit them. Smoke consists of very fine solid or liquid particles produced by the incomplete combustion of fuels. The size of many of these particles is much less than $1\,\mu$m ($1\,\mu$m $= 10^{-6}$ m). Pollutant gases include sulphur dioxide, carbon dioxide, nitrogen oxides, carbon monoxide and ozone. Halogens, hydrocarbons, ammonia and ammonium sulphate are among those gases usually considered to be minor pollutants. The smoke fumes and pollutant gases mix with the air and may travel some distance with it unless they are washed out by rain or fog formation occurs.

Industrial chimneys should be of such a height that pollution emission occurs well above the turbulent wind conditions caused by adjacent buildings. The effective chimney height and the efficiency of pollution dispersion are enhanced if the speed of the effluent smoke and gases carries them above the turbulent wind conditions around the chimney mouth. If the speed of the smoke and gases leaving a chimney is insufficient to prevent their being drawn into the turbulent wind conditions on the leeward side of the chimney they will be drawn down around the outside of the chimney before dispersal and the effective height of the chimney will be less than its real height. Intense pollution may occur during squally weather when gusts of wind may bring almost undiluted smoke and gases to ground level.

In residential areas the pollution concentration in the street is similar to that at the roof top since domestic chimneys are never high enough for their emissions of smoke and gases to avoid the turbulent wind currents at roof level. The installation of suitable heating and cooking appliances and the use of approved fuels (smokeless fuels) in Smoke Control Areas have benefited many cities and towns.

In cities and towns traffic congestion may lead to a high concentration of pollution at street level, especially where streets are narrow and sheltered from the wind by buildings of similar height. Wider streets and a mix of building heights provide a freer circulation of air and therefore some dilution of the exhaust fumes and gases.

The removal of pollutants from the point at which they are released into the air and their dilution by the air are encouraged by strong winds. Light wind conditions and a temperature inversion over an industrial city prevent their dispersion. A temperature inversion occurs when a layer of air warmer than the air at ground level hangs over the city at a height of about 100 m or more. This warm air layer prevents air adjacent to the ground diffusing into and mixing with the remainder of the atmosphere. Pollutants are trapped beneath this ceiling of warm air and accumulate near the ground to form smog.

Heavy industry is normally grouped together and sited downwind of a town so that grit and dust do not fall within the town and the smoke and gases are swept away by the wind. New towns are often located at fairly high windy sites where advantage is taken of the wind to clear fogs and mists and clean the atmosphere. Enclosed sheltered localities where pollution is likely to be trapped by temperature inversions are avoided.

Smoke particles are deposited upon and adhere to building surfaces, especially edges of window and door frames and any protrusions which increase air turbulence. Smoke contains carbon compounds such as tarry hydrocarbons and resins and is able to absorb much of the sulphur dioxide present in flue gases. Upon contact with moisture the sulphur dioxide forms sulphuric acid. In the presence of adherent smoke particles metal surfaces may experience accelerated attack by corrosive gases such as sulphur dioxide.

True control of man-made pollution must take place at source prior to the release of contaminants into the air, but the dispersion of this pollution is influenced by the weather, particularly the wind, and the local topography.

Exposure of a structure to the elements

Important physical parameters in the determination of climate are air temperature, precipitation (any moisture deposit from the atmosphere such as rain, dew, frost, hail or snow), humidity, wind speed and solar radiation. Because of the unpredictable nature of the weather, the average or the extreme values of these parameters required in building design are calculated from long-term meteorological records.

A building must provide shelter from wind, precipitation and temperature extremes. Wind and rain are the most important factors in the determination of the exposure of a building.

Wind

In the United Kingdom the highest wind speeds occur near coasts, particularly western coasts and over high ground. Inland areas are relatively sheltered from the wind. The structural frame of a building, the building elements such as walls and roof, and any cladding units should be able to withstand the pressure variations due to wind. *BRE Digest*, 119, 'The Assessment of Wind Loads', gives methods of assessing the maximum wind speed appropriate to the structure and site, and the wind loads which should be taken into account when designing buildings. The wind load on the roof of a structure is often more severe than that experienced by any other structural element. Suction over the roof is reinforced by increased air pressure beneath any overhanging roof edges (Fig. 12.4). Turbulence and suction also occur at corners, along the edges of walls and roofs and around projections such as chimney stacks. The wind loads on these vulnerable regions must be resisted by the firm fixing of roofs and cladding units.

Wind speed influences both thermal resistance and air infiltration rate, therefore reduced heating loads may be expected when a building is sheltered from high winds.

Precipitation

The transmission of water from the external environment through the roof and walls into the interior of a building must be prevented. Rain may be carried either at various angles by the wind or fall vertically

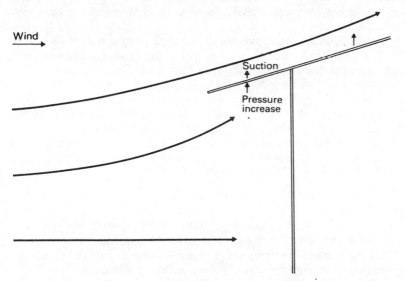

Fig. 12.4 Suction over the roof is reinforced by increased air pressure beneath the overhanging roof edge

upon a roof. The waterproof qualities of the traditional pitched roof depend upon:

(a) an impervious covering;
(b) satisfactory detailing of overlaps or jointing;
(c) the rapid run-off of water from the sloping roof to the gutters.

A flat roof with correct falls should also drain but even a minor defect in the waterproof membrane greatly increases the risk of moisture ingress.

In the absence of wind rain falls parallel to vertical surfaces, barely wetting them. Also the air pressure differences across the building shell are not great enough to force moisture through gaps into the interior of the building. Driving rain, rain blown against a wall by the wind, may penetrate the building shell because:

(a) absorption and movement of moisture through porous materials such as brick, stone and concrete occur under the influence of capillary forces;
(b) rainwater may be driven through joints and gaps such as those around windows and doors by wind forces.

Moisture migration through masonry material increases the thermal conductivity and may lead to dampness on the inner surface and a worsening of any condensation problems. It may be controlled by a barrier of impermeable material or by the provision of an air space within the wall. An unbridged air cavity (the build-up of mortar on wall ties is not recommended) is a very effective means of preventing moisture transfer.

While capillary forces move water through porous materials and very fine cracks, wind forces are responsible for driving rainwater through openings (cracks and joints) in excess of about 0.1 mm in width. Rainwater flowing along a surface beside a joint or bridging the gap across an opening may be forced into the interior by the wind. Small dry snowflakes with a much lower density than raindrops may be carried directly through joints by infiltrating air currents. Special care at joints is essential since even a small quantity of rain accompanied by a high enough wind speed may lead to rain penetration.

Rainwater run-off

Rain driven on to a wall by the wind will run off its surface if: (i) the rate of driving rain exceeds the rate at which the surface can absorb water; (ii) the surface becomes saturated with water; (iii) the surface is impervious. Restraint of water flow over the building surface reduces the risk of leakage at joints and promotes even weathering of the surface. Examples of some features by which control may be effected include overhanging eaves to protect walls beneath; gutters to prevent run-off from the roof reaching walls and foundations; traditional window sills to throw water run-off from the glass away from the wall; and since raised edges and protrusions create turbulent wind patterns, a window or any other element recessed into an opening in a wall has more shelter from the wind and is less susceptible to leaks. A wide variation in water flow from one part of the building surface to another should be avoided, because even if the risk of rain penetration and decay can be eliminated, differential weathering is unsightly.

Moisture movement

Many building materials absorb moisture, either from precipitation or a high humidity environment. A component absorbing or losing water or water vapour will usually experience a dimensional change (an increase in size as the moisture content increases). The order of magnitude of this change must be determined so that adequate provision may be made for moisture movement at the design stage.

Exposure grading

The likelihood of rain penetration through a masonry wall may be assessed from exposure gradings based upon the driving rain index. The annual mean driving rain index is proportional to the total rainfall driven during one year on to a vertical surface facing the wind.

Driving rain index = annual × average × $\frac{1}{1000}$
$\quad\quad$ rainfall $\quad$ wind speed
(m²/s) $\quad\quad$ (mm) $\quad\quad$ (m/s)

The three exposure gradings indicated on the map (Fig. 12.5) apply to a standard building occupying a site upon which the exposure is normal for the location. Adjustment to these exposure gradings may be necessary to take account of special factors (Table 12.1).

Table 12.1

Exposure grading	Driving rain index (m²/s)	Adjustment
Sheltered	up to 3	Regrade areas lying within 8 km of the sea or large estuaries as moderate
Moderate	3 to 7	Regrade areas having an index of 5 m²/s or more and lying within 8 km of the sea as severe
Severe	7 or more	

In areas of sheltered or moderate exposure, an exposure one grade more severe than indicated on the map should be applied to buildings on hilltops or hill slopes and to tall buildings which rise above their surroundings. Conversely if a site is particularly sheltered for its location the severity of its exposure may be reduced by one grade.

Local exposure may even vary over the surface of a building as corners, edges and projections cause wind turbulence and increased wind speeds.

Except in some eastern regions of the country where most rain is driven on to north and north-east walls, walls facing south-east, south and west experience most driving rain. However, the wind may blow from any direction during periods of intense driving rain.

Building materials susceptible to damage when frozen in a very wet condition (frost action) should not be used on exposed sites. The driving rain index is helpful in estimating the likely wetness of absorbent walls.

Temperature and solar radiation

The extremes of external temperature and solar radiation imposed upon a structure may be moderated by adequate thermal resistance, thermal capacity and ventilation so that a comfortable internal

480

Exposure gradings

Severe – Over 7 m²/s

Moderate – 3 – 7 m²/s

Sheltered – under 3 m²/s

Fig. 12.5 Simplified driving-rain index map (m²/s)

environment may be maintained without excessive energy usage. In addition the ratio of transparent to opaque material in the building envelope, building orientation, colour of surface and shading, control solar heat gain.

Thermal movements, dimensional changes in building components, are produced in response to temperature changes. Thermal expansion and contraction are greater in large units, in plastics which have high coefficients of thermal expansion compared to traditional building materials, and in membranes and thin sections where low thermal capacity leads to extreme surface temperatures. For example when a thin black covering on a flat roof is exposed to environmental extremes it may experience surface temperatures from a maximum of about 80 °C to a minimum of about -25 °C (R. E. Lacy). The maximum surface temperature may be reduced by embedding chippings in a layer of bitumen dressing compound to form a light-coloured reflective surface.

As with moisture movements, thermal movements must be located and estimated so that joints may be designed and unnecessary stresses and cracking prevented. The moisture content of a material tends to decrease as its temperature increases. These dimensional changes may compensate for one another, except in materials where one type of movement is dominant, such as moisture movement in wood and thermal movement in relatively impervious materials.

Deterioration of relatively impervious materials such as asphalt, paints and plastics is mainly caused by the effect of heating and cooling stresses and exposure to sunlight.

The microclimate

The exposure of a building is affected by its site position. While the ideal site would not be affected by unfavourable weather conditions such as strong winds, driving rain, prolonged fog or severe frost, a well-considered site should be arranged to ameliorate these climatic factors around the building envelope. The climate in the immediate vicinity of the building envelope is often called the microclimate. Trees, shrubs, hillocks and existing buildings may give useful shelter from strong winds and driving rain. The building should be orientated to take advantage of winter sunshine. Moderation of the microclimate provides a more comfortable environment and reduces the exposure of the structure to the elements.

Questions

1. Show how wind speed change with height above ground level is influenced by the nature of the terrain.
2. Describe the air flow around tall buildings.

3. Discuss the effects of weather, especially the wind, and local topography upon pollution dispersion.
4. Outline the problems associated with the exposure of a structure to wind and rain.
5. Explain when the driving rain index may be used in the assessment of the exposure of a building.
6. Discuss measures to control and minimise the effects of exposure to the elements upon the building envelope.

Answers

Chapter 1   Thermal transmission

1. (a) (i)   32 kW   (ii)   96 kW
   (b) (i)   14 W   (ii)   42 W
2. (a) $0.125 \, m^2 \, K/W$ (b) $0.5 \, m^2 \, K/W$ (c) $1.32 \, m^2 \, K/W$
3. (b)   144 W
4. 17.544 W
5. $0.03 \, m^2 \, K/W$

6. Heat flow	High emissivity $(m^2 \, K/W)$	Low emissivity $(m^2 \, K/W)$
horizontally	0.12	0.30
upwards	0.10	0.22
downwards	0.14	0.56

7. $3.3 \, W/m^2 \, K$
8. (a) $1.34 \, W/m^2 \, K$
   (b) $1.40 \, W/m^2 \, K$
9. $0.91 \, W/m \, K$
10. (b) $0.92 \, W/m^2 \, K$
    (c)   276 W
11. (b)   50 per cent
12. (b) $-0.3 \, °C$
13. (b)   100 mm

14. (i) $5.0 \, \text{W/m}^2 \text{K}$
  (ii) $5.6 \, \text{W/m}^2 \text{K}$
  (iii) $6.7 \, \text{W/m}^2 \text{K}$
15. (a)   11.8 per cent   (b)   27.3 per cent
17. (b)   (i)   12.2 kW   (ii)   10 per cent
18. 2.4 °C, 2.8 °C when the U value of the roof is improved to 0.4 W/m² K.

Chapter 2 Thermal comfort

4. (b)   45 per cent, 45 per cent
5. (a)   (i)   47 per cent   (ii)   74 per cent, 15.3 °C

Chapter 3 Condensation

1. (b)   14.7 °C, 2.6 °C
2. (b)   8.2 °C
  (c)   Yes (the dewpoint temperature is estimated as 8.2 °C).
4. (a)   Condensation occurs since the internal surface temperature is 12.0 °C.
  (b)   4.1 °C
  (c)   No (internal surface temperature = 14.0 °C).

5.

Surface (see Fig. A.1)	Structural temperature (°C)	Dewpoint temperature (°C)
1	18.9	8.5
2	18.6	7.5
3	14.0	3.5
4	1.6	2
5	0.5	−1

6.

Surface (see Fig. A.2)	Structural temperature (°C)	Dewpoint temperature (a) (°C)	(b) (°C)
1	17.4	8.5	8.5
2	16.0	8	1
3	7.0	4	−0.5
4	1.3	−1	−1

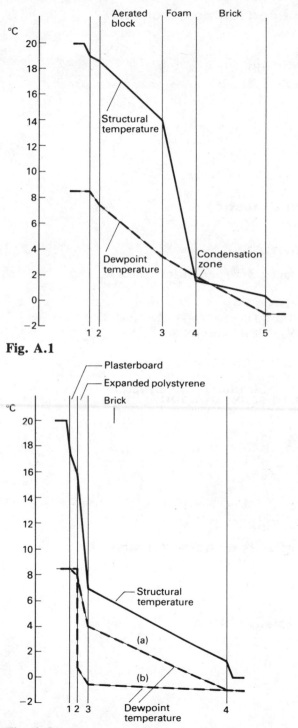

Fig. A.1

Fig. A.2

Chapter 4   Sound and its measurement

1. 73.3 dB
4. (b)   66.2 dB
5. 100 dB
6. 97 dB
7. 63.5 dB
8. $2.2 \times 10^{-6}\,W/m^2$
9. $2\,N/m^2$
10. 62 dB
11. 90 dB

Chapter 5   Room acoustics

2. (b)   1.633 s
    (c)   169 m^2
3. 1.8 s
4. Both 0.66 s (Assuming 3 m^3/person for speech)
5. 4789 (4800)
6. Actual 1.5 s = Optimum 1.4 s.
7. (a)   Volume 1800 m^3 (3 m^3/person) = Reverberation time 1 s.
    (c)   Optimum 1.9 s = 48 m^2 absorption.
9. 1.95 s.

Chapter 6   Transmission of sound

4. 1.9 m^2
5. 2.4 m^2
7. (a)   50 dB
    (b)   55 dB
8. 27.5 dB
9. 99 dB

Chapter 7   The effects of noise on man

3. 6 minutes

Chapter 8   Artificial lighting

1. 63.8 lux
2. (A)   22.8 lux
    (B)   12.03 lux
3. 60.3 lux, 63 lux

14. Room index = 1.59, utilisation factor = 0.63
number of lamps = 15, S/H_m = 0.88:1
15. 70 luminaires, 10×7 array.

Chapter 9   Daylighting

9. (a)   SC = 12.6 per cent corrected to 9.1 per cent
ERC = 0 per cent
Corrected average IRC = 0.41 per cent
DF = 9.5 per cent
(b)   SC = 13.3 per cent
As the position is close to the window allowance must be
made for the wall thickness which produces cut-off at the
sill. This is done by altering D to $(D - 0.25)$ in the ratios for
H/D for 'Windows' 2 and 4. The ratios W_1/D and W_2/D
are not altered.
Making this allowance:
SC = 12.1 per cent corrected to 8.7 per cent
DF = 9.1 per cent
10. Average daylight factor for the room = 2.2 per cent uncorrected
(1.54 per cent corrected).
This is well below the recommendations for an office and a
PSALI or a PAL scheme is needed.
(Correction:

$2.2 - 0.63 = 1.57$ per cent
$1.57 \times 0.8 \times 0.9 = 1.13$
$0.63 \times 0.8 \times 0.9 \times 0.9 = \underline{0.41}$

AV. DF = 1.54 per cent)

11. Approximately $6.5 \, \text{m}^2$

Chapter 10   Illumination for human comfort

8.   GI = 17.6
9. (a)   GI = 15.4
(b)   GI = 14.95
(c)   GI = 15.85

Bibliography

Section I Heat and thermal effects

British Standards

1. BS 5250:1975   Code of basic data for the design of buildings: the control of condensation in dwellings.
2. BS 5925:1980   Code of Practice for design of buildings: ventilation principles and designing for natural ventilation.

Building Research Establishment (BRE)

BRE Current Papers

3. CP 1/70   Pamela J. Arnold, *Thermal Conductivity of Masonry Materials*.
4. CP 9/78   M.A. Humphreys, *The Optimum Diameter for a Globe Thermometer for Use Indoors*.
5. CP 53/78   M.A. Humphreys, *Outdoor Temperatures and Comfort Indoors*.

BRE Digests

6. 108   *Standard U-values*.
7. 110   *Condensation*.
8. 145   *Heat Losses Through Ground Floors*.
9. 180   *Condensation in Roofs*.
10. 221   *Flat Roof Design: The Technical Options*.
11. 224   *Cellular Plastics for Building*.
12. 226   *Thermal, Visual and Acoustic Requirements in Buildings*.

13. 236   *Cavity Insulation.*
14. *Chartered Institution of Building Services (CIBS) Guide.*
    Section A1   Environmental Criteria for Design (1978).
    Section A3   Thermal Properties of Building Structures (1980).
    Until recently the *CIBS Guide* has been known as the *IHVE Guide* (Institution of Heating and Ventilating Engineers). The following sections were published as part of the *IHVE Guide.*
    Section A6   Solar Data (1970).
    Section A7   Casual Gains (1977).

Department of the Environment (DOE)

15. DOE Condensation in Dwellings, HMSO (1971).
16. DOE, *Thermal Insulation of Buildings*, HMSO (1971).
17. Chrenko, F.A. (Ed), *Bedford's Basic Principles of Ventilation and Heating* (3rd edn), H.K. Lewis (1974).
18. Fanger, P.O., *Thermal Comfort, Analysis and Applications in Environmental Engineering*, McGraw-Hill (1972).
19. Houses in the 80s, *Architects Journal*, 16 Jan. 1980.
20. Marsh, Paul, *Thermal Insulation and Condensation*, Construction Press (1979).
21. McIntyre, D.A., *Indoor Climate*, Applied Science Publishers (1980).
22. O'Callaghan, P.W., *Building for Energy Conservation*, Pergamon Press (1978).
23. Wilberforce, P.R., The effect of solar radiation on window energy balance. In: *Energy Conservation in the Built Environment*, Construction Press in conjunction with CIB (1976).

Section II   Sound

Building research Establishment (BRE)
BRE Digests

1. 192   *The Acoustics of Rooms for Speech.*
2. 143   *Sound Insulation: Basic Principles.*
3. 187   *Sound Insulation of Lightweight Dwellings.*
4. 228   *The Acoustic Requirements in Buildings.*
5. 203 & 204   *Noise Abatement Zones.*

British Standards

6. BS 661:1969   Glossary of Acoustical Terms.
7. BS 3638   Method for the Measurement of Sound Absorption Coefficients in a Reverberation Room.
8. BS 2750   Measurement of Airborne and Impact Sound Transmission in Buildings.
9. BS 3593:1963   Preferred Frequencies for Acoustical Measurements.
10. BS 3489:1962   Specification for Sound Level Meters (Industrial Grade).

11. BS 4197:1976   Specification for a Precision Sound Level Meter.
12. BS 4198:1967   Method for Calculating Loudness.
13. BS 5363:1967   Method for Measurement of Reverberation Time
in Auditoria.
14. BS 4142:1967   Method of Rating Industrial Noise Affecting
Mixed Residential and Industrial Areas.
15. BS 5228:1975   Code of Practice for Noise Control on
Construction and Demolition Sites.
16. BS 4078:1966   Cartridge Operated Fixing Tools.

General
17. Parkin, P.H., Humphreys, H.R. and Cowell, J.R., *Acoustics,
Noise and Buildings*, Faber (1979)
18. Smith, B.J., Peters, R.J. and Owen, S., *Acoustics and Noise
Control*, Longman (1982)
19. Burns, W., *Noise and Man*, John Murray (1973)

Section III   Lighting

The Chartered Institution of Building Services – Lighting Division

1. IES Code for interior lighting 1977.
2. IES Technical Reports.
3. IES Technical Report No. 2:   The Calculation of Utilisation
Factors. The BZ method.
4. IES Technical Report No. 4:   Daytime Lighting in Buildings.
Supplement: Control of discomfort sky glare from windows.
5. IES Technical Report No. 9:   Depreciation and Maintenance of
Interior Lighting.
6. IES Technical Report No. 10:   Evaluation of Discomfort Glare:
The IES Glare Index System for Artificial Lighting Installations.
Supplement: Additional data for asymmetric light distributions.
7. IES Lighting Guide: Building and Civil Engineering Sites.
Eyestrain: The Environmental Causes and their Prevention – J.H.
Goacher pre-print of papers CIBS National Lighting Conference
1980.

HMSO

BRE Digests
8. 41   *Estimating Daylight in Buildings* – 1.
9. 42   *Estimating Daylight in Buildings* – 2.
10. 226   *Thermal, Visual and Acoustic Requirements in Buildings.*

General
11. Hopkinson, R.G., *Architectural Physics: Lighting* (1963).
12. Longmore, J., *BRS Daylight Protractors* (1968).

13. Pritchard, D.C., *Lighting* (2nd edn), Longman (Environmental Physics Series) (1978).
14. Hopkinson, R.G., Petherbridge P. and Longmore, J., *Daylighting*, Heinemann (1966).
15. Henderson, S.T. and Marsden, A.M. (Eds), *Lamps and Lighting* (2nd edn), Edward Arnold (1972).
16. Weston, H.C., *Sight, Light and Work* (2nd edn.) H.K. Lewis (1962).
17. Bean, A.R. and Simons, R.H., *Lighting Fittings Performance and Design*, Pergamon Press, Oxford and London (1968).
18. Hopkinson, R.G. and Collins, J.B., *The Ergonomics of Lighting*, Macdonald Technical and Scientific, London (1970).
19. Tyler, H.A., *Environmental Science Level IV*, Van Nostrand Reinhold, London (1980).
20. *Photometric Data*, Vols 1 and 2, Thorn Lighting, London (1976).
21. *Interior Lighting Design Handbook*, Lighting Industry Federation.

British Standards Institution

22. Code of Practice 3   Chapter 1 Part 1 1964 – Daylighting
23. Code of Practice 3   Chapter 1 Part 2 1964 – Artificial Lighting

Section IV   Fire

British Standards

1. BS 476 Part 3:1975   External Fire Exposure Roof Test.
2. BS 476 Part 4:1970   Non-Combustibility Test for Materials.
3. BS 476 Part 5:1979   Method of Test for Ignitability.
4. BS 476 Part 6:1968   Fire Propagation Test for Materials.
5. BS 476 Part 7:1971   Surface Spread of Flame Tests for Materials.
6. BS 476 Part 8:1972   Test Methods and Criteria for the Fire Resistance of Elements of Building Construction.

Building Research Establishment (BRE)

BRE Digests
7. 214   *Cavity Barriers and Fire Stops*, Part 1.
8. 215   *Cavity Barriers and Fire Stops*, Part 2.
9. 220   *Timber Fire Doors*.

Section V   Climatic effects

Building Research Establishment (BRE)

BRE Digests
1. 119   *The Assessment of Wind Loads*.
2. 127   *An Index of Exposure to Driving Rain*.

492

3. 141 *Wind Environment Around Tall Buildings.*
4. 210 *Principles of Natural Ventilation.*
5. Lacey, R.E., *Climate and Building in Britain*, HMSO (1977).
6. Marsh, Paul, *Air and Rain Penetration of Buildings*, Construction Press (1977).

Appendix

Environmental temperature

In summer, solar radiation may raise the temperature of outside surfaces many degrees above the ambient air temperature, but on a winter night the outside surface temperature of roof cladding may fall several degrees below the air temperature because of radiation loss to a clear sky. Within a building radiant heat from heating appliances or solar radiation may lead to surface temperatures higher than the internal air temperature. During cold weather warm air heating in a poorly insulated building may result in internal surface temperatures noticeably lower than the internal air temperature. Environmental temperature is introduced to account for these aspects in heat transfer calculations in buildings.

Environmental temperature (t_e) is defined as the temperature of a hypothetical uniform environment which would give the same rate of heat transfer through a building element as occurs under the existing conditions. The hypothetical uniform environment has surroundings and air at equal temperatures. Internally, the environmental temperature is t_{ei}, externally it is often referred to as the *sol-air temperature*, t_{eo}.

Approximately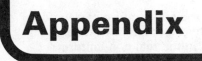
$$t_{ei} = \tfrac{2}{3}t_r + \tfrac{1}{3}t_{ai}$$
$$(°C) \quad (°C) \quad (°C)$$

where t_{ei} = internal environmental temperature

   t_r = mean radiant temperature

and $\quad\quad\quad\quad\quad t_{ai}$ = internal air temperature

Its exact value depends upon room configuration and the convective and radiant heat transfer coefficients.

Mean radiant temperature

Radiation exchange between a body in a room and the surfaces of that room depends upon the nature, shape, size and temperature of the surfaces and the solid angles subtended at the body by the various surfaces.

Determination of precise values is so complex that it is convenient to replace the room surfaces by a hypothetical uniform enclosure which would give the same net radiation exchange between body and room surfaces and to define the mean radiant temperature (t_r).

Mean radiant temperature (t_r) is the temperature of a uniform black enclosure in which a solid body or occupant would exchange the same quantity of radiant heat as in the real non-uniform environment. Mean radiant temperature is closely related to the mean surface temperature.

Mean radiant temperature is a function of position, but within a room with unheated surfaces an average value, sufficiently accurate in most circumstances, may be obtained from the mean of the surface temperatures weighted by area

$$t_r = \frac{a_1 t_1 + a_2 t_2 + a_3 t_3 + \ldots}{a_1 + a_2 + a_3 + \ldots}$$

where $\quad t_1, t_2, t_3 \ldots$ are surface temperatures °C and

$\quad\quad\quad a_1, a_2, a_3 \ldots$ are surface areas (m^2)

This approximation should not be made when large areas of metal or high surface temperatures are present.

Index

498